AF903489

NONDESTRUCTIVE TESTING METHODS AND NEW APPLICATIONS

Edited by **Mohammed Omar**

Nondestructive Testing Methods and New Applications
Edited by Mohammed Omar

Published by InTech
Janeza Trdine 9, 51000 Rijeka, Croatia

Copyright © 2012 InTech
All chapters are distributed under the Creative Commons Attribution 3.0 license, which allows users to download, copy and build upon published articles even for commercial purposes, as long as the author and publisher are properly credited, which ensures maximum dissemination and a wider impact of our publications. After this work has been published by InTech, authors have the right to republish it, in whole or part, in any publication of which they are the author, and to make other personal use of the work. Any republication, referencing or personal use of the work must explicitly identify the original source.

As for readers, this license allows users to download, copy and build upon published chapters even for commercial purposes, as long as the author and publisher are properly credited, which ensures maximum dissemination and a wider impact of our publications.

Notice
Statements and opinions expressed in the chapters are these of the individual contributors and not necessarily those of the editors or publisher. No responsibility is accepted for the accuracy of information contained in the published chapters. The publisher assumes no responsibility for any damage or injury to persons or property arising out of the use of any materials, instructions, methods or ideas contained in the book.

Publishing Process Manager Ivona Lovric

Technical Editor Teodora Smiljanic

Cover Designer InTech Design Team

Additional hard copies can be obtained from orders@intechopen.com

Nondestructive Testing Methods and New Applications, Edited by Mohammed Omar
p. cm.
ISBN 978-953-51-0108-6

We are IntechOpen, the world's leading publisher of Open Access books

Built by scientists, for scientists

4,000+
Open access books available

116,000+
International authors and editors

120M+
Downloads

151
Countries delivered to

Our authors are among the
Top 1%
most cited scientists

12.2%
Contributors from top 500 universities

CLARIVATE ANALYTICS BOOK CITATION INDEX INDEXED

WEB OF SCIENCE™

Selection of our books indexed in the Book Citation Index in Web of Science™ Core Collection (BKCI)

Interested in publishing with us?
Contact book.department@intechopen.com

Numbers displayed above are based on latest data collected.
For more information visit www.intechopen.com

For EU product safety concerns:
IN TECH d.o.o., Prolaz Marije Krucifikse Kozulić 3, 51000 Rijeka, Croatia,
info@intechopen.com or visit our website at intechopen.com.

Meet the editor

Dr. Omar is an associate professor at the Clemson University International Center for Automotive Research CU-ICAR. His research and teaching interests are in the area of Manufacturing, Materials and, specifically, knowledge based manufacturing systems and lightweight design, non-destructive testing of materials and structures. He is the recipient of many prestigious awards including: the SME Richard L. Kegg Manufacturing award, the SAE Foundation Leadership in Manufacturing award, and the Murray Stokely Excellence in Engineering Education award. Dr. Omar is an active participant in several engineering societies: he hosted and chaired the ASNT topical conference series in 2009, was on the organizing committee of the NAMRC 37, and the SAE World Congress in 2009 and 2010. Prior to joining Clemson, he served as a visiting scholar at the Toyota Motor Company Measurement and Instrumentation Engineering Division, in addition to serving as a Post-Doctoral scholar at the University of Kentucky Center for Manufacturing.

Contents

Preface

The Nondestructive testing science is a broad field that covers variety of testing methods and applications, in addition to the associated pre and post processing mathematics. In terms of methods and techniques the Nondestructive testing modalities rely on different physical phenomena such as the electromagnetism, the acoustic emission, the thermal emission and the penetration of high-energy radiation through materials and structures. This diversity in the Nondestructive testing tools is only matched by its fields of application, which covers the testing of civil and mechanical structures and components, the online monitoring of manufacturing processes and products, and a wide array of medical applications that include dental and veterinary medicine.

This book will seek to introduce several Nondestructive testing embodiments to address different testing techniques that rely on several physical phenomena while addressing the wide range of its applications. This is done in an effort to highlight several types of the Nondestructive evaluations and its ability to accommodate multitudes of fields and tests. Also the manuscript will explain the different mathematical and statistical processing techniques used in pre-processing the acquired data in terms of noise reduction, data compression and signal conditioning; in addition to processing the signals and correlating it with the properties of the materials or structures that are being tested. Sections of this book will be solely dedicated to new applications or to using innovative NDT technologies.

The specific Nondestructive techniques addressed in this book include; the magnetic adaptive testing, the ultrasonic testing methods, the Neutron Radio-graphy, the Superconducting Quantum Interference Device SQUID sensor based testing routines. The text will also include chapters to discuss the testing reliability and validation studies. This book is structured in three main sections; mainly a section on the General Nondestructive Testing Methods and its Specific Considerations, a section on Innovative Nondestructive Testing Systems and Applications, and finally a section on the Concrete Nondestructive Testing Methods.

Dr. Mohammed Omar
Clemson University,
International Center for Automotive Research CU-ICAR, Greenville, SC
USA

Part 1

General Nondestructive Testing Methods and Considerations

1

Nondestructive Inspection Reliability: State of the Art

Romeu R. da Silva[1] and Germano X. de Padua[2]
[1]Federal University of Rio de Janeiro,
[2]Petróleo Brasileiro S.A. (PETROBRAS),
Brazil

1. Introduction

In health, there are numerous types of tests for the identification of pathologies in patients. Some questions that can be brought up: How accurate are these tests? What are the "losses" of a medical report error if the patient has a serious health problem and it cannot be detected by the examination chosen? On the other hand, if the patient has no problem and the medical report shows positive? What consequences are there in a medical report error?

If we imagine that the medical risks assumed in inaccurate reports may lead to serious consequences, which can happen with the result of an inspection of equipment without reliability? Unlike the medical field instead of a fatal case, there may be multiple fatalities, environmental damage, irreparable financial losses, etc.

There is several non-destructive inspection methods used to evaluate the integrity of industrial equipment and thus raise several questions. What are the most reliable? Which ones provide lower risk of decision? There is an ideal method for a given type of equipment? A more reliable inspection method also costs more? Some of these questions are answered in the study of methods for estimating the reliability of Nondestructive Testing (NDT), area of scientific research that has been the focus of many investments in recent decades, aiming mainly to provide greater operational reliability of equipments from different branches of industries.

PoD curves may become a powerful tool for quantifying the performance of inspection techniques, as well as inspectors and can be used to:

- Establish criteria for project acceptance;
- Set up maintenance inspection intervals;
- Qualification of NDT procedure;
- Performance verification of qualification of persons;
- Qualify improvements in NDT procedures.

Considering the thematic importance and the increasing trend of investment projects aimed at better understanding the reliability of NDT methods, this chapter has the main objective of making an approach on the state of the art studies of the reliability of non-destructive inspection to be used as the first bibliographic guidance to future researches. Firstly, it covers topics of major theoretical techniques used in the estimation of reliability curves. Then, some of the most relevant research publications in the area of reliability of NDT are commented in their main results. It must be noted that this work does not exhaust all the literature produced; there are other references that can be studied to obtain detailed information on this research topic.

2. Methods for reliability assessment

2.1 PoD - Probability of Detection curves

It's supposed that the first PoD (Probability of Detection) studies arose by the end of 60's and beginning of 70's, when most of studies were from aeronautic industry. At that time, it was realized that the question "what is the smaller detectable discontinuity with NDT methods?" was less appropriate than "what is the larger not detectable discontinuity?".

Currently, the most used method to determine the reliability and sensitivity of a NDT technique is through the assessment of probability of detection curves. A PoD curve estimates the capacity of detection of an inspection technique in regard to discontinuity size. In the ideal technique, the PoD for discontinuities smaller than established critical size would be zero. In the other hand, discontinuities greater than critical size would have PoD equal 1, or 100% of probability of detection. In such ideal technique, would not happen what we know as False Positive (rejection of acceptable components) or False Negative (approval of defective components). However, in real situation, PoD curves do not have an ideal behavior, presenting regions of False Positive and False Negative. Figure 1 illustrates a real and ideal PoD curve [2].

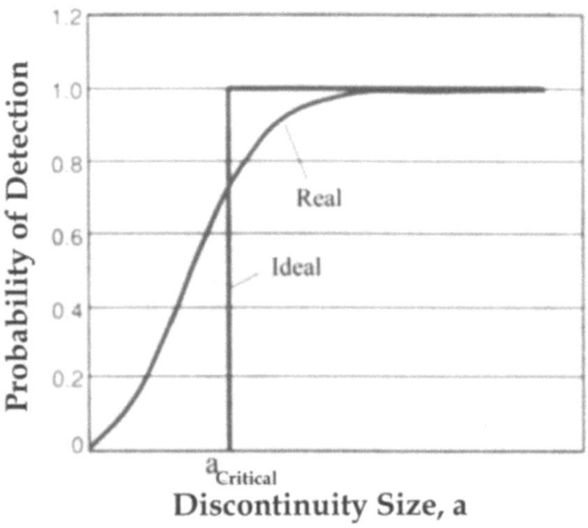

Fig. 1. Pattern of real and ideal PoD curves [2].

These curves are commonly constructed empirically. The most known method is *Round Robin Testing* (RRT), where a group of inspectors proceed a nondestructive examination of test pieces with artificial defects, simulating real defects that may be found in welded joints, for example. Artificial defects are fabricated in various dimensions. PoD curves may be drawn from results of one inspector or based on a group of inspectors [2-4]. Two issues need to be highlighted in this RRT methodology, the first is the amount of test pieces necessary to guarantee statistic reliability of the estimated curve, and second is the complexity of obtaining artificial defects in dimension, location and characteristics as similar as real defects. In welding, for instance, only skilled, experienced and well trained welders are able to produce defective welds in such way that simulate real situations of inspections that provide representative results of PoD.

At *First European-American Workshop* of reliability (Berlin, June 1997), a model of reliability was proposed, which recognize three functions connected to the reliability of a nondestructive testing technique: intrinsic capacity of the system, characteristics of specific applications and human factor. Thus, it's suggested that reliability of a NDT technique will never be higher than that idealized. The reliability of a technique, when applied to a specific type of defect, may be represented by following concept:

$$Re = f(IC) - g\,(PA) - h\,(HF) \tag{1}$$

Where,

Re is the total reliability of the system.
f (IC) is function of intrinsic capacity of the NDT system;
g (PA) is function of parameters applied (access, surface finishing etc.);
h (HF) is function of human factor (skills, training, experience etc.).

By this concept, the function *f* is associated to intrinsic capacity of the specific inspection technology in ideal conditions. In case of any noise (deviation of ideal conditions), the ideal reliability is going to be reduced as function of *g* nature. When there are human factors associated to manual inspection, reliability is reduced, according to function *h*. Automatic inspections are free of these factors, due to this fact, often provide higher probability of detection [2].

The PoD of a discontinuity sizing "a" is determined as the average of probability of detection for all discontinuities sizing "a". A PoD curve is constructed from the average of PoD for each dimension of discontinuity. Normally, a confidence level is associated, since it is estimated in function of a finite sample space. The length is the dimension commonly used, although the height (internal defect) or depth (surface defect) may be used as well [2-4].

Difficulties in fabricating a number of test pieces high enough, frequently provide a poor sample space. Due to this, there are various statistic models used to estimate PoD curves [2-5]. These models run data obtained from two types of analysis: $\hat{a}$ versus a and *hit/miss* [1-5]. According to Carvalho [2], some NDT techniques connect a signal with response "$\hat{a}$" to a real dimension a of the discontinuity. Nevertheless, some inspection techniques do not size the defects, the response is only detected or not. The analysis *hit/miss* get useful due to its simplicity. Both methods may be used to implement PoD curves, however, different results are obtained when applied to the same data set.

Figure 2 shows a scheme presented by Carvalho [2] to describe the methodology of analysis $\hat{a}$ versus a. Observe that a defect sizing a in a welded test piece cause a signal with magnitude $\hat{a}$ on the ultrasonic apparel during examination.

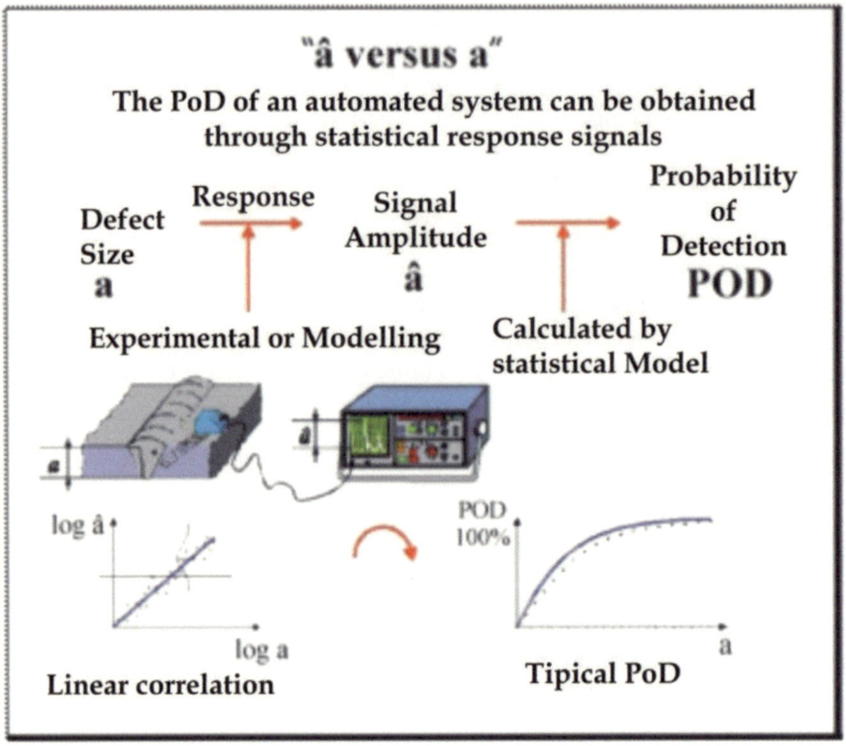

Fig. 2. Scheme of method $\hat{a}$ versus $\boldsymbol{a}$ to implement PoD curves [2].

An inspection procedure may be prepared with two purposes:

1. Detecting defects with any dimension, or detecting defects within specific dimension, or even detecting a specific type of defect;
2. Ratify the inspected part is free of defect, or if the inspected part is free of defects larger than specific dimension, or even if the part is free of specific type of defect.

A practical procedure to prepare PoD curves, from aerospace industry, may be summarized as follows:

1. Fabrication of test pieces containing high amount and various types of defects;
2. Proceed inspection of test pieces using proper technique;
3. Record the results as function of defect dimensions;
4. Plot PoD curve as function of defect dimension.

Nevertheless, prior fabrication of test pieces, it is necessary to have the answer to the questions: which defect dimension will be used, length, width or depth? What is the range of defect dimension will be investigated, 1 to 9 millimeters for example? How many intervals are necessary within the range of dimension? [5].

To stipulate the number of test pieces, two important issues must be considered. First, the amount of test pieces shall be great enough to estimate PoD curve and the limit of confidence interval. Second, the sample space shall be great enough to determinate the statistic parameters of PoD curve that provide better data adjustment.

2.1.1 Statistic model for hit/miss

For analysis of *Hit/Miss* cases, various statistic distributions have been proposed. Distribution *log-logistics* or *log-probability* was found to be more suitable and function PoD(a) may be written as follows [5]:

$$PoD = \frac{e^{\frac{\pi}{\sqrt{3}}\left(\frac{\ln a - \mu}{\sigma}\right)}}{1 + e^{\frac{\pi}{\sqrt{3}}\left(\frac{\ln a - \mu}{\sigma}\right)}} \tag{2}$$

Where a is a defect dimension, and μ e σ are average and standard deviation, respectively [5]. Equation 2 can be written as follows:

$$PoD = \frac{e^{(\alpha+\beta \ln a)}}{1 + e^{(\alpha+ \beta \ln a)}} \tag{3}$$

It is simple to reach the equation 4:

$$\ln\left(\frac{PoD(a)}{1-PoD(a)}\right) = \alpha + \beta \ln a \tag{4}$$

Where $\mu = -\frac{\alpha}{\beta}$ and $\sigma = \frac{\pi}{\beta\sqrt{3}}$.

Thus

$$ln(probability) \propto \ln\,(a). \tag{5}$$

2.1.2 Statistic model for data of response signal

Concerning to response signal of the inspection technique, it is considered a linear relation between $\ln \hat{a}$ and $\ln a$, where a is the dimension established of the defect [5]. This relation may be represented by equation 6:

$$ln\,(\hat{a}) = \alpha_1 + \beta_1 \ln(a) + \gamma \tag{6}$$

Where γ is the error with normal distribution, presenting average equal zero and standard deviation constant and equal σ_1. Equation 6 represent normal distribution of $ln\,(\hat{a})$ centered at $\mu\,(a)$ and deviation σ_γ^2, where,

$$\mu\,(a) = \alpha_1 + \beta_1 \ln\,(a) \tag{7}$$

PoD (a) function for the NDT response signal ($ln\,(\hat{a})$) may be presented as follows:

$$PoD\,(a) = Probability(ln\,(\hat{a}) > \ln\,(\hat{a}_{th})) \tag{8}$$

Where $\ln\,(\hat{a}_{th})$ is the limit of defect evaluation [3].

Using statistic pattern simbology, the PoD function for the response signal of NDT may be represented by equation 9:

$$PoD(a) = 1 - F\left[\frac{\ln(\hat{a}_{th}) - (\alpha_1 + \beta_1 \ln(a))}{\sigma_\gamma}\right] \tag{9}$$

Where F is a continuous cumulative distribution function.

Using the symmetric property of normal distribution:

$$PoD\ (a) = F\left[\frac{\ln(a) - \mu}{\sigma}\right] \tag{10}$$

Which is a cumulative log-normal distribution, where $\mu(a) = \frac{\ln \hat{a}_{th} - \alpha_1}{\beta_1}$ and the standard deviation $\sigma = \frac{\sigma_\gamma}{\beta_1}$. The parameters α_1, β_1 e σ_γ are estimated through the maximum verisimilitude method. Such function is often used on analysis Hit/Miss as well [5].

2.1.3 Estimation of PoD curve parameters

To estimate PoD curve parameters using *hit/miss* method, it is recommended that dimension of defects being uniformly distributed from the smallest to largest dimension of interest, containing at least 60 defects. For signal response analysis, it is recommended, at least, 30 defects [5].

2.1.4 Confidence interval of PoD curve

For a hit/miss analysis, a confidence interval of 95% is usually applied, it is necessary a minimum of 29 defects on each dimension range of study, taking into account that the number of discontinuities detected follows a binomial distribution. It can be interpreted as 29 test pieces containing one defect each. Thus, as an example, if an analysis requires 6 ranges of dimensions, it is going to be necessary, at least, 174 test pieces, increasing costs for fabrication of test pieces to estimate PoD curve and confidence intervals correctly [5].

As stated previously, a confidence interval may be calculated, assuming it follows a normal distribution, through the equations 11 and 12.

$$P\left[-z\left(\frac{\alpha}{2}\right) \leq \frac{\bar{x} - \mu}{\sigma/\sqrt{n}} \leq z\left(\frac{\alpha}{2}\right)\right] = 1 - \alpha \tag{11}$$

$$\left[\bar{x} - z\left(\frac{\alpha}{2}\right)\frac{\sigma}{\sqrt{n}},\ \bar{x} + z\left(\frac{\alpha}{2}\right)\frac{\sigma}{\sqrt{n}}\right] \tag{12}$$

Where α is the significance level, μ is is average and σ is standard deviation.

Figure 3 shows a didactic example of 95% confidence interval (α=5%).

2.1.5 General aspects of experimental PoD curves

Experimental PoD curves are plotted when a high volume of inspection data were obtained experimentally. They can be applied in projects that include fabrication of test pieces containing defects with controlled characteristics, such as type, dimensions and location. Another application is to equipments which inspection history is fully recorded from the same reference block containing well known defects.

For fabrication of test pieces, a significative number of artificial defects is necessary to provide a sample space that enable estimation of the curves. To reproduce the field situation as feasible as possible, many inspectors and defects characteristics shall be used.

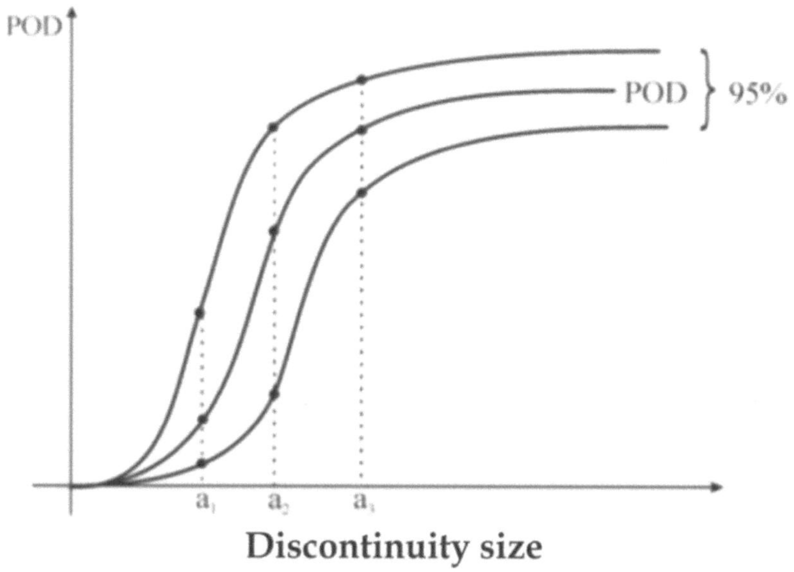

Fig. 3. Example of PoD curve with 95% confidence level [2].

The main advantage, in this case, is obtaining the curves without application of mathematic models. Based only on the detection rates obtained, which result is the closest to the field inspection. On the other hand, the disadvantage is the high number of experimental tests required, what increases the cost of project and may extend it a lot.

2.1.6 General aspects of PoD curves modeled through experimental data

When only a few numbers of experimental data is available, due to insufficient number of test pieces or inspection data, it is possible to plot a PoD curve through a mathematic model. Thus, we can, for example, extrapolate defects out of the dimension scale inspected. The main advantages of this methodology are low cost, easiness and readiness. The disadvantage is that in case of extrapolation of larger defects inspection data to smaller defects, the PoD obtained may be too low, what do not represent real situation. But, this is the most employed method.

2.1.7 Mathematic simulation of PoD curves

Recently, modeling of PoD curves has increased considerably. The low computational cost of simulation, compared to fabrication of test pieces, acquisition of resources for inspection and use of equipments, is driving to amplify the use of this methodology [2, 5]. Furthermore, modeling of a PoD may provide a study of inspection parameters before its execution, and enable an evaluation of False Positive rates. In this chapter, only Monte Carlo Simulation Method will be approached, despite other methodologies to simulate PoD curves are available.

2.1.8 Monte Carlo simulation method

The Method Monte Carlo (MMC) is a statistic method applicable to stochastic simulations, suitable to other areas such as physics, mathematics and biology. MMC has been used a long time in order to obtain numeric approaches of complex functions. This method is typically used to generate observations of any distribution of probabilities and use of sampling to approximate the interest function.

The application of Monte Carlo simulation to estimate PoD of a defect may be obtained through the equation 13.

$$PoD(D) = E[I(x,y)] = \int_0^{da}\int_0^{dt} I(x,y) f_{x,y}(x,y)dxdy \tag{13}$$

Where "D" is the diameter of defect, x and y are random variables associated to the position of the center of circular defect, $f_{x,y}(x,y)$ is the density function of probability for both variables, $E[\_]$ is the expected or average value. In an ultrasonic inspection for example, the elements dt and da are distance between each probe and distance between two data acquisitions, respectively. As the center of defect may be located randomly in a rectangle (inspected area), the function $f_{x,y}(x,y)$ is given by two normal distributions, one for each coordinate of center of defect, as follows:

$$f_{x,y}(x,y) = \frac{1}{da}\frac{1}{dt} \tag{14}$$

$I(x, y)$ is an indicator of inspection function, it assumes value 1 if the defect was detected and 0 if not detected. In case of ultrasonic examinations [2], detection is considered successful when an overlap between defect and ultrasonic beam occur and the amplitude of the echo produced by defect is larger than reference curve. The simulation of test pieces is accomplished by random definition of the center of defect (x, y), according to equation 14. Then, results of inspection of simulated defects, detected or not detected, shall be considered for the study. Equation 13 may be rewritten as follows:

$$PoD(D) = \sum_{i=1}^{N} \frac{I(x_i,y_i)}{N} \tag{15}$$

Where N means the number of simulations (simulated test pieces), it must be great enough to provide statistic reliability of the results. According to literature [6], the error rate of results obtained through equation 15, considering a confidence level 95%, is given by equation 16.

$$error = 200\sqrt{\frac{1-PoD}{N.PoD}}(\%) \tag{16}$$

From equation 16, it is possible to determinate the number of simulations necessary to reach the error level wished. Details of Method Monte Carlo are provided by Carvalho [2] and Ang [6].

3. ROC curves (Receiver ou Relative Operating Characteristic)

The ROC curves are well known in theory of signal detection and accessed on technical referenced of pattern recognition [7-10]. These curves are result of relation between number of false positives (FP), abscissas axis, and number of true positives (TP), ordinates axis. Alike PoD, reliability is given by area under the curve. Reliability of technique is better as much as higher values of TP and lower values of FP. Ideal reliability is encompassed in a 100% of a square area, according to didactic example of figure 4.

The probability of detection, or in other word, the probability of True Positive is:

$$PoD = P(TP) = \frac{TP}{TP+FN} \tag{17}$$

where *FN* is the value of False Negative.

The probability of False Alarm or False Positive (FP):

$$P(FP) = \frac{FP}{TN+FP} \tag{18}$$

where *TN* is the value of True Negative [7, 8].

The ROC curves have some advantages compared to PoD curves. One of these advantages is the evaluation of false positive index, which are not taking into account when PoD curves are plotted. Certainly, these indexes are very important for nondestructive testing. Just supposing a situation when false positive may imply in an unnecessary emergency shut down of an equipment or operating unit. In the other hand, a worse situation would be a false negative that may start up defective equipment, elevating the risk of a catastrophic occurrence, causing damages to facilities, environment and human deaths.

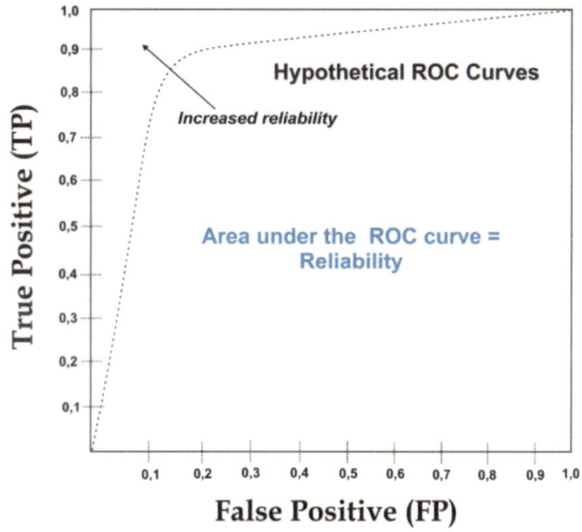

Fig. 4. Example of ROC curve.

4. Bibliographic review

4.1 Experimental PoD

4.1.1 Manufacture of specimens

The manufacture of specimens with artificial defects can be considered an art, as they should be induced in order to represent, in location, size and shape defects that occur in the reality of manufacturing processes and equipment operation. At this point, the main focus is to address some techniques of manufacturing well-done defects in materials in order to produce specimens for estimation of inspection reliability, which can also be used for training and certification of NDT personnel.

a. Fatigue cracks

For metal alloys, fatigue cracks are initiated and grown under controlled conditions with the purpose of construction of PoD curves. Fatigue cracks have particular characteristics, they are economical to produce and constitute a challenge for detection. These cracks can be initiated, for instance, through a notch. The controlled growth of the crack can be accomplished by loading constant at approximately 70% of the yield strength of the material, or by fatigue test, monitoring its growth with methods such as ultrasound by TOFD (*Time of Flight Diffraction*). The notch should be removed from the original specimen before the inspection process to allow a correct measurement of the crack only [11].

b. Welding defects

Silk's book [1], in chapter 3, contains an item devoted to description of the main causes of welding defects, which do is not scope of the proposed work.

Another interesting work is that of Bullough et al [12] describes a model for estimating the distribution of defects in submerged arc welding in equipment of nuclear industry.Potential defects are mainly: lack of fusion, solidification cracking in weld metal hydrogen cracking in the HAZ and weld metal. It also estimated the probability of the presence of defects in weld inspection after manufacturing by calculating the probability of defect formation versus the probability of not detecting [12].

The table 1 below contains some recommendations to produce controlled defects in welded plates resulting from the experience of welders of the SENAI RJ Technology Welding Center.

Type of Weld Defect	Recommendations
Lack of root fusion	Welding with lower amperage/Welding only one side of root face.
Lack of fusion in the wall groove	Place a piece of graphite on the wall and make the filling.
Lack of Penetration	Use a rod thicker than the root gap/ Put a piece of material (carbon steel) at the root gap.
Excessive Penetration	Welding with higher amperage.
Surface crack	Add copper, aluminum or cobalt before welding the face reinforcement.
Internal Crack	During the filling step of the weld beam, create a notch through a cutting disc or saw blade thinness. Afterwards, finish the face reinforcement.
Crack in the Root	The same procedure for internal crack, but carried out in the root step.
Porosity	Lower gas flow (Ar for GTAW). For a 7 mm diameter nozzle, the flow of gas is recommended 8 l/minute. You can use a flow 3l/minute to generate porosity. To generate porous in Shielded Metal Arc Weld, the best practice is to weld in direct polarity.
Face Undercutting	Apply high amperage / increase the speed of welding. For GTAW and SMAW, weld with angle ≤15° or ≥ 30°.

Table 1. General Recommendations to produce typical welding defects.

4.1.2 Estimation of experimental PoD

One of the first projects of reliability of NDT method was the Program for Inspection of Steel Components (PISC) in mid-70s, which has been initiated in order to assess the ability of defect detection by method of ultrasound in the walls of pressure vessels of up to 250mm in the nuclear industry [13-15]. Several ultrasound procedures existing at that time were strictly applied to the results of inspections, which resulted in PoD with low values [13-15]. However, some inspectors could also use the procedures they wanted, thus achieving much better results in terms of detection for the same defects analyzed. In the PISC II and III programs, the project drew on more flexible procedures. The results showed that characteristics of defects such as shape and geometry are more relevant to the POD when compared to other physical parameters. They also concluded that there were some mistakes in the ASME code, however, the most relevant contribution was made to a detailed evaluation of NDT techniques for detecting and sizing of defects [13-15].

Another important program was titled NORDTEST, which was developed in Scandinavia by the Netherlands Institute of Welding (NIL), the ICON (Inter Calibration of Offshore Nondestructive Examination) and TIP (Topside Inspection Project). The main object of this project was to compare the manual method of ultrasonic with the method applied to X-ray inspection of welded plates carbon-manganese steel with thickness smaller than 25 mm. The results also were used to establish acceptance curves (curves (1-POD) versus height of the defect) [13]. This project also compared the technique of inspection by manual ultrasonic with the automated inspection assisted by processing techniques (such as focusing system), certifying that the computerized inspection result in a PoD significantly higher than the manual inspection [16].

The UCL (University College London) conducted a project in the offshore area for the preparation of PoD from fatigue cracks in tubular joints in the mid 90s. The aim was to compare the probability of defect detection of these cracks by the method of magnetic particles with the method of eddy currents, as well as the method of using ultrasonic Creeping waves. Which has reached PoD between 90 and 95% for cracks larger than 100mm[5, 17]?

In the 90's, the Netherlands Institute of Welding (NIL) has issued a report with the results of a project to study the reliability of the method of mechanized ultrasonic, among other methods, to detect defects in welded plate of 6 to 15mm thick. The results proved that the mechanized method and TOFD (Time of Flight Diffraction) technique have probability of detection much higher than the manual method (60-80% of PoD compared to about 50% of the manual). The mechanized method is also more effective in sizing of defects [2, 18].

Carvalho [2] employed the method of ultrasonic pulse-echo both in manual and in automated form, as well as the automatic method of TOFD. More information about inspection procedures can be obtained in [2]. The inspection by the pulse-echo manual technique was carried out by five (5) inspectors duly certified by ABENDI - Brazilian Association of Nondestructive Tests and Inspection, recognized by SNQC - National System of Qualification and Certification, in accordance with ISO 9712. Thus, the PoD curves were constructed from an average of 75 defects (samples), since each length was repeated 15 times. Each set was replicated by this bootstrap technique [19] in 1500 a new set containing 75 samples. The average probability of detection of each length was estimated for each of these sets. The 1500 PoD values were arranged in ascending order by choosing a 95% confidence interval [2].

By figures 5(a) and (b), it is possible to certify that it reaches close to 100% detection for defects larger than 20 mm for both LP (Lack of Penetration) and LF (Lack of Fusion) classes. Figure 5 (c) shows that for defects in lengths of about 12 mm, the class LP has a higher value of PoD, whilst that from it value, the opposite happens. The integrated curve shows that the class LP has higher PoD (77%) than LF (63%). Carvalho [2] concluded that it must be explained by the fact that the defect LP is usually located in the root of weld and can therefore be detected by both sides of the beam. As for high dimensional defects, LP can be confused with the background echo, which does not happen with the LF, this may justify the higher PoD of LF from a given value of length.

Fauske et al [20] and Verkooijem et al [21] also concluded in their work that automatic inspection allows PoD values higher than manual inspection by ultrasonic. The first reached the value of 80% PoD for automatic detection of cracks of 10mm with a depth of 1 mm, while the manual inspection resulted in only 60%. Verkooijem [21], who worked with the classes LP, inclusion of slag and porosity, reached 83.6% probability of detection with automated pulse-echo ultrasound, 52.3% for manual technique and 82.4% for TOFD technique.

Carvalho [2] also concluded that automatic inspection provide PoD values much higher than manual inspection by ultrasonic. The graphs in Figure 5, which also include PoD curves for each inspector used in the tests show that the PoD of automated systems (pulse-echo and TOFD) is a typical case of ideal PoD, where there is a critical size defect below which there is no detection. Discussions on the performance of the inspectors can be found in [2].

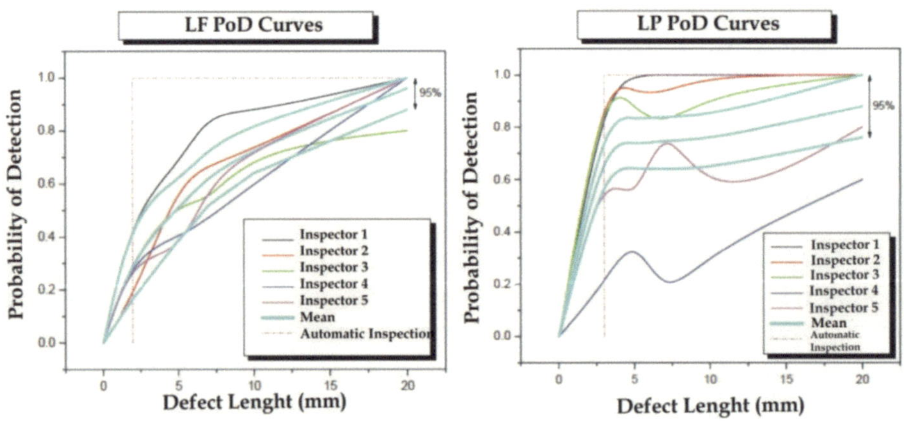

(a) (b)

Fig. 5. PoD curves for defects LF and LP, respectively [2].

Figure 6 below shows the results of Carvalho [2] for what is called PoS (Probability of Sizing), which is a graph where the x-axis represents the expected size (projected) of the defect, and the y-axis represents the size found by the inspector. Thus, a point located at y = x, if the scales are the same in Cartesian axes, means accuracy in sizing of the defect (discontinuity). By the result obtained, it became evident that there was overestimation in most of the results. It is relevant to emphasize that this overestimation was very significant, because defects with 3, 5, 7 and 10mm have been scaled up to 20mm. Carvalho [2] discusses the cost-benefit of this behavior, from the point of view of the operational safety it is good, on the other side it can cause an unnecessary shut down of equipment operation, which will result in interrupted profit [2].

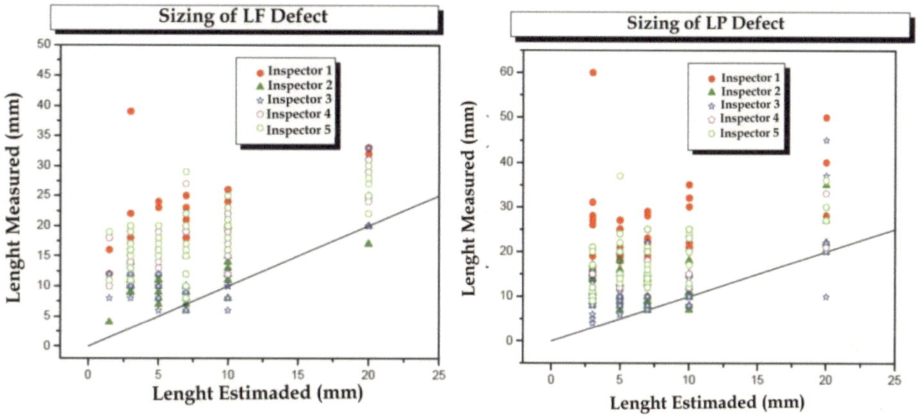

(a) (b)

Fig. 6. Probability of sizing for the LF and LP defect classes [2].

4.2 Simulation of PoD curves

The sonic intensity of a divergent ultrasonic beam decreases in relation to the center. Carvalho [2] emphasizes that the sonic distribution of beam divergence follows the equation 19, which actually describes a bell-shaped function, as illustrated in Figure 7.

$$s(x,y) = S_0.exp\left(-a\left(\sqrt{x^2+y^2}\right)^b\right) \tag{19}$$

S_0 is the intensity at the center of the sonic beam and "a" and "b" constants determined by data supplied by the manufacturers of transducers [2].

In addition to the decrease in intensity due to the sonic beam divergence, there is the attenuation caused by the absorption and dispersion of the wave, thus, considering a distance "d" of the transducer, the equation that models the distribution is:

$$S_d(x,y,d) = e^{-\alpha.d}.S_0.k(d).exp\left(-a\left(\sqrt{x^2+y^2}\right)^b\right) \tag{20}$$

where,

$s_d\,(x,y,d)$ = sound intensity encountering a point (x,y) at a distance d from the transducer;
α = material attenuation coefficient;
d = distance between the transducer and the point of interest (see Figure 7);
$k\ (d)$ = factor to maintain total sound intensity of an ideal material (α = 0) constant at any depth considering the variation of the ultrasonic beam aperture.

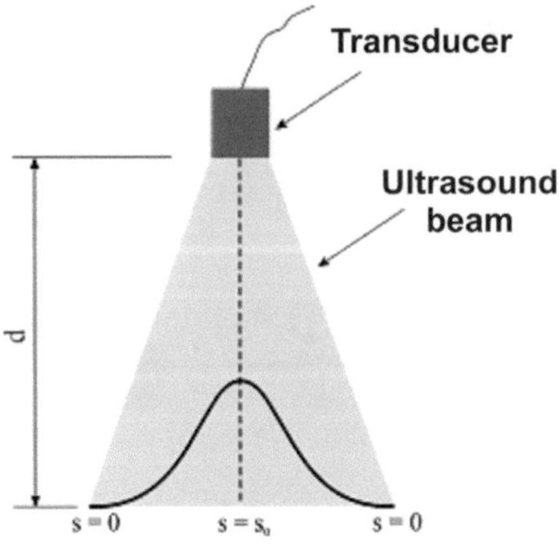

Fig. 7. Sonic intensity distribution as a function of beam divergence [2].

In conventional ultrasonic method, the reference curves are constructed with standard blocks containing defects of known dimensions to determine whether there is a defect in the material examined. It is also possible to estimate this reference curve numerically by using equation 21 below, which is the integral of the equation 20 over the area of the defect.

$$S_i = \iint_{\sqrt{x^2+y^2}\leq\frac{\phi c}{2}} e^{-\alpha.d}.S_0.k(d_i).\exp\left(-a\left(\sqrt{x^2+y^2}\right)^b\right) \tag{21}$$

In relation to a situation of inspection, as shown in Figure 8, there is an incidence onto a defect ϕ diameter at a distance d from the transducer, the total intensity sonic inspection, depending on the position of the sonic beam in relation to the defect, is obtained by equation 22, which is compared with equation 21 to determine whether or not detection.

$$S_\phi = \iint_A e^{-\alpha.d}.S_0.k(d_i).\exp\left(-a\left(\sqrt{x^2+y^2}\right)^b\right) \tag{22}$$

Where A is the area of the defect reached by the ultrasound beam (Figure 8).

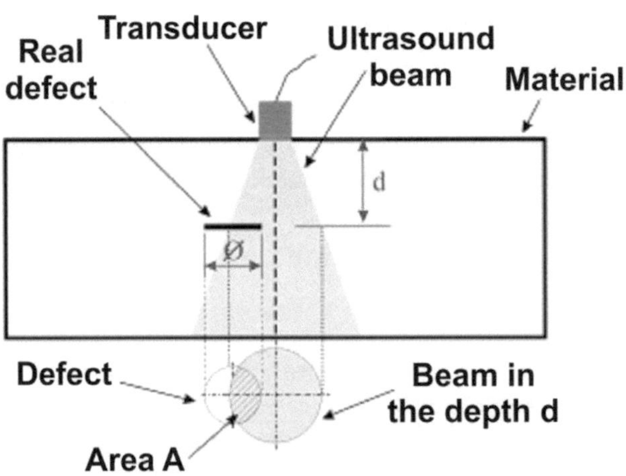

Fig. 8. Illustration showing an ultrasound beam focusing on a defect of area A.

After the normalization process in relation to the pattern flaw ϕ c at d position, the R valued is:

$$R = \frac{\iint_A exp\left(-a\left(\sqrt{x^2+y^2}\right)^b\right)dxdy}{\iint_{\sqrt{x^2+y^2}\leq\frac{\emptyset c}{2}} exp\left(-a\left(\sqrt{x^2+b^2}\right)^b\right)dxdy} \tag{23}$$

Then, if $R \geq R_{curve}$ the flaw is detected, where R_{curve} is the value of the employed reference curve, that is, $R_{curve} = 1$ for the 100% reference curve, and so on. After the simplification R becomes independent of s_0 and $k(d)$ [2].

As a practical result, Carvalho [2] obtained the curves shown in Figure 9. Figure 9 (a) represents the result of PoD for a flat plate considering various distances between transducers

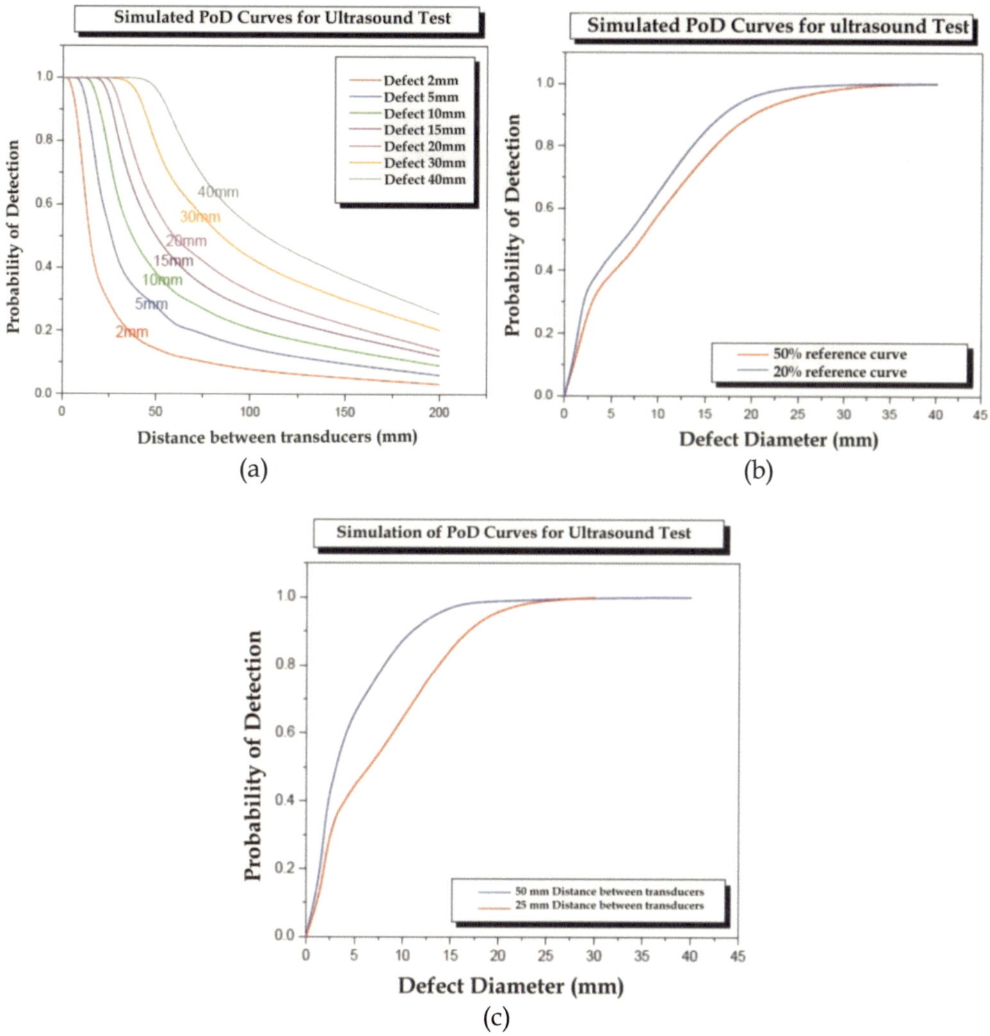

Fig. 9. PoD curves obtained by numerical simulation [2].

of different diameters and defects. The results show that as greater the distance between the transducers, the PoD is lower. In contrast, as larger the diameter of the defect, PoD is higher. The reference curve used was 50%. Figure 9 (b) presents the results obtained for evaluation by considering two conditions of reference curve: 20% and 50%. The PoD curve obtained for 20% was higher than 50%, what was expected since it is a condition of more conservative detection. Since the figure 9 (c) represents the results obtained for variation of the distance from the transducer to the defect: 25mm and 50mm. The distance of 50mm resulted in higher PoD, according to Carvalho [2], due to greater divergence of the beam at this distance. However, Carvalho [2] points out that this simulation was not considered the fact that as greater the distance, greater the attenuation that will be the sonic intensity. Another important aspect is that the practice of ultrasonic inspection by the inspector is based on amplitude curves (measured in dB) for the detection of defects.

4.3 ROC curves

There are few works of development of ROC curves on nondestructive testing. One of the most important work is Fücsök's [7, 8] for the construction of ROC curves of radiographic films. This study aimed to estimate the reliability of human factors in the evaluation of radiographs, using the RRT method (Round-Robin Test) with inspectors of laboratories in Croatia, Hungary and Poland. A set of films was selected at BAM (Federal Institute of Materials Berlin) who provided technical and scientific support to the implementation of the project. Thirty eight radiographic films were used, containing 206 weld defects of different types and sizes. These films were scanned with the scanner LS85 SDR Lumisys. The digitized radiographic images were used as the basis for preparation of templates of the reports. With the digital images, series of identical films were printed in a laser printer AGFA Scopix so that each inspector could review the same films. The inspectors received instructions, a procedure of reports and special forms to fill in the result. The identity of each inspector was preserved through identification codes. At the end, a total of 60 inspectors from different laboratories were used. The best result of detection was 88.3% (probability of true positive) and 12.9% probability of false positive, when all defects have been considered (even below the lower acceptance criterion). In an assessment that considered only major defects that the level 1 of acceptance of EN 12517:1998, the PoD was 85.2%, and only 2.1% for false alarms (positive). Fücsök [7] concluded that the results in general were not good, with high values of false alarms, indicating a need for better training and qualification of inspectors.

4.4 Other references

In addition to the publications already mentioned, it is relevant highlight some other very interesting, as the publication of the Department of Defense [23] which is a complete theoretical approach of the reliability of NDT, including chapters on method and not just PoD curves, but also the construction of ROC curves [23].

Another interesting publication is the Meeker [24] that presents a "step by step" how to implement a POD curve by Hit / Miss and $\hat{a}$ x a methods, which is a practical and fast tutorial. In addition to the MATLAB tools of programming (Mathworks), there is the STATUS program (Statistical Analysis Tools for Ultrasonics Test System) to prepare PoD curves by the two methods of analyzing, as well as analysis of PoS (Probability of Sizing). Another program is the "mh1823" that also allows the construction of curves and performing reliability analysis (freely available).

There are several other publications which are listed below, which although not directly addressed in this work, especially not to make the technical discussion of the subject too repetitive, since many are similar to the methodology presented by the work described, they can be analyzed because present other applications of experimental situations [25-45].

4.5 General discussion

In general, the human factor is usually the cause of most failures to detect discontinuities, and that even well qualified and experienced inspectors make mistakes. Thus, automatic inspection methods allow significantly higher probability of detection as the ultrasonic method. However, it must be noted that sometimes the automatic inspection is still economically impracticable. Another important issue that a PoD analysis can support a choice when there are two possible methods of inspection of a given equipment, for example, if a method "A" cost 50% of a method "B" and if your PoD is 30% of "B" PoD , your actual cost is greater than "B".

The experimental methods require high costs to be carried out, as well as time, as they demand the production an enough quantity of specimens to allow the correct estimation of relevant statistical parameters of the reliability curves and their confidence intervals. Another important point here is that the preparation of specimens with defects properly controlled, as in the case of welding defects, is an art a bit rare to find skilled professionals to do so. The defects made should reproduce the maximum of reality in terms of size, morphology and location in the weld. These methods employ mathematical tools for estimation of curves and log-normal function is most often employed for this purpose, although other functions can also be used. Considering the lack of samples in many studies, some researchers also resort to the techniques of simulation data, as the bootstrap technique to generate new data sets and produce the needed confidence intervals.

There is now a series of works focusing on numerical simulation of reliability curves as a solution to the disadvantages presented by the method of preparation of specimens. Nevertheless, despite the cost advantages and speed, it is extremely difficult to reproduce in a simulation all variables that are embedded in a real inspection and which are not easily quantified, such as those related to the human factor. In Mores' study [46], there are descriptions of 39 variables that are relevant to an inspection by ultrasonic. Considering advances in tools and simulation programs, this methodology should gain more space compared to the experimental methods.

The probability of detection of defects has been used to meet the requirements for equipment design, and has worked as a tool for comparing the performance capabilities of various NDT methods, procedures and guidance of qualified professionals.

5. Conclusions

The reliability of nondestructive testing has been the subject of important research projects over the past 40 years, many of them highlighted in this work, which can reach the main conclusions:

Most of the works present experimental methodologies of specimens manufacture. In this case, one should always look for the minimum number of defects generated in order to get the better estimation of PoD curves and their confidence intervals.

- The preparation of samples requires extreme skill, in particular, the welded specimens must contain defects that are the most realistic possible with ideal dimensions to regulatory acceptance criteria used.
- Some functions are usually employed to model mathematically the PoD curve obtained with some experimental data, and the Log-Normal function is the normally chosen by the researchers.
- Due to the disadvantages of experimental methods related to the time, cost and logistics, the simulation of PoD and ROC curves can be a solution, having already carried out several experiments in this direction. However, it is necessary to emphasize that is very difficult to reproduce in simulation, either physical or numerical (simulations of ultrasonic or radiographic tests, for example), several factors that occur in a real inspection, especially those related to the human factors, which generally make a major influence on PoD. But with the advances of the simulators, it is possible that this disadvantage be mitigated in a near future.
- The methodologies for estimating the reliability of NDT are extremely important for modern industry and will increasingly focus investments.

6. References

[1] Silk, M G. The Reliability of Non-destructive Inspection. Assessing the Assessment of Structures under Stresse. Adam Hilger, Bristol. 1987.

[2] Carvalho, A. A. Confiabilidade de Técnicas de Ensaios Não Destrutivos na Inspeção de Dutos Utilizados na Indústria do Petróleo. Tese de Doutorado. COPPE. Rio de Janeiro, 2006. (In portuguese).

[3] Carvalho, A. A., Sagrilo, L. V. S., Silva, I. C., et al. The POD Curve for the Detection of Planar Defects Using a Multi-Channel Ultrasonic System. Insight, Vol. 44, N°. 10, p. 689-693, 2002.

[4] Carvalho, A.A. Rebello, J.M.A. Silva, R.R. Sagrilo, L.V.S. Reliability of the Manual and Automatic Ultrasonic Technique in the Detection of Pipe Weld Defects. Insight, vol. 48, n.11, November 2006.

[5] Georgiou, G.A. PoD Curves, their Derivation, Application and Limitations. Vol. 49, n.7. Insight, p. 409-413, 2007.

[6] Ang, A.H.S. Tang, W.H. Proabability Concepts in Engineering Planing and Design – Decision, Risk an Realibility, vol. 3, New York, Editora Jonh Wiley & Sons, 1984.

[7] Fücsök, F. Muller, C. Reliability of Routine Radiographic Film Evaluation – An Extended ROC Study of the Human Factor. 8th ECNDT, Barcelona 2002.

[8] Fücsök, F. Muller, C, Scharmach, M. Measuring of The Reliability of NDE. 8th International Conference of the Slovenian Society for Nondestrutive Testing. Application of Contemporary Nondestrutive Testing in Engineering. September 1-3, Slovenia, pp. 173-180.

[9] Schoefs, F., Clément, A., Nouy, A. Assessment of ROC curves for inspection of random fields. Structural Safety. Vol. 31, p.409-419, 2009.

[10] Nockemann, C. Heidt, H. H. Thomsen, N. Reliability in NDT: ROC Study of Radiographic Weld Inspections. C. Nockemann, H. Heidt and N. Thomsen. NDT&E International, vol. 24, n.5, October 1991.
[11] http://www.asnt.org/publications/materialseval/basics/jan98basics/jan98basics.htm (acesso em 10/05/2011).
[12] Bullough, R., Dolby, R.E., Beardsmore, D.W., Burdekin, F.M., Schneider, C.R.A. The Probability of Formation and Detection of Large Flaws in Welds. International Journal of Pressure Vessels and Piping. Vol. 84, p.730-738.
[13] Lamaitre, P., Koblé, T.D. Summary of the PISC Round Robin Results on Wrought and Cast Austenitic Steel Weldments, Part I: Wrought to Wrought Capability Study. International Journal of Pressure Vessels & Piping. Vol. 69, pp. 5-19, 1996.
[14] Lamaitre, P., Koblé, T.D. Summary of the PISC Round Robin Results on Wrought and Cast Austenitic Steel Weldments, Part II: Wrought to Cast Capability Study. International Journal of Pressure Vessels & Piping. Vol. 69, pp. 21-32, 1996.
[15] Lamaitre, P., Koblé, T.D. Summary of the PISC Round Robin Results on Wrought and Cast Austenitic Steel Weldments, Part III: Cast to Castt Capability Study. International Journal of Pressure Vessels & Piping. Vol. 69, pp. 33-44, 1996.
[16] PoD/PoS Curves for Nondestrutive Examination - Offshore Technology Report, 2000/018. Prepared by Visser Consultancy Limited for the Health and Safety Executive (HSE).
[17] NDT of Thin Plates - Evaluation of Results. Nil Report. NDP 93-38 Rev.1. (in Dutch), 1995.
[18] Dover, W.D., Rudlin, J.R. Results of Probability of Detection Trials. Proc. IOCE 91, Aberdeen, 13-16 October 1992.
[19] Efron, B., Tibshirani, R. J., "AN Introduction to the Bootstrap", 1a Edição,New York, Editora Chapman & Hall/CRC, 1993.
[20] Fauske, T. H., Dalberg, P., Hansen, A. Ultrasonic Crack Detection Trials. 15th WCNDT - World Conference on Nondestructive Testing. Roma, October, 2000.
[21] Verkooijem, J. The Need for Reliable NDT Measurements in Plant Management Systems. 7th ECNDT - European Conference on Nondestructive Testing. Copenhagen, May, 1998.
[22] Bartholo, P.U. Modelagem da Probabilidade de Detecção do Ensaio Ultrassônico e Avaliação da Influência de Inspetores e Tipos de Defeitos. Projeto Final de Curso. Universidade Federal do Rio de Janeiro, RJ, Brasil, 2008. (In portuguese).
[23] MIL-HDBK-1823. Non-Destructive Evaluation System Reliability Assessment, 1999.
[24] Meeker, W. Q. and Escobar, L. A. Statistical Methods for Reliability Data, John Wiley and Sons, New York.
[25] Forsyth, D. S., Fahr A., "On the Independence of Multiple Inspections and the Resulting Probability of Detection", *Quantitative Nondestructive Evaluation*, Iowa, July, 2000.
[26] Wall, M. Wedgwood, F. A. Burch, S. Modeling of NDT Reliability (POD) and Applying Corrections for Human Factors. *7th* ECNDT European Conference on Nondestructive Testing. Copenhagen, May, 1998.
[27] Forsyth, D. S. Fahr A. Leemans, D. V. et al. Development of POD from In-Service NDI Data. Quantitative Nondestructive Evaluation. Iowa, July, 2000.
[28] Topp, D. A. Dover, W. D. Reliability of Non-Destructive Inspection Test. Análisis de Riesgo y confiabilidad Estructural de Instalaciones Marinas. México, diciembre, 2001.

[29] Olin, B.D, Meeku, Willian Q. Application of Statistical Methods to Nondestructive Evaluation. Technometrics, vol. 38, n.2, 1996, p. 95-112.

[30] Wall, M. Burch, S. Lilley, J. Human Factors in POD Modelling and Use of Trial Data. Insight, Vol. 51, N. 10, October 2009.

[31] Rummel, W D. Recommended practice for a demonstration of non-destructive evaluation (NDE) reliability on aircraft production parts'. Materials Evaluation Vol. 40. August 1982.

[32] Generazio, E.R. Directed Design of Experiments for Validating Probability of Detection Capability of NDE Systems. Review of Quantitative Nondestructive Evaluation. Vol.27, p. 1693-1700, 2008.

[33] Burch, S.F., Stow, B.A., Wall, M. Computer Modelling for the Prediction of the Probability of Detection of Ultrasonic Corrosion Mapping. Vol. 4, n.12, Insight, pp.761-765, December 2005.

[34] Gandossi, L., Simola, K. Derivation and Use of Probability of Detection Curves in the Nuclear Industry. Insight, Vol. 52, n. 12, December, 2010.

[35] Sweeting, Trevor. Statistical Models for Nondestructive Evaluation. International Statistical Review. Vol. 63, n.2, p. 199-214, 1995.

[36] Wall, M., Burch, S.F., Lilley, J. Review of Models and Simulators for NDT. Vol. 51, n.11, Insight, November 2009.

[37] Hong, H.P. Reliability Analysis with Nondestructive Inspection. Structural Safety, Vol. 19. N.14, pp. 383-395, 1997.

[38] Bullough, R., Burdekin, F.M., Chapman, O.V.J., Green, V.R., Lidbury, D.P.G, Pisarski, H., Warnick, R.G., Wintle, J.B. The Probabibility of "Large" Defects in Thick-Section Butt Welds in Nuclear Components. International Journal of Pressure Vessels and Piping, Vol. 78, p. 553-565, 2001.

[39] Kazantsev, I.G., Lemakien, I., Salov, G.I., Denys, R. Statistical Detection of Defects in Radiographic Images in Nondestructive Testing. Signal Processing. Vol. 82, p. 791-801, 2002.

[40] Wall, M., Burch, S.F., Lilley, J. Simulators for NDT Reliability (PoD). Insight, Vol. 51, p. 612-619, 2009.

[41] Wall, M.,Wedgwood, F.A., Burch, S. Modeling of NDT Reliability (PoD) and Applying Correction for Human Factors. 7th European Conference on Nondestructive Testing. Copenhagen, 1998.

[42] Forsyth, D.S., A., Leemans, D.V., et al. Development of PoD from In-service NDI Data. Quantitative Nondestructive Evaluation, EUA, 2000.

[43] Topp, D.A., Dover, W.D. Reliability of Nondestructive Inspection Test. Análisis de Riesgo y Confiabilidad Estructural de Instalaciones Marinas, México, 2001.

[44] Simola, K., Pulkkinen, U. Models for Nondestructive Inspection Data. Reliability Engineering and System Safety. Vol. 60, p. 1-12, 1998.

[45] Serabian, S. Ultrasonic Probability of Detection of Subsurface Flaws. Materials Evaluation. Vol. 40, p. 294-298, 1982.

[46] Moré, J.D. Aplicação de Lógica Fuzzy para Avaliação da Confiabilidade Humana nos Ensaios Não Destrutivos Tipo Ultrassom. Tese de Doutorado. COPPE/UFRJ, Rio de Janeiro, 2004.

Part 2

Innovative Nondestructive Testing Systems and Applications

2

SQUID Based Nondestructive Evaluation

Nagendran Ramasamy and Madhukar Janawadkar
Indira Gandhi Centre for Atomic Research, Kalpakkam
India

1. Introduction

Nondestructive testing (NDT) or Nondestructive evaluation (NDE) refers to techniques, which are used to detect, locate and assess defects or flaws in materials or structures or fabricated components without affecting in any way their continued usefulness or serviceability. The defects may either be intrinsically present as a result of manufacturing process or may result from stress, corrosion etc. to which a material or a component may be subjected during actual use. It is evident that techniques to detect critical flaws before they have grown unacceptably large are of vital importance in the industry for in-service inspection, quality control and failure analysis. There are several NDE techniques of which one of the widely used techniques is based on eddy currents. However, the conventional eddy current technique has the drawback that it can detect flaws upto a certain depth under the surface of the conducting specimen under investigation and is not suitable for locating deep subsurface defects. Such limitations can often be overcome with the use of high sensitivity SQUID sensor. Nondestructive evaluation of materials and structures using low temperature SQUID (LTS) as well as high temperature SQUID (HTS) has been proposed and the potential of the technique has been demonstrated during the last two decades (H. Weinstock, 1991 & G.B. Donaldson et al, 1996). The SQUID based NDE offers many advantages such as high sensitivity (~ 10 to 100 $fT/\sqrt{Hz}$), wide bandwidth (from dc to 10 kHz), broad dynamic range (>100 dB) and its intrinsically quantitative nature; disadvantage of this technique is that the SQUID sensor operates only at cryogenic temperatures which makes it relatively expensive. However, despite the expensive cryogen and associated inconvenience in handling, SQUID sensors find a niche in areas where other NDE sensors fail to achieve the required performance (H.-J. Krause & M.V.Kreutzbruck, 2002).

The SQUID based NDE systems have been developed and utilized in several areas of application. At university of Strathclyde, system based on SQUID sensor has been used for the detection of flaws in steel plates (R.J.P. Bain et al., 1985, 1987; S. Evanson et al., 1989). Weinstock and Nisenoff were the first to demonstrate the possible use of SQUID sensors for the study of stress - strain behaviour in a ferromagnetic material (H. Weinstock & M. Nisenoff, 1985, 1986). The SQUID sensors have been used for the detection of tendon rupture in pre-stressed steel tendons of concrete bridges through magnetic flux leakage method (G. Sawade et al., 1995; J. Krieger et al., 1999). Marco Lang et al evaluated the fatigue damage of austenitic steel by characterizing the formation of martensite due to quasi-static and cyclic loading with the help of the SQUID based measuring instrument (M. Lang et al., 2000). SQUID sensors have been successfully used for the detection of ferrous inclusions in aircraft turbine discs by Tavrin et al. (Y.Tavrin et al., 1999). The presence of such inclusions may initiate cracks in these critical parts and may eventually lead to failure. The magnetic inclusions in the nonmagnetic alloy of the turbine discs have been investigated by pre-magnetizing the turbine discs and probing their remanent field using the SQUID sensor. Since the sensitivity of the SQUID is very high (under 5 fT/√Hz) and remains constant down to frequencies as low as 1 Hz, these sensors are widely utilized for the detection of deep sub-surface flaws in conducting materials through low frequency eddy current excitation to take advantage of increased skin depth at low frequencies. The SQUID based eddy current NDE technique plays a major role in detecting deep subsurface defects embedded in thick multilayered aluminum structures used in aircraft lap joints.

In this chapter we describe the working principle of the DC SQUID sensor, schemes generally used to couple an external signal to the SQUID sensor, construction of a SQUID-based NDE system by using a low temperature DC SQUID sensor and its associated readout electronics developed in our laboratory. The SQUID system in our laboratory has been used for the measurement of deep subsurface flaws by inducing eddy currents in conducting materials at relatively low frequencies. Detailed experimental studies have been carried out for the determination of optimum eddy current excitation frequencies for the flaws located at different depths below the top surface of an aluminum plate. This system has also been used for the measurement of extremely low content of magnetic δ - ferrite in the 316L(N) stainless steel weldment specimens subjected to high temperature low cycle fatigue (LCF).

2. SQUID sensor

The SQUID (Superconducting Quantum Interference Device) is an extremely sensitive sensor for magnetic flux and its output voltage is a periodic function of applied magnetic flux with the periodicity of one flux quantum Φ_0 (= $h/2e$ = 2.07 x 10^{-15} Wb). To utilize the SQUID sensors for real applications, Flux Locked Loop (FLL) readout electronics has been developed in our laboratory to linearise the periodic output voltage of the SQUID. By using SQUID sensor and its associated readout electronics, it is possible to detect a change in applied magnetic flux whose magnitude is much less than one flux quantum. SQUID sensor can measure any physical quantity that can be converted into magnetic flux, and has been used, for example, for the measurement of magnetic field, magnetic field gradient, magnetic susceptibility, electric current, voltage, pressure, mechanical displacement etc. with an unprecedented sensitivity. The systems based on SQUID sensors offer a wide bandwidth (from near dc to hundreds of kHz), wide dynamic range (>100dB) and an intrinsically quantitative response. The unprecedented sensitivity of the SQUID sensor together with the use of superconducting pickup loops (used as input circuits) enables one to realize practical measuring instruments for the measurement of extremely weak magnetic signals with a high sensitivity.

2.1 Principle of operation

The SQUID is basically a superconducting sensor, which operates below the superconducting transition temperature (T_c) of the superconducting materials used for the fabrication of the device. The basic phenomena governing the operation of SQUID devices are flux quantization in superconducting loops and the Josephson effect (fig.1.). While detailed descriptions are available in the literature (J. Clarke, 1993; H. Koch, 1989), a brief description of the working principle of the SQUID sensor is included here to make this chapter self-contained.

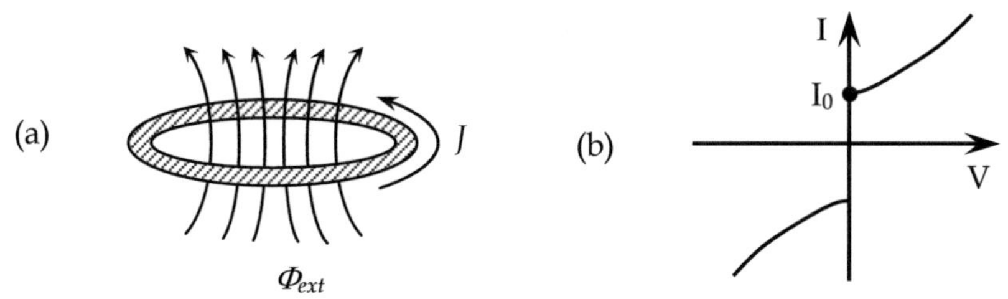

Fig. 1. (a) Flux quantization. (b) Non hysteretic I-V characteristic of a resistively shunted Josephson junction

Flux quantization refers to the fact that the total flux linked with a superconducting loop is always constrained to be an integral multiple of a flux quantum (Φ_0).

$$\Phi_{tot} = \Phi_{ext} + LJ = n\Phi_0 \tag{1}$$

where Φ_{ext} is the externally applied magnetic flux, L is the self inductance of the superconducting loop, J is the screening current induced in the superconducting loop because of the application of the external magnetic flux and n is an integer. The Josephson effect refers to the ability of two weakly coupled superconductors to sustain at zero voltage a supercurrent associated with the transport of Cooper pairs, whose magnitude depends on the phase difference between the two superconductors.

$$I = I_0 \sin \delta\varphi \tag{2}$$

where I_0 is the maximum current the junction can sustain without developing any voltage and is known as critical current of the Josephson junction and $\delta\varphi$ is the phase difference between the two weakly coupled superconductors. When two superconductors are separated by a very thin oxide barrier (tunnel junction), the establishment of tunneling assisted phase coherence leads to Josephson effect; I-V characteristic of such a tunnel junction shows hysteretic behaviour due to non-negligible value of junction capacitance. This hysteresis is, however, undesirable and can be eliminated by shunting the junction with appropriate on-chip thin film resistor to provide sufficient damping of the phase dynamics. A DC SQUID consists of two such non-hysteretic Josephson junctions connected by a superconducting loop.

To describe the operation of the dc SQUID, we assume that the bias current I_b is swept from zero to a value above the critical current ($2I_0$) of the two junctions. An external magnetic flux varying slowly in time is applied perpendicular to the plane of the loop. When the external applied magnetic flux is zero (or $n\Phi_0$, n is an integer), there is no screening current induced in the superconducting loop and the bias current I_b simply divides equally between the two junctions assuming the SQUID to be symmetric. When external magnetic flux Φ_{ext} is applied, the requirement of flux quantization generates a screening current $J = -(\Phi_{ext} - n\Phi_0)/L_s$, where L_s is the inductance of the SQUID loop and n is an integer which makes the value of $n\Phi_0$ to be nearest to the applied flux Φ_{ext}. The screening current induced in the SQUID loop adds to the bias current flowing through junction 1 and subtracts from that flowing through junction 2. When the junction 1 reaches its critical current $I_0 = I_b/2+J$, the current flowing in junction 2 is $I_b/2-J$ (that is, I_0 -2J) and the total current flowing in the SQUID is $2I_0$-2J. At this point the SQUID switches to the non-zero voltage state. When the applied flux is increased to $\Phi_0/2$, the screening current J reaches a value of $\Phi_0/2L_s$ and the critical current falls from $2I_0$ to $2I_0$ – (Φ_0/L_s) as shown in fig.2 (b). When the flux Φ_{ext} is increased further the SQUID makes a transition from the flux state n = 0 to n = 1, J changes its sign and reaches zero again when Φ_{ext} becomes equal to Φ_0. At this point, the critical current of the SQUID is restored to its maximum value of $2I_0$. In this way the critical current oscillates as a function of Φ_{ext}. If we bias the SQUID with a dc current greater than the critical current of the two Josephson junctions, the voltage developed across the SQUID oscillates with a period of Φ_0 when the input magnetic flux steadily increases. Thus, the SQUID produces output voltage in response to a small input flux $\delta\Phi$ (<< Φ_0), and is effectively a flux to voltage transducer. The voltage swing δV produced at the output of the SQUID when the flux changes from $n\Phi_0$ to $(n + 1/2)\Phi_0$ is known as the modulation depth of the SQUID. The modulation depth of a typical low T_c DC SQUID based on Nb/AlO_x/Nb Josephson junctions is ~ 20 to 30 μV. The modulation depth δV is maximum for bias currents a little above the maximum critical current of the SQUID, and during operation the SQUID is tuned to the bias current at which the modulation depth is a maximum.

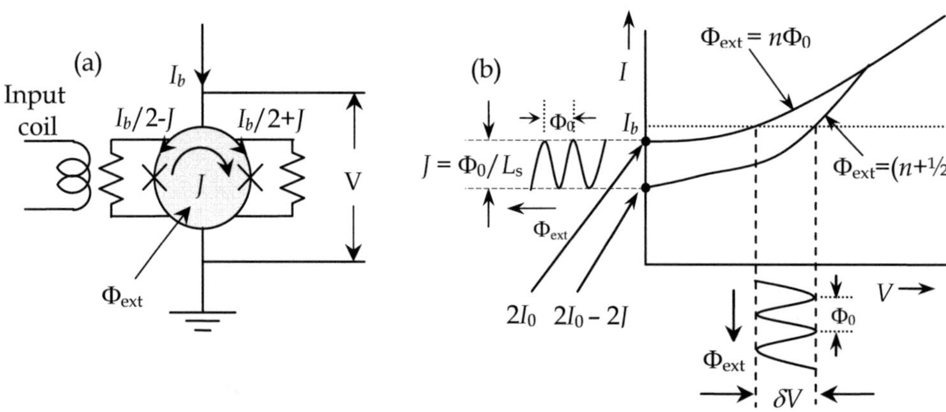

Fig. 2. (a) dc SQUID (b) I-V characteristic of the dc SQUID with the application of input magnetic flux.

3. Flux Locked Loop (FLL) readout electronics

The periodic output voltage of the SQUID allows the SQUID to be operated in a small signal mode around the optimum working point but the linearity of response is limited to flux range much less than $\Phi_0/2$. The small signal readout can only be used when the amplitude of variation of magnetic flux signal is limited to the linear range around the working point (<$\Phi_0/4$). However, in most of the applications, the signal flux, which is required to be measured, varies from a fraction of a flux quantum to several hundreds of flux quanta.

Therefore, the measurement system based on the SQUID sensor should be designed to provide a wide dynamic range as required in any application. In order to linearise the periodic output voltage, the SQUID should be operated in a feedback loop as a null detector of magnetic flux; the voltage at the output of the readout electronics will then be proportional to the input signal flux. In order to suppress the $1/f$ noise and dc drifts in the preamplifier the signal of interest is shifted to frequencies well above the threshold of $1/f$ noise by using high frequency flux modulation scheme. The flux modulation scheme is illustrated in fig.3. In this scheme, the signal flux, $\delta\Phi_{sig}$ which is to be measured is modulated by a high frequency carrier flux $\Phi_m(t)$. The sinusoidal modulation flux, $\Phi_m(t)$ of frequency f_m with a peak-to-peak value of nearly $\Phi_0/2$ is applied to the SQUID. When there is no applied signal flux or the applied input flux is $n\Phi_0$, the SQUID produces output voltage with a frequency twice the frequency of the modulation flux $\Phi_m(t)$ and there is no component at the modulation frequency present in this output. When this voltage output is fed to a lock-in detector referenced to f_m, the output of the lock-in detector is zero. On the other hand, when the applied signal flux is $(n+1/4)\ \Phi_0$, the SQUID output voltage has a component at frequency f_m, which is in-phase with the carrier frequency and the output of the lock-in detector is a maximum. Similarly, when the signal flux is $(n-1/4)\ \Phi_0$, the SQUID output voltage has a component at frequency f_m, which is out of phase with the carrier frequency and the output of the lock-in detector has a maximum negative value. Thus, as one increases the flux from $n\Phi_0$ to $(n+1/4)\ \Phi_0$, the output from the lock-in detector referenced to f_m steadily increases from zero to a maximum positive value; if instead the flux is decreased from $n\Phi_0$ to $(n-1/4)\ \Phi_0$, the output from the lock-in detector referenced to f_m decreases from zero to a maximum negative value. The lock-in output is integrated and fed back to the same coil as that used for flux modulation via a feedback resistor. The signal flux $\delta\Phi_{sig}$ applied to the SQUID produces an output of $-\delta\Phi_{sig}$ from the feedback loop to maintain a constant flux in the SQUID, while producing an output voltage across the feedback resistor, which is proportional to $\delta\Phi_{sig}$.

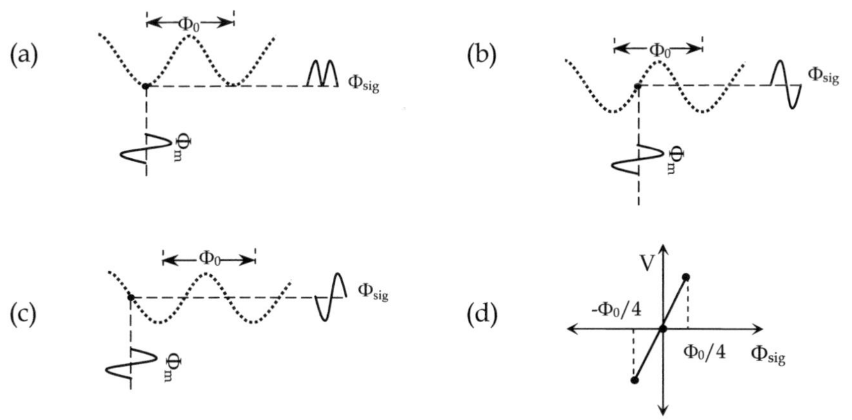

Fig. 3. Schematic representation of flux modulation scheme when the signal flux is at (a) $n\Phi_0$, (b) $(n+1/4)\ \Phi_0$ and (c) $(n-1/4)\ \Phi_0$. (d) Linearized output as the signal flux varies from $(n-1/4)\ \Phi_0$ to $(n+1/4)\ \Phi_0$.

The schematic diagram of the FLL readout electronics is shown in fig.4. In order to achieve a low system noise, it is necessary to match the output impedance of the SQUID device and input impedance of the preamplifier. LT1028 (Linear Technology) was chosen to construct the preamplifier. The spectral density of the voltage noise (e_n) and current noise (i_n) of the preamplifier at a frequency of 100 kHz are specified by the manufacturer to be about 0.9 $n\text{V}/\sqrt{\text{Hz}}$ and 1 $p\text{A}/\sqrt{\text{Hz}}$ respectively. The optimum input impedance (e_n/i_n) of the preamplifier is, therefore, about 900 Ω. Since the dynamic resistance of the SQUID at its optimum bias point is about 1 Ω, one requires a coupling circuit with an impedance transformation of about 900 for effective signal extraction from the SQUID. In addition, the bandwidth of the coupling circuit should be as large as possible to extract the 100 kHz modulation signal without attenuation. For this, a room temperature step up transformer with a turns ratio of 30 has been fabricated and used as an impedance matching circuit. The room temperature transformer consists of 10 turns of 24 SWG copper wire as primary coil and 300 turns of 28 SWG copper wire as secondary coil wound on a toroidal ferrite core. The inner and outer diameters of the toroidal core are 10 mm and 18 mm respectively. The step up transformer is housed in the preamplifier box and the preamplifier is mounted at the top of SQUID insert. The SQUID is biased with an optimum dc bias current, I_b to get maximum voltage modulation from the SQUID sensor. The magnetic flux, which is to be measured (in actual measurements) is applied to the input coil of the SQUID, which is inductively coupled to the SQUID loop via the mutual inductance, $M_i \propto \sqrt{L_i L_s}$, where L_i is the self inductance of the input coil and L_s is the self inductance of the SQUID loop. The signal flux is modulated by a 100 kHz sinusoidal flux whose peak-to-peak amplitude is less than $\Phi_0/2$. The modulated output voltage from the SQUID is stepped up by the impedance matching transformer and further amplified by a two stage amplifier with sufficient gain and fed to the signal input channel of the analog multiplier. The modulated output is phase sensitively detected with respect to the reference signal supplied from the same 100 kHz oscillator to

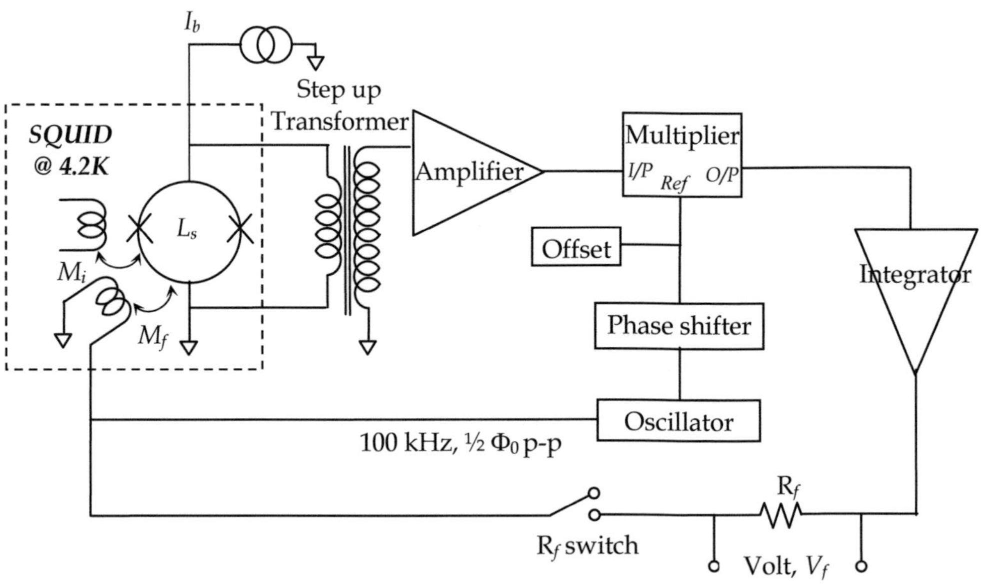

Fig. 4. Schematic diagram of the flux locked loop electronics with flux modulation scheme.

the reference input channel of the analog multiplier. The output of the analog multiplier is integrated and fed back as a current to the feedback coil in order to counterbalance the signal flux applied to the SQUID. The voltage, V_f developed across the feedback resistor R_f is proportional to the input flux and the transfer function of the system ($\partial V_{FLL}/\partial \Phi_{sig}$) is given by ($R_f / M_f$), where R_f is the feedback resistance and M_f is the mutual inductance between the feedback coil and SQUID (J. Clarke et al., 1976). When the feedback switch is closed, the feedback flux at the SQUID cancels the input flux and the system uses the SQUID as a null detector of magnetic flux. The voltage developed across the feedback resistor is proportional to the applied input flux and the time variation of this voltage is an exact replica of the time variation of the input signal flux coupled to the SQUID. The system gain can be varied by simply varying the value of the feedback resistor. The figures of merit of the SQUID system with the FLL readout electronics are the flux noise, slew rate and bandwidth; these values are listed in table 1 for the system developed in our laboratory.

Flux noise ($\mu\Phi_0/\sqrt{Hz}$)	Slew rate (Φ_0/s)	Bandwidth (kHz)
10	$5x10^5$	10

Table 1. Benchmarks of the SQUID system

4. Important issues in building SQUID based measuring systems

There are some important issues to be considered while designing and building measuring instruments based on SQUID sensors. Since the sensing area of the SQUID is very small (typically 10^{-2} mm^2), direct sensing of the signal flux leads to poor field resolution. The field resolution can be improved to some extent by increasing the effective area of the SQUID sensor's square washer (J.M. Jaycox & M.B. Ketchen, 1981). A better way to increase the field resolution, however, involves the use of superconducting pickup loop with larger area in the form of a magnetometer or a gradiometer connected to the on-chip integrated multi-turn input coil which is magnetically tightly coupled to the SQUID. The noise energy coupled to the SQUID is minimum only when $L_i = L_p$, where L_i and L_p are the inductances of the input coil and the pickup loop respectively. For a typical low-T_c dc SQUID and radius of the pickup loop r_p ~ 10 mm, one can expect the white magnetic field noise of the order of 1 $fT/\sqrt{Hz}$. Such sensitivities can only be utilized fully inside a superconducting shield or in a magnetically shielded room. In unshielded environments, fluctuations in the earth's magnetic field, local fields at power line frequency and disturbances due to strong sources such as rotating machinery will dominate the applied magnetic field to the SQUID. For sensitive measurements using SQUID sensors, like the measurement of biomagnetic fields and high resolution measurement of magnetic susceptibility, the use of gradiometers for rejection of distant sources of magnetic noise is almost inevitable. Fig.5(b) shows the principle of a first derivative, axial gradiometer consisting of two pickup loops of equal turns areas connected in series; the two loops are wound in opposition and are separated by the baseline, b. A uniform magnetic field B_z ideally induces zero net screening current into the two loops, and hence couples no net flux to the SQUID. A gradient $\partial B_z/\partial z$, on the other hand, induces a net screening current in the gradiometer, which flows through the input coil and a corresponding flux is inductively coupled to the SQUID. Fig.5(c) shows the configuration of the second derivative axial gradiometer consisting of two first derivative gradiometers wound in opposition. For uniform magnetic field and first order field gradient, the net screening current induced in the second order gradiometer is zero. The second derivative of the magnetic field $\partial^2 B_z/\partial z^2$ induces screening current in such a gradiometer and couples a net flux to the SQUID.

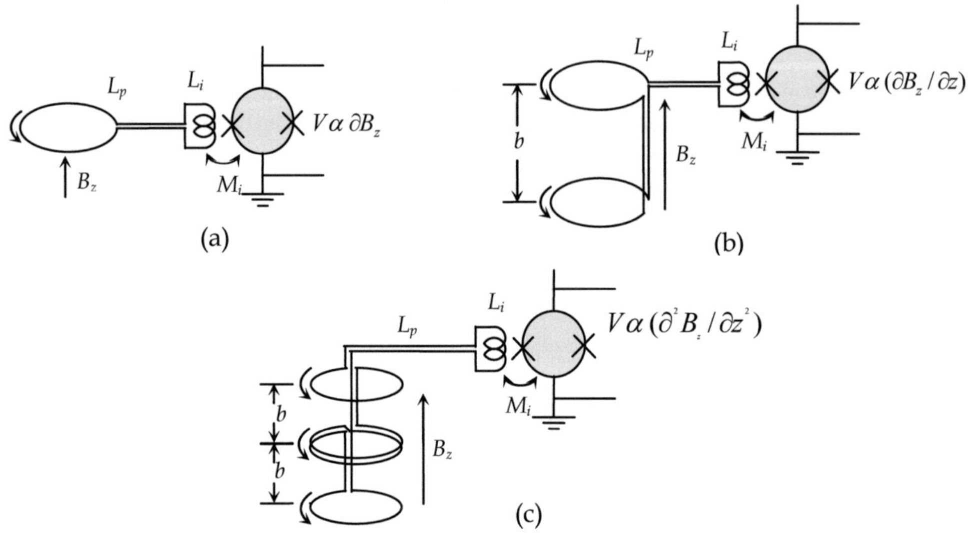

Fig. 5. Pickup loop configuration with different geometries connected to the input coil of the SQUID (a) magnetometer, (b) first order gradiometer and (c) second order gradiometer

The SQUID based measuring system has to be designed based on the nature of the signal source of interest. In some cases the signal source may be exposed to high temperature, high magnetic fields like magnetization measurements and in some other cases like nondestructive testing, biomagnetism, geophysics, SQUID microscope etc. the signal source may have to be kept at room temperature and is often outside the cryostat.

5. SQUID based NDE system

The SQUID based NDE system designed and constructed in our laboratory consists of a SQUID probe with superconducting pickup loop in the form of first order axial gradiometer housed in a nonmagnetic liquid helium cryostat, stepper motor driven precision XY scanner with non-magnetic platform for the sample movement and the data acquisition module to acquire the SQUID output signal with respect to the positional coordinates of the sample which is under investigation.

5.1 SQUID probe

The SQUID probe comprises of a SQUID sensor and a superconducting first order gradiometer connected to the input coil of the SQUID. The SQUID sensor is enclosed by a superconducting magnetic shield while the superconducting gradiometer is exposed to detect the magnetic signal of interest. The local source of interest generates a much larger field gradient at detector (superconducting pickup loop) than that generated by a more distant noise source.

A holder to mount the SQUID sensor together with a former to wind the pickup loop of the gradiometer has been fabricated as a single piece and is connected at one end of the 13 mm diameter thin walled stainless steel (SS) tube designed to be insertible in a liquid helium cryostat. The electrical leads in the form of twisted pairs as required for the SQUID sensor are routed through the thin walled SS tube and terminated at the top with appropriate electrical feedthroughs. A first order gradiometer made of superconducting NbTi wire with diameter of 0.1 mm is wound over the former and is inductively coupled to the SQUID device. The gradiometer consists of two loops of 4-mm diameter wound in opposition and separated by a baseline of 40 mm. This configuration enables discrimination against distant sources of magnetic noise which produce equal and opposite response from the two loops constituting the gradiometer. The SQUID sensor is shielded with a lead cylinder with an inner diameter of 16 mm and a length of about 120 mm.

The SQUID probe is housed in a FRP liquid helium cryostat having a capacity of 11.5 litres of liquid helium. The cryostat has a low boil-off rate of under 2.2 litres per day and is designed to have a minimum warm-to-cold distance of 6 mm. The cryostat is equipped with a top loading clear access of 25 mm diameter for the insertion of SQUID probe. The cryostat was constructed with fiberglass-reinforced epoxy, which has been found to have suitable structural and thermal properties without introducing a high level of magnetic noise and distorting noise fields. The setup was calibrated by measuring the system response to the magnetic field produced by a large circular coil and the calibration constant was inferred to be 20 nT/cm per flux quantum coupled to the SQUID. Fig.6. shows the cross sectional view of the SQUID probe housed in a liquid helium cryostat.

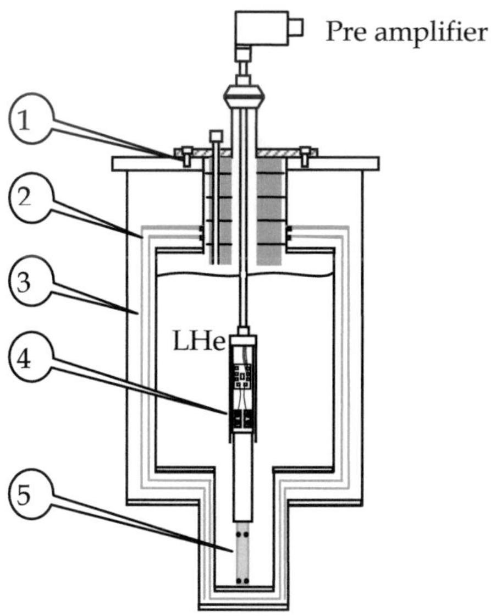

Fig. 6. Cross sectional view of the SQUID probe housed in liquid helium cryostat. 1. LHe cryostat, 2. Radiation shields, 3. Vacuum space, 4. SQUID, 5. Pickup loop

5.2 XY scanner

One of the most important requirements for SQUID based NDE system is the scanner to move the specimen in order to scan it under the stationary liquid helium cryostat. In most SQUID based NDE systems, the specimen is moved under a rigidly fixed stationary SQUID system. Commercial general purpose scanners are not suitable for the SQUID NDE applications because of the use of magnetic materials in their construction and the consequent generation of heavy electromagnetic noise. Therefore a stepper motor driven nonmagnetic precision XY scanner has been specially designed and fabricated for SQUID NDE system operating in our laboratory to enable precision movement of the sample.

The XY scanner has been designed for the scanning of flat plate samples with a positional accuracy of 0.025 mm and repeatability of 0.1 mm. The detailed specifications of the XY scanner have been listed in Table 2. The major components of the XY scanner are computer controlled XY stage, supporting platform which moves smoothly over a frictionless table and a non-metallic and non-magnetic sample holder. The whole assembly has been mounted on a single frame. Necessary vibration isolation has been provided to minimize the transmission of floor vibrations. Care has been taken to isolate the SQUID system from the magnetic noise generated by the high torque stepper motors. The details of the XY scanner are discussed below.

Sample	Flat plates
Scanning area	300 mm x 300 mm
Step size	0.25 mm to 5 mm
Speed	1 to 50 mm/sec

Table 2. Specifications of X-Y scanner

5.2.1 XY scanner: Computer controlled XY stage

The main component of the XY scanner is a computer controlled stepper motor driven XY stage. For flat samples the XY table is activated to scan the sample in either X or Y directions. The XY table has a traverse of 300 mm in each direction and is driven by two stepper motors (model KML092F13, Warner SLOSYN). Each stepper motor is connected to a micro-stepping driver (SS2000 MD808, Warner SLO-SYN) controlled by a programmable stepper motor controller (model SS2000PCi, SLO-SYN). The motion is achieved by driving the lead screws through the stepper motors to achieve the desired position of the table. Each axis of the stage is equipped with proximity switches which set the travel end-limits of the stage and also define the home position. The home position is a reference point from which all distances are measured, and is always approached from the positive end of travel limit with a preset velocity and acceleration, to ensure the reproducibility of scans to better than 0.1 mm. In addition, there are safety limit switches at both ends for over travel protection.

5.2.2 Supporting platform

Since the SQUID system is extremely sensitive to any magnetic flux changes, it is necessary to keep this system far away from the stepper motors. The high frequency current pulses applied to the high torque stepper motors generate undesirable magnetic field noise, which could disturb the locking of the sensitive SQUID system. To avoid such magnetic noise, the stepper motors have been located 2.5 m away from the SQUID system. The sample holder has been mounted at one end of the supporting platform which is made of non-magnetic and non-metallic materials. Since the platform has a large length, two pairs of roller tables have been provided. The central one is made of a stainless steel case with brass ball bearings and the other one close to the sample holder is made of a nylon case with glass ball bearings. These rollers smoothly move over the frictionless tables. The supporting frame is made of fiber glass square pipes. The fasteners used in the region close to the sample holder are made of nylon.

5.2.3 Sample holder

The sample holder is fabricated using non-metallic and non-magnetic materials such as fiberglass and polypropylene. Care has been taken to avoid the use of any magnetic components around the sample holder up to a radial distance of about 500 mm.

5.2.4 Shielding

Even though the stepper motors are kept far away from the SQUID system, it was found that the SQUID picks the magnetic field noise associated with the stepper motors causing the SQUID system to frequently unlock. The noise level has been reduced to some extent by covering the stepper motors by two layers of μ - metal and by wrapping all the cables coming from the control panel to XY stage by a flat copper braid. The flat copper braid has been properly grounded with the main power ground. The electrical leads coming from the SQUID to the preamplifier mounted at the top of the liquid helium cryostat and the leads that connect the preamplifier to the flux locked loop electronics module kept in the instrument rack have also been shielded and the shield is properly grounded.

5.3 Data acquisition system

Fig.7 shows the schematic diagram of the SQUID based NDE measurement system. The specimen or the source of magnetic signal is mounted over the sample holder of the XY scanner. Whenever there is a change of magnetic flux in the vicinity of the pick-up loop, a screening current is induced in the pickup loop, which is inductively coupled to the SQUID. The SQUID in conjunction with the flux locked loop readout electronics generates a voltage output which is proportional to the magnetic field gradient, ($\partial B_z/\partial z$) over the baseline of the pick-up loop. The oscilloscope is used to display the necessary signals for tuning the SQUID readout electronics by setting optimum bias current, amplitude of modulation flux, etc. A complete data acquisition system has been developed based on Visual Basic. The user can select the initial and the final positions of the sample coordinates, step size and scanning speed for each scan axis. The software selects grid points within the selected area depending upon the specified step size. The sample is scanned under the SQUID system and the SQUID output is recorded in the computer as a function of position coordinates. For every specified step size, the intelligent micro-stepping drive unit sends TTL pulse to the data logger (model 34970A, Agilent), which stores the data in its buffer whenever a trigger pulse is received from the drive unit; the first data is collected by a software trigger. After each line scan (say y-axis) the sample is moved one step with the specified step size perpendicular to y-axis (say x-axis). During this time the data is transferred from the data logger to the computer and stored for further analysis. This wait time is user selectable before the start of scan. In addition to this, a linear scale has been provided along each scan axis and the actual position coordinates displayed in a digital readout. Provision has been made for manual operation of each scan axis, and for selection of home position for each scan axis. Fig.8 shows the photograph of the SQUID based NDE system with XY scanner.

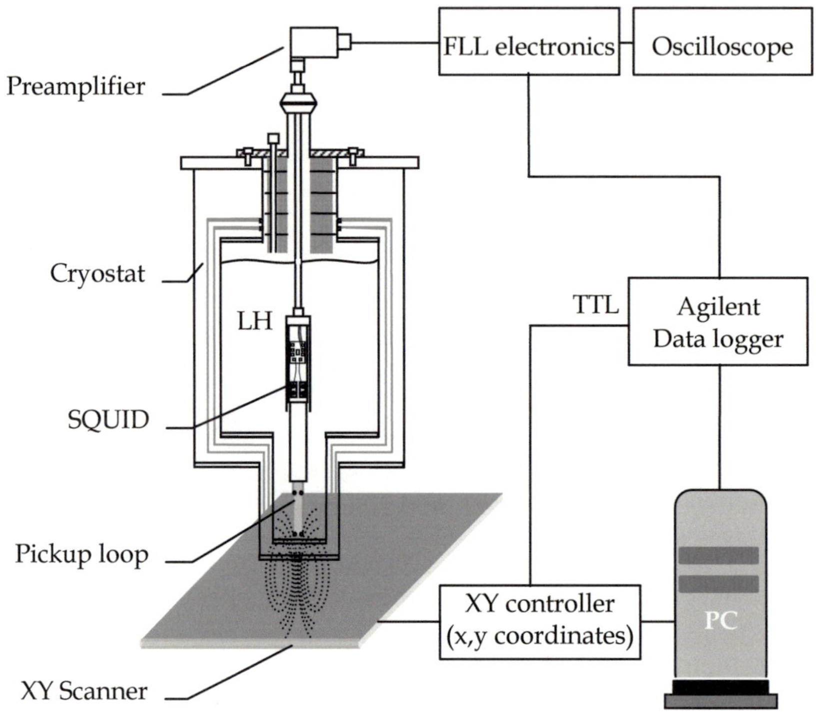

Fig. 7. Block diagram of the SQUID based NDE system

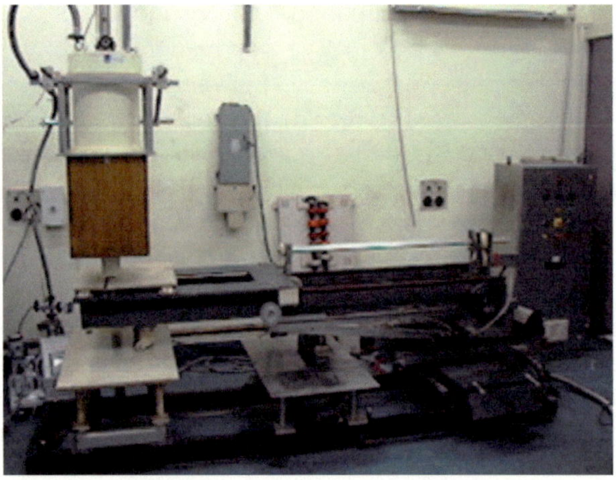

Fig. 8. Photograph of the SQUID based NDE system

To check the performance of the system a steel ball weighing 2.2 mg has been magnetized by applying a magnetic field of about 35 Oe for a preset time of about 5 minutes. The magnetized steel ball has been scanned under the SQUID system to measure the remanent magnetization. After scanning, the steel ball has been removed and the background noise has been recorded. It may be noted that even though the sample platform is made of non-magnetic material, the Y table which moves over the X table in the XY scanner is constructed by using lead screws and linear guides which are magnetic. The contribution of the background has been eliminated by subtracting the background from the signal. Fig 9 shows the recorded SQUID output voltage for the magnetized steel ball while it is scanned under the SQUID system. The amplitude of the magnetic anomaly associated with the 2.2 mg steel ball is about 80 millivolts while the width of the magnetic anomaly arises on account of the stand-off distance of about 10 mm between the steel ball and the sensing loop. It may also be noted that based on the noise floor, one can estimate that the SQUID based NDE system can detect the presence of a steel particle weighing as low as 10μg at signal-to-noise ratio unity.

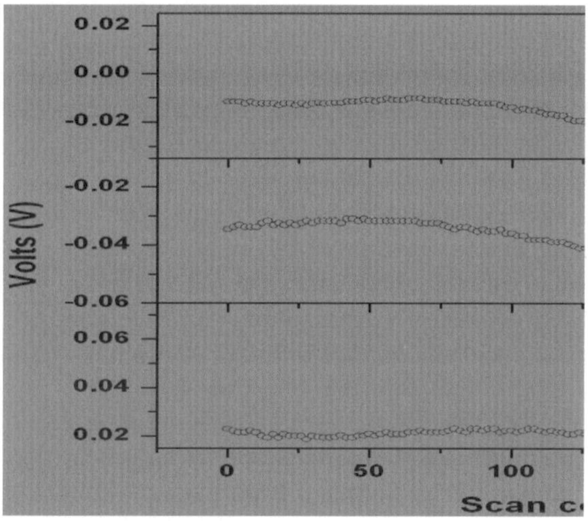

Fig. 9. Recorded SQUID output voltage for the steel ball scanning under the SQUID. (a) Steel ball with background noise (b) background noise and (c) steel ball alone

6. Utilization of SQUID based NDE system

This SQUID system has been used for the measurement of subsurface defects by inducing eddy currents in conducting materials at relatively low frequencies and also for the measurement of extremely low content of δ-ferrite in the 316L(N) stainless steel weldment specimens which are subjected to low cycle fatigue at high temperatures.

6.1 SQUID based eddy current NDE

Conventional eddy current testing is the popular NDE technique for detecting subsurface defects in conducting structures like aluminum, stainless steel etc. In this technique, eddy currents are excited in a conductor using a magnetic field varying in time at an excitation frequency *f*.

A suitable coil carrying AC current is used to induce eddy currents in the specimen by producing a time varying magnetic field. The physical basis for the eddy current induction in conducting material is contained in the Faraday's law $\nabla \times \vec{E} = -\frac{\partial \vec{B}}{\partial t}$ which can be rewritten as

$$\nabla \times \vec{J} = -\mu\sigma \frac{\partial \vec{H}}{\partial t} \tag{3}$$

where μ is the magnetic permeability and σ is the electrical conductivity of the material. The time varying applied magnetic field will induce a current distribution inside the conductor in accordance with equation 3 and these currents are referred to as eddy currents.

If the applied magnetic field varies sinusoidally with respect to time, $H = H_0 e^{i\omega t}$, then eq. (1) becomes

$$\nabla \times J = -i\omega\sigma\mu H \tag{4}$$

In qualitative terms, the curl of the eddy currents is out of phase with the applied time varying magnetic field, is proportional to the frequency of the applied magnetic field and opposes the changes in the applied magnetic field with time. In conventional eddy current NDE, sinusoidal current with the frequency ranging from 10 kHz to 10 MHz is passed through an excitation coil having hundreds of turns and the voltage induced across the coil is measured to determine the impedance of the coil. The presence of a conductor near the coil affects its impedance and the changes in the coil impedance are measured which are associated with changes in some characteristics of the conductor (variation in local thickness, conductivity or permeability due to the presence of defects). In the case of SQUID based eddy current NDE the change of the magnetic field due to the presence of a defect is directly measured by the SQUID. The defects perturb the flow pattern of the induced eddy currents in the conductor and hence they manifest as localized magnetic anomalies when the conducting plate is scanned under the SQUID sensor. The amplitude and shape of the magnetic anomaly corresponding to a defect depend on the amplitude of the eddy current flowing in the immediate vicinity of the defect and also on the defect characteristics such as size, orientation etc. The use of higher excitation frequencies tends to increase the amplitude of the eddy currents induced on the surface; however, the strength of the eddy current induced exponentially decreases with depth d measured from the surface as $e^{-d/\delta}$, where δ is the skin depth defined as

$$\delta = \left(\frac{1}{\pi f \mu\sigma} \right)^{1/2} \tag{5}$$

Here f is the frequency of the excitation current used to induce eddy current in the conducting material. In the conventional eddy current testing, excitation frequencies higher than 10 kHz are generally used to increase the voltage induced in the pick-up loop; therefore, the depth of the defect detection is limited by the skin depth (δ) of the material, which is rather low at such high frequencies. If the excitation frequency is reduced in order to increase the skin depth, the induced voltage in the pick-up loop also decreases, often reaching the system noise floor, and thus severely limiting the depth of defect detection. The system based on the use of SQUID as a detector promises better signal-to-noise ratio at lower operating frequencies. Since the sensitivity of the SQUID device is independent of the operating frequency from near dc to several kHz (white noise regime), it is possible to use a SQUID based system for the detection of deep subsurface defects, which are not detectable in the conventional eddy current testing owing to skin depth limitations. In the SQUID based system, pickup loop can be wound using superconducting wire; unlike normal metal pickup loops for which induced voltage is proportional to the rate of change of flux, superconducting pickup loop directly senses the change in magnetic flux and hence its use is crucial in retaining the requisite low frequency sensitivity. Several groups have successfully utilized this feature to detect subsurface defects not detectable by conventional eddy current techniques (W.G. Jenks et al., 1991; P. Chen et al., 2002; H. Weinstock, 1991; M. Mück et al., 2005; U. Klein et al., 1997).

The excitation coil used to induce eddy currents in the SQUID based eddy current NDE is also often different from that used in the conventional eddy current methods. In conventional testing, the excitation coil is in the form circular coil and in some cases a set of differential coils are used. To enhance the eddy current in the specimen, it is required to apply a large time varying magnetic field. If we use circular excitation coil for the SQUID based eddy current NDE system the direct magnetic field coupled to the SQUID is very large and the output voltage of the flux locked loop readout electronics may reach saturation levels. Under these conditions, the changes of magnetic field due to the presence of a defect cannot be detected by the SQUID. To overcome this, a double "D" excitation coil at room temperature, placed just below the cryostat, is often used to excite eddy currents in the specimen at room temperature. This double "D" configuration for the excitation coil enables realization of a wide dynamic range since the direct coupling of magnetic flux to the pick-up loop is a minimum when the axis of the excitation coil coincides with that of the pick-up loop. The schematic view of the double "D" excitation coil and the location of the superconducting pick-up loop are shown in fig.10. The central axes of the excitation coil and the pick-up loop are adjusted to coincide and held stationary during the measurement. The current in the double "D" coil flows through the central straight segment and then around the perimeter of the semicircular loops. Thus the field through one D section of the coil is always 180 degrees out of phase with the field through the other D section of the coil. The specimen, which is to be evaluated, is kept at room temperature at a stand-off distance of about 14 mm from the bottom loop of the superconducting gradiometer located inside the cryostat and is scanned under the stationary liquid helium cryostat. Eddy currents were excited in the specimen under investigation using the double "D" coil with 120 turns and an overall diameter of 40 mm carrying a current of 150 mA(rms). Changes in the induced eddy current flow associated with the presence of a defect manifest as changes of the flux signal detected by the SQUID device as the specimen is scanned. The SQUID, in turn, produces an output voltage corresponding to these changes in magnetic flux. This output is phase sensitively detected by a lock-in-amplifier whose reference channel is fed by the oscillator, which is used to supply the necessary excitation current to the double-D coil. The schematic diagram of the experimental set-up for the SQUID based eddy current NDE is shown in fig.11.

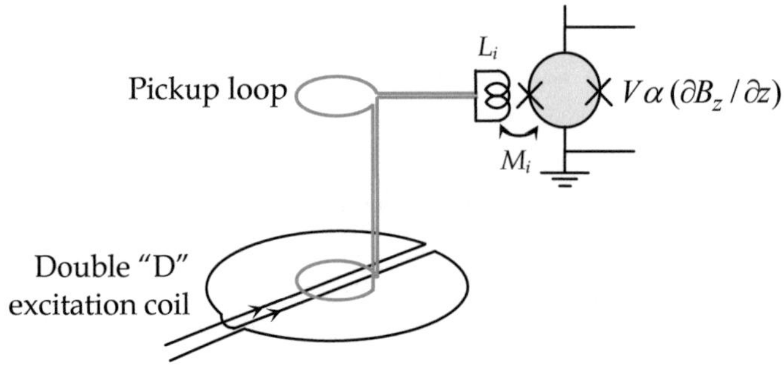

Fig. 10. Schematic view of the double "D" excitation coil and pickup loop arrangement

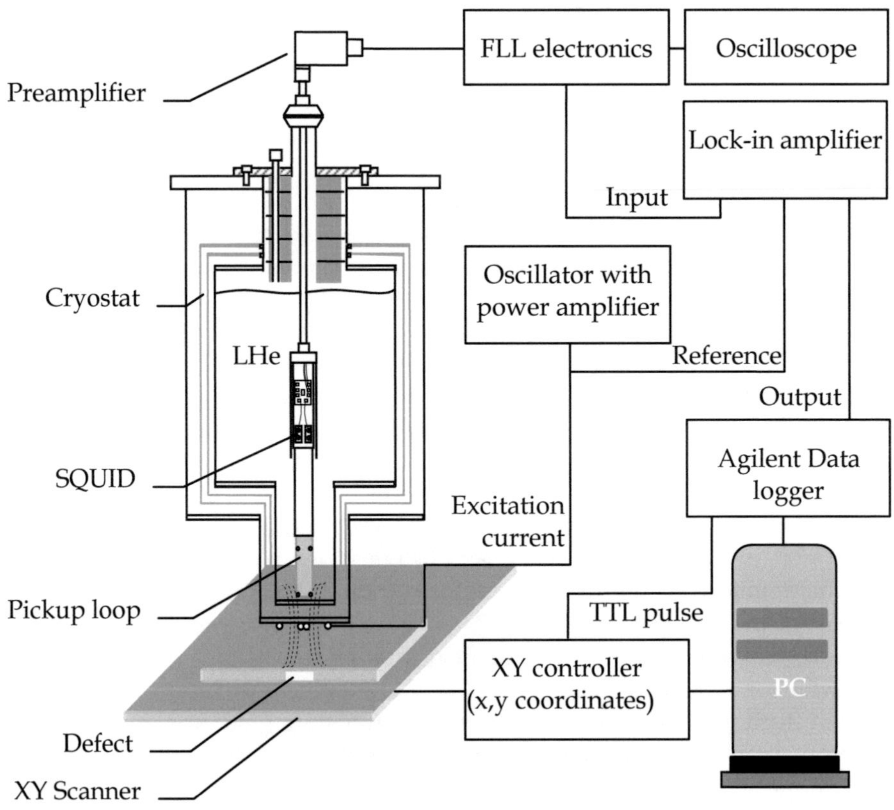

Fig. 11. The SQUID based eddy current NDE system

Both the real (in-phase with the excitation current) and imaginary (90° out-of-phase with the excitation current) components of the SQUID output voltage were measured using the lock-in-amplifier and were simultaneously recorded by a computer with respect to the positional coordinates of the defect as the plate was scanned under the stationary cryostat using a precision X-Y scanner.

6.2 Remnant magnetization measurements

Fatigue damage and effect of thermal ageing are the most important problems for a number of high temperature components in power plants particularly in welded regions. Austenitic stainless steels are the preferred materials for sodium cooled fast breeder reactors as they possess excellent high temperature mechanical properties for nuclear reactor service. A nitrogen-added low carbon version of SS 316 austenitic stainless steel (316 L(N)) has been chosen for the high temperature components of Prototype Fast Breeder Reactor (PFBR) in India. A major consideration during the welding of 316 L(N) stainless steel is its resistance to hot cracking. The presence of an optimum amount of delta ferrite in the austenitic weld metal is desirable to prevent hot cracking in the weldment. However the δ-ferrite structure is highly unstable during high temperature service and transforms to carbides and brittle intermetallic phases, e.g. sigma phase. Small quantities of these phases would cause large variations in the mechanical properties and influence corrosion behaviour of the material. Hence the knowledge of these phases such as type, amount etc is essential in designing the optimal operational parameters to achieve the desired lifetime for welded stainless steel components. Since the δ-ferrite is a magnetic phase, one can evaluate the δ-ferrite content by measuring remanent magnetization of the weldment sample. Since the SQUID device has an extremely high sensitivity for magnetic flux signals, one can detect an extremely low content of δ-ferrite in the weldment samples through remanent magnetization measurements. The experimental set up for the remanent magnetization measurements is similar to that shown in fig.11 except that the double "D"excitation coil is not used. In this case, the SQUID output is directly read by data logger and recorded by the computer with respect to the coordinates of the sample as it is scanned under the stationary cryostat.

7. Results of experimental investigations

We have carried out two different kinds of studies such as detection of artificially engineered subsurface flaws in aluminum plates by using the SQUID based eddy current technique and NDE of fatigue cycled stainless steel weldment specimens to detect the δ-ferrite content through remanent magnetization measurement by the SQUID system.

7.1 Detection of subsurface flaws in aluminum plates

The SQUID based NDE system has been initially tested with an aluminum plate with a length of 300 mm, width of 100 mm, and thickness of 10 mm with artificially engineered defects, as shown in fig. 12 (a). Two rectangular defects have been engineered with a separation of 150 mm. One defect has a length of 50 mm, width of 1 mm, and height of 1 mm, and another defect has a length of 50 mm, width of 1 mm, and height of 0.5 mm and the magnetic anomalies corresponding to these two defects were studied as part of initial exploratory studies on subsurface defect characterization. Eddy currents were excited at a relatively low frequency of about 200 Hz using double "D" excitation coil. The aluminum plate with defects on its bottom surface was scanned under the SQUID probe and the changes of magnetic field associated with the defects were recorded with respect to the positional coordinates. Fig. 12 (b) shows the magnetic anomalies associated with these defects recorded using a SQUID. The amplitude of the magnetic anomaly corresponding to a defect having a height of 1 mm was 2.3 nT/cm. Since the noise floor of the system is few tens of pT/cm, it is estimated that even a localized loss of conductor thickness as low as 0.1 mm can be detected at a depth of 9 mm at signal to noise ratio unity. The potential of the system for detection of subsurface defects using low-frequency eddy current excitation is evident from this data.

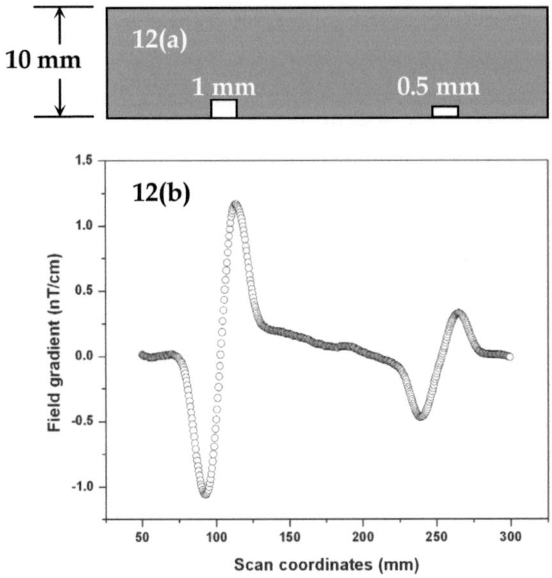

Fig. 12. (a) Aluminum plate with artificially engineered defects. (b) Magnetic anomalies associated with the defects buried at a depth of 9 mm below the surface recorded by the SQUID.

Subsequently detailed experimental studies have been carried out for the determination of optimum eddy current excitation frequencies for the defects located at different depths below the top surface of an aluminum plate. This optimum excitation frequency corresponds to the frequency at which the amplitude of the magnetic anomaly associated with a defect at a certain depth reaches a maximum value. The optimum excitation frequency has been theoretically investigated by Sikora et al (R. Sikora et al., 2003) and by Baskaran and Janawadkar (R.Baskaran and M.P. Janawadkar, 2004) using the analytical solution (C.V.Dodd & W.E. Deeds, 1968) to the distribution of eddy currents excited by a circular coil in a defect free semi-infinite conductor. It may, however, be noted that the results of these theoretical investigations cannot be directly compared with the experimental results reported here since in the present studies, a non-axisymmetric double-D coil was used as an excitation coil to induce eddy currents in a conductor stack of finite thickness with a defect embedded within the stack.

A set of aluminum plates with a length of 300 mm, width of 200 mm and with different thicknesses ranging between 2mm to 12 mm in steps of 2 mm were fabricated for this study. An artificial defect is engineered in a separate 1 mm thick aluminum plate with a defect length of 60 mm and width of 0.75 mm to simulate a localized loss of conductor volume in the plate. The length and width of the plate carrying the defect are identical to those of the plates without defects. SQUID based eddy current NDE measurements have been carried out for the stack of aluminum plates with the defect plate located at different depths ranging from 2 mm to 14 mm (in steps of 2 mm) while keeping the total thickness of the conductor stack to have a constant value of 15 mm throughout the experimental investigations. As an example, to simulate the defect located at a depth of 2 mm below the top surface, the defect plate was sandwiched between a 2 mm thick aluminum plate at the top and a 12 mm thick aluminum plate at the bottom. An interchange of the top and bottom plates shifts the defect location to 12 mm below the top surface. In this way, the location of the defect was shifted to different depths from 2 mm to 14 mm below the surface. A photograph of the aluminum plates used in this study in order to vary the defect depth is shown in fig. 13. The stack of plates was scanned under the SQUID in a direction such that the defect length is perpendicular to the scanning direction. To avoid edge effects, the scanning region was limited to a total length of 120 mm (60 mm on each side of the defect). For each scan, both the real and imaginary components of the SQUID output voltage were measured using the lock-in-amplifier and were recorded with respect to the positional coordinates of the defect. For each defect depth, the experiments were repeated at different excitation frequencies ranging between 33Hz and 853Hz in steps of 20 Hz.

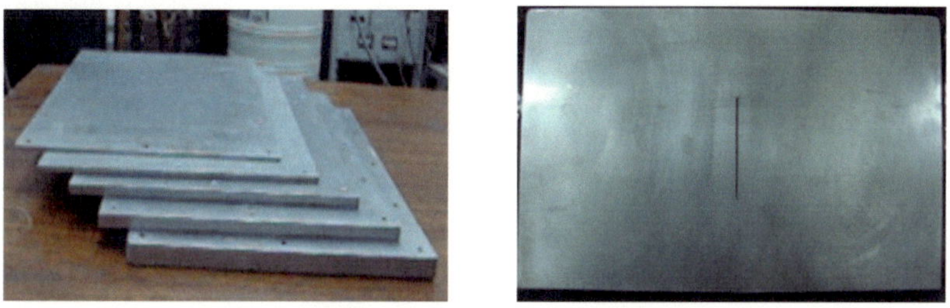

Fig. 13. Aluminum plates used in the study (plate with the defect is also shown).

As an illustration of the magnetic anomaly detected by the SQUID system, fig. 14 shows the measurements corresponding to defect located at a depth of 14 mm below the top surface when the eddy currents are excited at the frequency of 33 Hz, 103 Hz and 243 Hz. At the lower excitation frequency of 33 Hz, the skin depth is high and the resultant amplitude of the change of magnetic field gradient due to the defect is 0.657 nT/cm. At the higher excitation frequency of 103 Hz, the skin depth is lower but the induced eddy current amplitude is higher and thus the detected magnetic anomaly has a higher amplitude of 1.8 nT/cm. When the excitation frequency is further increased to 243 Hz, the exponential attenuation of the eddy current amplitude at the location of the defect more than compensates for the small linear increase resulting from the use of higher frequency of excitation; the amplitude of the detected magnetic anomaly therefore decreases to 1.356 nT/cm. In this case, as shown in fig.14 (b) one can also clearly see the signal changes due to the small variations in lift-off as the sample moves relative to the gradiometric pickup loop. At the optimum excitation frequency of 103 Hz, the signal to noise ratio is evidently better. It may also be noted that the phase of the magnetic anomaly changes systematically when the excitation frequency is varied as shown in fig.14. The change in the resultant amplitude of the magnetic anomaly corresponding to a defect located at a depth of 14 mm below the surface when eddy currents are excited at different frequencies is shown in fig.14(d). At low frequencies, the eddy current induced in the conductor stack has low amplitude and hence the amplitude of the magnetic anomaly associated with the defect is also low. As the excitation frequency increases, the amplitude of the eddy current induced on the surface increases monotonically; however, owing to the skin effect, the amplitude of the eddy currents at a depth of 14 mm increases initially, reaches a maximum at the optimum excitation frequency of 103 Hz and decreases steadily thereafter. At higher frequencies, the eddy current tends to be concentrated at the upper surface of the material due to the skin effect. Therefore the amplitude of the magnetic anomaly due to the presence of the defect at a depth of 14 mm decreases when excitation frequencies higher than 103 Hz are used. In the present experimental investigations, the excitation frequency was varied from 33 Hz to 253 Hz in steps of 10 Hz and the optimum frequency was determined to be 103 Hz for the defect located at a depth of 14 mm below the surface. The experiments were repeated for defects located at different depths of 12 mm, 10 mm, 8 mm, 6 mm, 4 mm and 2 mm from the top surface and a comprehensive characterization of the magnetic anomalies was carried out as a function of defect depth and frequency. Fig.15 shows the magnetic anomaly detected by the SQUID system as the sample with defect located 10 mm below the surface is scanned. These plots show the variations in the magnetic anomalies resulting from the use of different excitation frequencies on the same scale. Fig. 16 shows the changes in the resultant amplitudes of the magnetic anomalies corresponding to defects located at different depths from the surface when eddy currents are excited at different frequencies. The excitation frequencies were varied in steps of 10 Hz for the defects at greater depths and 20 Hz for the defects at lower depths. The experimental parameters such as standoff distance, excitation

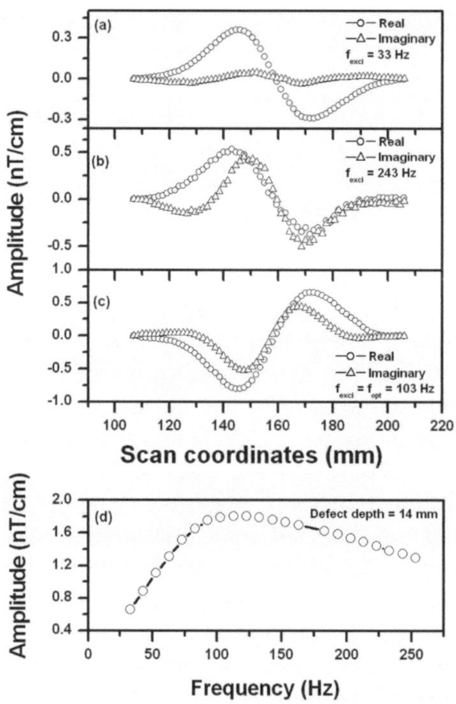

Fig. 14. (a),(b) and (c) show the changes in the magnetic anomaly corresponding to the defect buried at a depth of 14 mm below the surface with eddy current excitation at a frequency of 33 Hz, 243Hz and 103 Hz respectively. (d) The resultant amplitude of the magnetic anomaly due to the defect at different excitation frequencies.

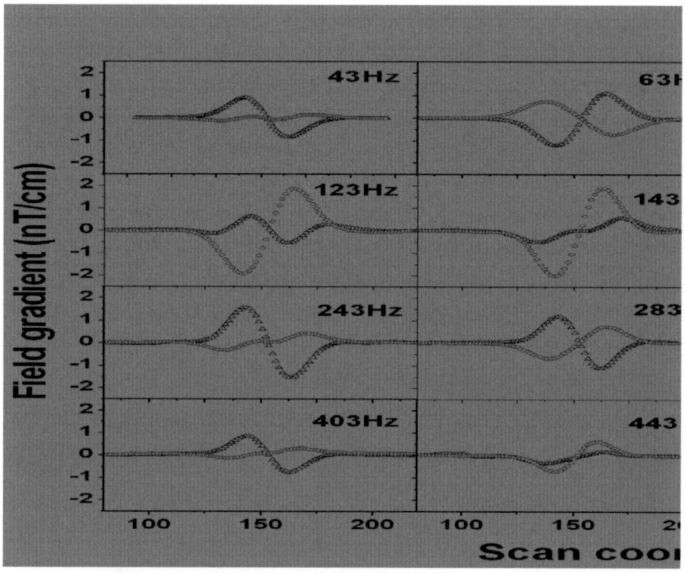

Fig. 15. The magnetic anomalies detected by the SQUID system for different frequencies when the defect was located 10 mm below the surface.

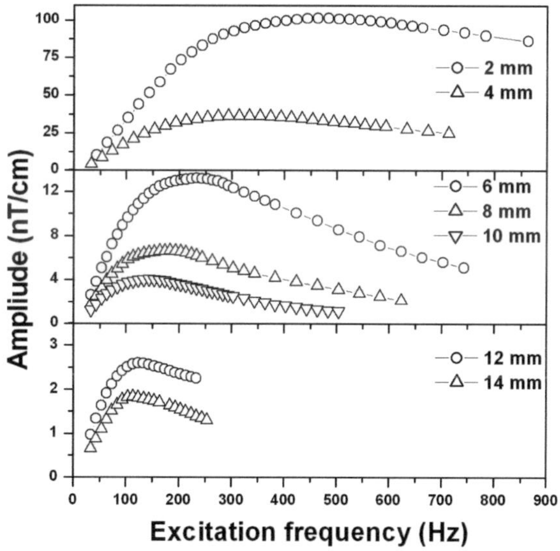

Fig. 16. The resultant amplitude of the magnetic anomalies corresponding to defects located at different depths when excited at different frequencies

coil current, flux locked loop electronics gain, scanning speed, step size, time constant of the lock in amplifier etc were kept constant for the entire series of experiments. The frequency at which the amplitude of the measured magnetic anomaly reaches a maximum value is taken to be the optimum frequency of excitation for defect located at a particular depth. As a result of these detailed experimental investigations, the optimum excitation frequencies have been estimated for the defects located at different depths from the surface and are listed in table 3.

Depth (mm)	2	4	6	8	10	12	14
f_{opt} (Hz)	453	313	233	183	143	123	103

Table 3. Optimum excitation frequencies for the defect at different depths

As shown in fig.17, the square root of the optimum excitation frequency is found to be inversely proportional to the defect depth. For the defects at lower depths from the surface, the signal to noise ratio is excellent for a wide range of frequencies; for the defects at greater depths from the surface, however, the range of frequencies which can be used for defect detection becomes narrow due to the poor signal to noise ratio and the skin effect of the conducting materials. It may also be noted that such defects can be detected only by using the SQUID based system for nondestructive evaluation.

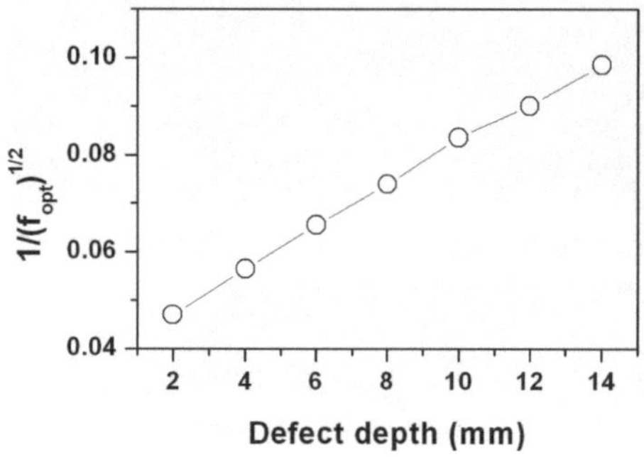

Fig. 17. The variation of the optimum excitation frequency with defect depth

7.2 Remanent magnetization measurements on fatigue cycled stainless steel weldments

The 316L(N) stainless steel weldment specimens were prepared by welding 316L(N) base metal with 316N electrodes by a manual metal-arc welding process. Welding was carried out on a 25 mm thick plate with a double-V configuration with an included angle of 70° (Fig. 18). Welds were radio-graphed and only sound joints were taken for the fabrication of the specimens. The low cycle fatigue (LCF) tests were conducted at a strain amplitude of ±0.6% using an Instron servohydraulic fatigue testing machine under total axial strain control mode at 600 °C. A strain rate of 3 x $10^{-3}s^{-1}$ was employed for the test. The magnetic flux signal arising from the remanent magnetization of the sample was coupled to the SQUID device via a superconducting flux-transformer. As the sample was scanned under the cryostat, SQUID output revealed a characteristic magnetic anomaly when the center of the gauge length of the sample passed under the pickup loop indicating the presence of a magnetic phase at the location of the weld. To show the repeatability of the XY scanner a virgin weldment sample was scanned under the SQUID system for measuring the remanent magnetization. The sample was placed at room temperature; about 25 mm below the pickup loop and scanned using the XY scanner. The scanning started far away from the sample and as the sample approaches the pick up loop, SQUID output increased and it subsequently decreased as the sample moved away from the pick up loop. The experiment has also been repeated by rotating the sample by 90°, 180° and 270° about its central axis. As the sample was rotated about its axis, depending on the orientation of the magnetization of the sample relative to the axis of the pick-up loop, the shape of the magnetic anomaly changed from bell shaped to dipolar The magnetic flux profile measured near the center of the weldment sample for different angles of rotation is shown in Fig. 19. It may be noted that the center of the measured magnetic anomaly coincides for all scans showing excellent repeatability of the XY scanner. To show the system capability in measuring the virgin sample which was strongly magnetic and fractured sample which was magnetically very weak, the two samples were kept at a separation of 150 mm and scanned under the SQUID. Fig. 20 shows the magnetic flux profile for the virgin and the fatigue fractured sample. For the virgin sample, the amplitude of the magnetic anomaly was measured to be ~1.78 Φ_0, whereas for the sample which was subjected to fatigue testing at 600 °C until fracture, the amplitude of the magnetic anomaly was measured to be as small as ~0.05 Φ_0 under identical measurement conditions. To complete the SQUID based measurements, measurement of δ - ferrite was also carried out using a magnagage along the 25 mm gauge length of both the virgin and LCF tested samples. Figs. 21 (a) and (b) show the δ - ferrite distribution profiles in the virgin and LCF tested specimens respectively. Measurements were made in two

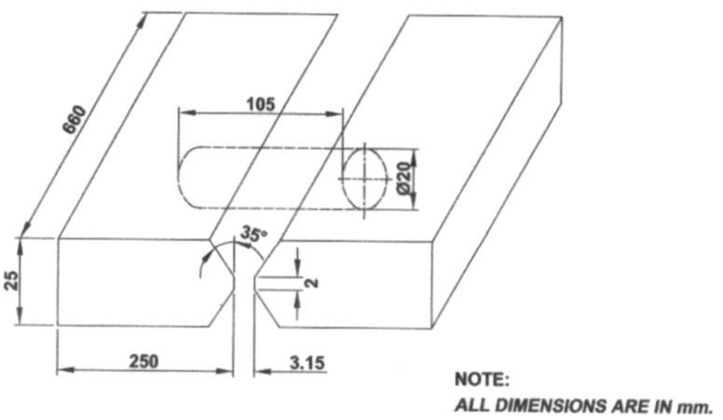

Fig. 18. Weld pad geometry used for the weld joint specimens.

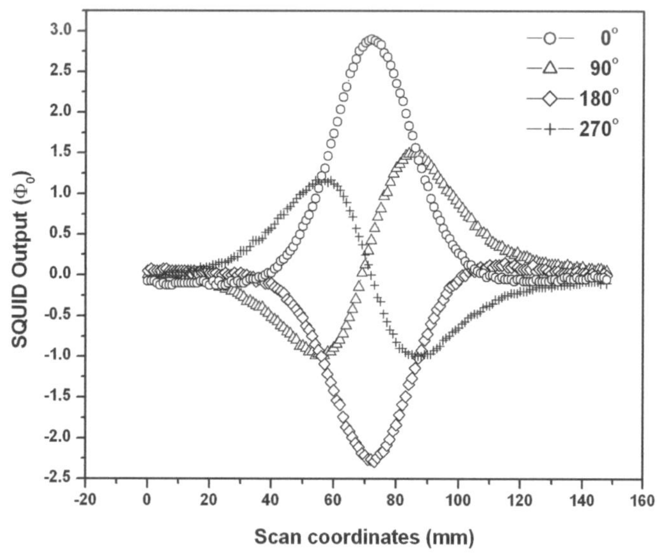

Fig. 19. Magnetic line profiles perpendicular to the gauge length of the weldment samples for different angles of rotation.

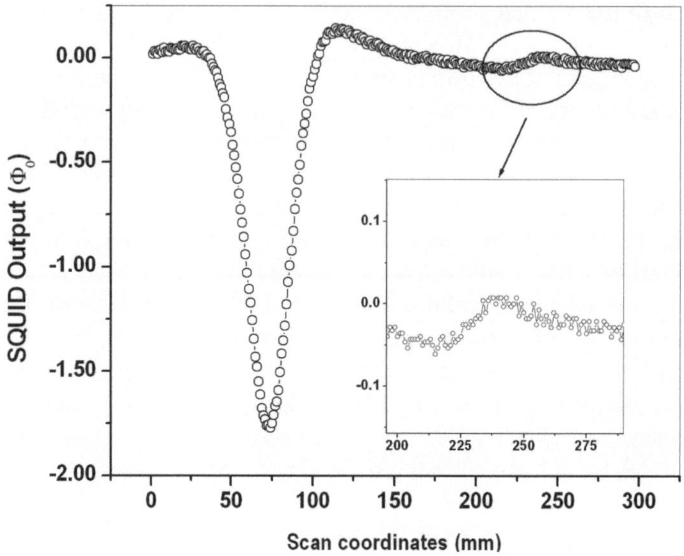

Fig. 20. Comparison of the magnetic flux profiles for the virgin and the fractured samples.

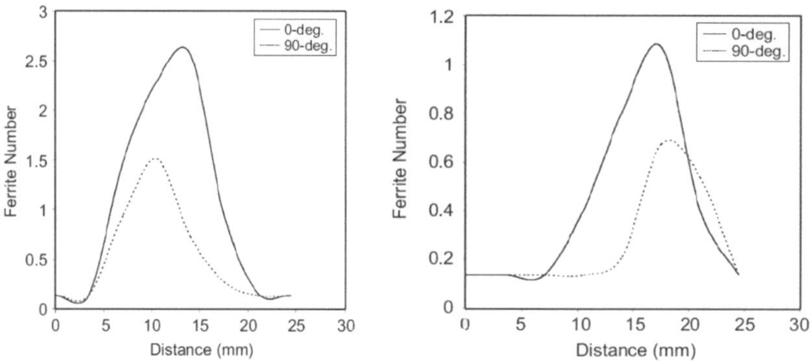

Fig. 21. (a) Delta ferrite distribution profile in the virgin sample. (b) Delta ferrite distribution profile in the LCF tested sample.

different orientations 90° apart from the central axis. The δ - ferrite was seen to peak around the mid-gage portion consisting of the joint proper. A considerable reduction in the amount of δ - ferrite after the LCF cycling could be seen. The experiments clearly show that the magnetic phase present in the virgin samples transforms to a non-magnetic phase (σ or carbides, etc.) during the fatigue deformation at 600 °C. The transformation of δ - ferrite to brittle σ - phase has been known to influence the cracking behavior and thereby the fatigue life in the weld metals and weld joints of 316 and 316 L(N) stainless steels (M.Valsan et al., 1995, 2002; A. Nagesha et al., 1999). The amount of transformation of the δ - ferrite into brittle phases has also been found to be a strong function of the temperature and frequency of testing (A. Nagesha et al., 1999). The results reported here demonstrate the potential of SQUID based measurements in tracking the magnetic-nonmagnetic (ex.δ-σ) transformation in stainless steel welds when subjected to high temperature fatigue loading. A series of experiments has been carried out in the stainless steel weldment specimens to evaluate the transformation of magnetic phase to nonmagnetic phases at different levels of fatigue deformation. In this study, a single weld joint was selected for remanent magnetization measurements in the virgin state and subsequently after every 50 fatigue cycles. At every stage prior to fatigue cycling and remanent magnetization measurement, the weld joint was properly demagnetized by subjecting it to low-frequency alternating magnetic field to eliminate the influence of the past history and then remagnetized by applying a preset dc magnetizing field (35 Oe) for a preset time (300 s) before the measurement of remanent magnetization using the SQUID-based setup commenced. The peak value of demagnetization field was kept slightly higher than the magnetizing field. To evaluate the relative changes in the magnetic content of the weldment specimen, the parameters of the measurement setup such as stand-off distance, FLL gain, etc., were maintained constant throughout the whole series of measurements. The virgin weldment specimen gave a maximum SQUID signal of 15 Φ_0, which decreased rapidly to 6.67 Φ_0 when the sample was subjected to LCF for 50 cycles. Thereafter, no significant changes in the maximum SQUID signal could be observed up to 150 cycles. However, a marked decrease in the SQUID signal was noticed when the sample was subjected to LCF for 200 cycles accompanied by the initiation of a crack at the boundary of the weldment specimen. Micro cracks were seen at the boundary of the weldment specimen when the specimen was examined through a microscope. The magnetic profile of the weldment specimen scanned under the SQUID probe is shown in fig. 22 in the virgin state as well as after subjecting the specimen to different levels of fatigue deformation. Fig. 23 shows the variation in the maximum SQUID signal as a function of the number of cycles of fatigue loading and portrays the transformation of magnetic δ-ferrite to nonmagnetic phases when the weldment specimen is subjected to LCF at 600 ^{0}C. It may be noted that although the initial signals were large, SQUID-based measurements were intended to look for small magnetic anomalies, if any, which could be correlated with the residual life of the weldment specimen. Further experiments are underway to clarify these issues.

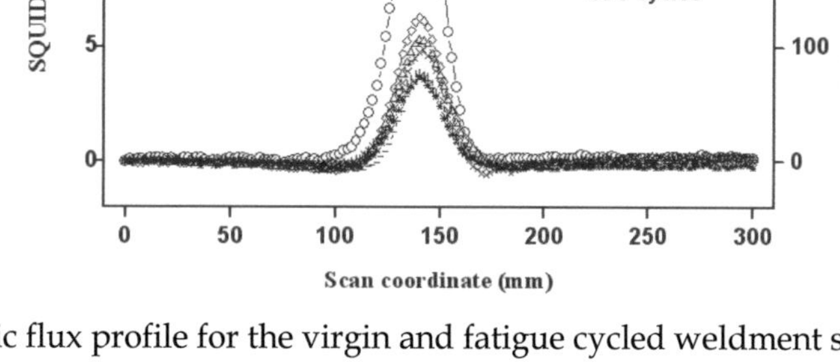

Fig. 22. Magnetic flux profile for the virgin and fatigue cycled weldment specimen

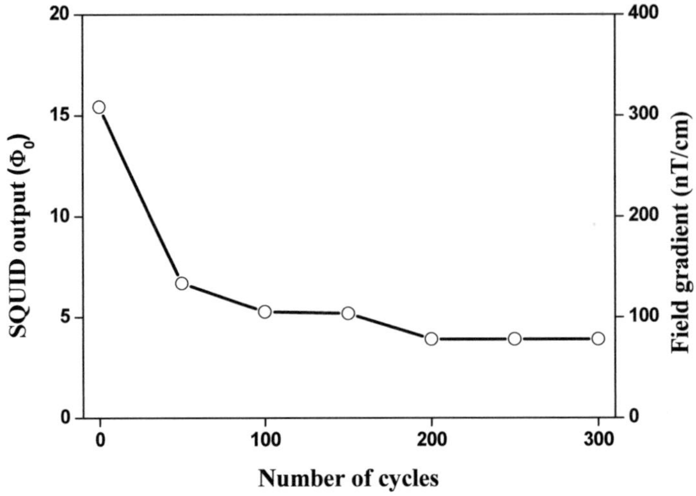

Fig. 23. Change of SQUID output vs number of fatigue cycles for the weldment specimen subjected to fatigue deformation

8. Conclusions

It is evident that the high magnetic field sensitivity of the SQUID sensors at relatively low frequencies can be usefully harnessed for a number of applications in nondestructive evaluation. In this chapter, we have attempted to illustrate the wide spectrum of possibilities by describing only a few representative studies and the reader is directed to the vast published literature in this emerging area. With the advent of HTS SQUID sensors operating at liquid nitrogen temperatures, SQUID based NDE is poised to be increasingly used in critical applications where the requirement of high sensitivity, not attainable by any other sensor technology, overrides the cost of cryogenic operating temperatures.

9. References

Bain, R.J.P; Donaldson, G.B.; Evanson, S. & Hayward, G. (1985) SQUID gradiometric detection of defects in ferromagnetic structures. In *SQUID '85 - Superconducting Quantum Interference Devices and their Applications,* (eds.) Hahlbohm, H.D. & L¨ubbig, H., pp. 841-846, ISBN 0899251447, Walter de Gruyter & Co., Berlin, New York

Bain, R.J.P.; Donaldson, G.B.; Evanson, S. & Hayward, G. (1987). Design and operation of SQUID-based planar gradiometers for non-destructive testing of ferromagnetic plates. *IEEE Trans. Magn.* Vol. 23, No. 2, pp. 473-476, ISSN: 0018-9464

Baskaran, R. & Janawadkar, M.P. (2004). Universal dependence of optimum frequency on defect depth for SQUID based eddy current evaluation. *Proc. National Seminar on NDE*, Pune, India, December 2004

Chen,P.; Chen, L.; Li, J. & Ong, C.K. (2002). Novel excitation coils for non-destructive evaluation of non-magnetic metallic structures by high-T_c dc SQUID. *Superconductor Science and Technology*, Vol. 15, No.6, pp. 855-858, ISSN 0953-2048

Clarke, J.; Goubau, W.M. & Ketchen, M.B. (1976). Tunnel junction d.c SQUIDs: fabrication, operation and performance. *J. Low Temp. Phys.*, Vol. 25, No. 1-2, pp. 99-144, ISSN 0022-2291

Clarke,J. (1993). SQUIDs: theory and practice. in *The New Superconducting Electronics* (eds) Weinstock, H. & Ralston, R. W., Kluwer, Dordrecht, pp.123–180, ISBN 0-7923-2515-X

Dodd, C.V. & Deeds, W.E. (1968). Analytical Solutions to Eddy-Current Probe-Coil Problems, *J. App. Phys.*, Vol. 39, pp. 2829-2838, ISSN 0021-8979.

Donaldson, G.B.; Cochran, A. & McKirdy, D.A. (1996) in: H. Weinstock (Ed.), SQUID sensors: Fundamentals, Fabrication and Applications, Kluwer, Dordrecht, pp. 599-628, ISBN 0-7923-4350-6

Evanson, S.; Bain, R.J.P.; Donaldson, G.B.; Stirling, G. & Hayward, G. (1989) A comparison of the performance of planar and conventional second-order gradiometers coupled to a SQUID for the NDT of steel plates, *IEEE Trans. Magn.* Vol. 25, No. 2, pp. 1200-1203, ISSN: 0018-9464

Hans Koch, H. (1989) in *Sensors – A Comprehensive Survey* (eds Boll, R. & Overshott, K. J.), VCH Publishers, NY, Vol. 5, pp. 381–445, ISBN 978-3-527-62058-6

Jaycox, J. M. & Ketchen, M. B. (1981) Planar coupling scheme for ultra low noise DC SQUIDs *IEEE Trans. Magn.*, Vol. 17, pp. 401–403, ISSN 0018-9464

Jenks, W.G.; Sadeghi, S.S.H. & Wikswo, J.P. (1997) SQUIDs for nondestructive evaluation. *J.Phys.D.* Vol. 30, pp. 293-323, ISSN: 0022-3727

Klein,U.; Walker, M.E.; Carr, C.; McKirdy, D.M.; Pegrum, C.M.; Donaldson, G.B.; Cochran, A. & Nakane, H. (1997). Integrated low-temperature superconductor SQUID gradiometers for nondestructive evaluation. *IEEE Trans. Appl. Supercond.*, Vol. 7, No. 2, pp. 3037-3039, ISSN 1051-8223

Krause, H.-J. & Kreutzbruck, M.v. (2002) Recent developments in SQUID NDE, *Physica C*, Vol. 368, No. 1, pp. 70 - 79, ISSN 0921-4534

Krieger, J.; Krause, H.-J.; Gampe, U. & Sawade, G.(1999) . Magnetic field measurements on bridges and development of a mobile SQUID system. in: *Proc. Intl. Conf. on NDE Tech. Aging Infrastr.*, pp. 229-239, Newport Beach, USA, (March 1999)

Lang, M.; Johnson,J.; Screiber, J.; Dobmann, G.; Bassler, H.J.; Eifler,D.; Ehrlich, R. & Gampe, U. (2000). Cyclic deformation behaviour of AISI 321 austenitic steel and its characterization by means of HTC-SQUID. *Nuclear Engineering and Design*, Vol. 198, No.1, pp. 185-191, ISSN 0029-5493

Mück, M.; Korn, M.; Welzel, C.; Grawunder, S. & Schölz, F. (2005). Nondestructive evaluation of various materials using a SQUID-based eddy-current system. *IEEE Trans.Appl.Supercond.* Vol. 15, No. 2, pp. 733-736, ISSN 1051-8223

Nagesha, A.; Valsan, M.; Bhanu Sankara Rao, K. & Mannan. S.L.(1999) Strain rate effect on the low cycle fatigue behaviour of type 316L(N) SS base metal and 316 SS weld metal, Proceedings of the international welding conference IWC'99, Vol. 2, February 15-17, 1999, New Delhi, pp. 696-703

Sawade, G.; Straub, J.; Krause, H.J.; H. Bousack, Neudert, G. & Ehrlich, R. (1995). Signal analysis methods for remote magnetic examination of pre-stressed elements. in: Schckert, G. & Wiggenhauser, H. (eds) *Proc.Intl. Sympos.on NDT in Civil Eng. (NDT-CE)*, vol. II, DGZfP, Berlin, pp. 1077-1084

Sikora,R.; Chady, T.; Gratkowski, S.; Komorowski, M. & Stawicki, K. (2003). Eddy Current Testing of Thick Aluminum Plates with Hidden Cracks. *Review of progress in Quantitative nondestructive evaluation*, Vol. 22, AIP Conference Proceedings, Vol. 657, pp. 427-434, ISBN 0-7354-0117-9, Bellingham, Washington, July 2002

Tavrin, Y.; Siegel, M. & Hinken, J.-H. (1999). Standard method for detection of magnetic defects in aircraft engine discs using a HTS SQUID gradiometer. *IEEE Trans. Appl. Supercond.* Vol. 9, No.2, pp. 3809-3812, ISSN 1051-8223

Valsan, M.; Sundararaman, D.; Bhanu Sankara Rao, K. & Mannan, S.L. (1995). High Temperature Low Cycle Fatigue of Steels and their Welds, *Metall. Mater. Trans.*, Vol. 26A, pp. 1207-1219, ISSN 1073 - 5623

Valsan, M.; Nagesha, A.; Bhanu Sankara Rao, K. & Mannan, S.L. (2002). High temperature low cycle fatigue and creep-fatigue interaction behaviour of 316 and 316(N) weld metals and their weld joints. *Trans Ind Inst Met.* Vol. 55, No. 5 pp. 341-348, ISSN 0019-493X

Weinstock. H. & Nisenoff, M. (1985) Nondestructive evaluation of metallic structures using a SQUID gradiometer, *SQUID '85, Proc. 3rd Int. Conf. On Superconducting Quantum Interference Devices and their Applications* (eds) Hahlbohm, H.D. & L¨ubbig, H. (Berlin: deGruyter) pp. 843–847 ISBN 0899251447

Weinstock, H. & Nisenoff, M. (1986) Defect detection with a SQUID magnetometer *Review of Progress in QNDE 6* (ed) Thompson, D.O. & Chimenti, D. (New York: Plenum) 669-704

Weinstock, H. (1991). A review of SQUID magnetometry applied to nondestructive evaluation. *IEEE Trans. Mag.,* Vol. 27, No. 2, (March 1991), pp. 3231-3236, ISSN 0018-9464

Weinstock, H. (Ed.), (1995) *SQUID sensors: Fundamentals, Fabrication and Applications*, Kluwer, ISBN 0-7923-4350-6, Dordrecht

3

Applications of Current Technologies for Nondestructive Testing of Dental Biomaterials

Youssef S. Al Jabbari[1,2,*] and Spiros Zinelis[3,4]
[1]*Director, Dental Biomaterials Research and Development Chair, School of Dentistry, King Saud University, Riyadh,*
[2]*Prosthetic Dental Sciences Department, School of Dentistry, King Saud University, Riyadh,*
[3]*Department of Biomaterials, School of Dentistry, University of Athens, Athens*
[4]*Dental Biomaterials Research and Development Chair, School of Dentistry, King Saud University, Riyadh*
[1,2,4]*Saudi Arabia*
[3]*Greece*

1. Introduction

Dental biomaterials consist of a wide range of synthetic products that are normally used to restore patients' oral health, function and aesthetic appearance. Generally, dental biomaterials are classified on the basis of their atomic bonding into metallic, ceramic, polymer and composite materials. In addition to this classification, biomaterials can be classified according to their interactions with the surrounding oral tissues (Mano et al. 2004). Based on the tissue responses, biomaterials can be divided into three different categories commonly known as bioinert, bioactive and bioresorbable materials.

- Bioinert materials are materials that have minimal interactions with the surrounding tissues, such as partially stabilized zirconia, alumina, pure titanium and some of its alloys, some grades of stainless steel and ultra high molecular weight polyethylene.
- Bioactive materials are materials that start to interact positively with the surrounding hard and sometimes soft tissues after placement inside a biological body. Synthetic hydroxyapatite [$Ca_{10}(PO_4)_6(OH)_2$] and bioglass are the best representative examples of commonly used bioactive materials.
- Bioresorbable materials are materials that begin to progressively dissolve and slowly become replaced by the adjacent biological tissues after placement or implantation inside a biological body. Polylactic-polyglycolic acid copolymers and tricalcium phosphate [$Ca_3(PO_4)_2$] are two examples of the most commonly used bioresorbable materials.

Further to these previous classifications of dental biomaterials, it is important to mention that the final products normally placed in the oral cavity are produced in two different ways. The first involves industrial line production, wherein thousands of identical dental

* Corresponding Author

devices are manufactured, such as dental implants, endodontic files and orthodontic wires/brackets. The second is the production of custom-made devices that are produced in dental laboratories, such as crowns, fixed partial dentures, removable partial dentures, complete dentures and orthodontic devices (Figure 1).

As with any industrial application, the structural integrity of dental devices has to be tested to preserve the reliability of function and avoid premature failures. This testing starts with quality control aiming to discard any defective items, while the end products should comply with safety regulations and materials specifications (i.e. ISO standards). Information for assuring proper quality is acquired through destructive and non-destructive testing. These techniques are easily distinguished from each other, as the latter leaves the tested devices undamaged and suitable for additional testing and/or permanent usage in patients. Non-destructive testing (NDT) involves a broad spectrum of analytical techniques, including macroscopic/microscopic visual inspection, eddy current testing, radiography, X-ray computed tomography (XCT), resonance frequency analysis (RFA) and many others. NDT aims to qualitatively and/or quantitatively characterize the tested devices by detecting,

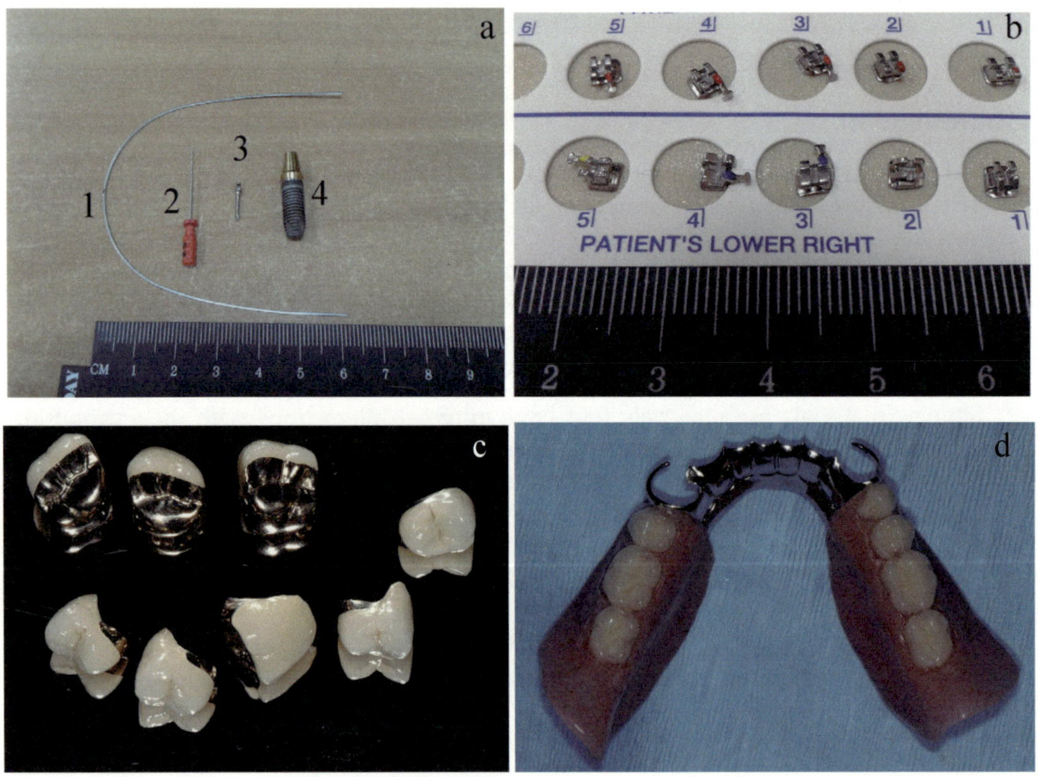

Fig. 1. a) Representative examples of industrially-made dental devices: 1) orthodontic archwire made of Ni-Ti alloy, 2) endodontic file made from Ni-Ti alloy, 3) Mini-orthodontic implant made of cp-Ti and 4) dental implant made of Ti. b) Industrially-made orthodontic brackets fabricated from stainless steel alloy. c) Custom-made ceramo-metal crowns. d) Custom made removable partial denture made of Cr-Co alloy and acrylic resin.

locating and sizing any external/internal structural defects. The acquired information is then assessed to evaluate whether the tested devices meet acceptable criteria or whether they should be rejected. Taking advantage of the NDT capabilities, the quality of the end products is increased while the level of reliability is enhanced. The application of these sophisticated NDT methods can easily be conducted industrially to produce acceptable dental devices. However, conducting extensive systematic quality control for custom-made dental devices is restricted owing to 1) the tremendous increase in relative costs, 2) the need for availability of specifically designed testing equipments in dental offices and laboratories and 3) the need for well-trained dentists and dental technicians to operate these equipments. Besides all these factors, if proper NDT is applied routinely to custom-made dental devices, it will dramatically increase the total cost of already expensive dental services.

Although there are a number of limiting factors for routine application of NDT in dentistry, NDT has a variety of important applications in the dental field. The best example is X-ray radiography, which is used for the detection of pores that are typically located in thin regions of cast custom-made dental devices (i.e. clasp shoulders of a cast metal removable partial denture framework). The early detection of pores at these critical regions and stage will allow the dentist and/or dental technician to correct the problem before the final delivery of the prostheses to the patients. The existence of such pores will lead to catastrophic premature failure in the oral cavity. In research applications, NDT contributes significantly to the determination of adverse effects of the oral environment on dental biomaterials/devices and vice versa. In addition, NDT is very helpful in studies of the interaction mechanisms between the oral environment and dental biomaterials as a result of intraoral aging. A variety of NDT methods can be used to fully characterize dental devices preoperatively (as received) and after retrieval from the oral cavity by tracking the occurrence of any changes during intraoral aging. Stereomicroscopy, variable pressure SEM (VPSEM) without the need for a conductive coating and X-ray microanalysis have been used to monitor the morphological and elemental alterations after intraoral aging of dental devices. In addition, computed X-ray micro-tomography (micro-XCT) has been used to correlate the locations and sizes of internal defects found preoperatively in all ceramic bridges with the fracture origin under clinical conditions. Such information acquired from micro-XCT regarding the defect locations will provide tremendous insights into the underlying mechanism of intraoral degradation. Other examples for applications of NDT in dentistry are 1) noninvasive and *in situ* testing of intraoral implant stability utilizing RFA and 2) detection of hidden or sub-surface caries (tooth decay) by utilizing X-radiography and laser technology measuring laser fluorescence within the tooth structure.

The aim of this chapter is to present and discuss the applications of current technologies for NDT of dental biomaterials covering a wide range of NDT techniques and their impacts on daily dental practice, noninvasive diagnosis and dental biomaterials research.

2. Applications of NDT for quality assurance purposes during dental treatment

Commonly, NDT techniques are performed by dentists and/or dental technicians to ensure the proper quality of custom-made dental devices. For example, they carry out routine checks by the naked eye or using a stereoscope for the marginal integrity of dental restorations (crowns) for precision before the restorations are delivered permanently to the patients. However, it is important to mention that precision in dentistry is a subjective term and varies from one dentist to another. This means that proper precision and acceptability are dependent on a dentist's personal experience and skills. Therefore, this type of testing/inspection is optional and in many situations is performed without specific requirements to accept or reject custom-made dental devices (e.g. crowns). Internal defects in cast and welded metallic custom-made devices are not uncommon and X-ray testing is performed for early detection of such defects. This section will present the dental applications of stereoscopic and X-ray examinations and testing.

2.1 Stereomicroscopic examination and testing

Stereomicroscopic examination is performed routinely by dental technicians and/or dentists to evaluate the quality of recently fabricated custom-made dental devices (e.g. crowns). Detection of any problems at this stage will allow for proper correction before the final prosthesis insertion in the patient's mouth. A recent study found that implant retaining screws deteriorate over a period of time when they are used to hold a dental prosthesis in a patient's mouth (Al Jabbari et. al, 2007a). Therefore, the authors encouraged dentists who provide extensive implant treatment in their practice to equip their offices with a stereomicroscope to enable regular evaluation of the quality of the tiny retaining implant screws at follow-up appointments. A low power stereomicroscope was found to be a powerful and useful tool for evaluating the quality of the external structures/surfaces of tiny dental devices (e.g. prosthetic retaining screws) (Al Jabbari et. al, 2007a). As illustrated in Figure 2, the stereomicroscope was able to clearly reveal the damage and deterioration of an implant screw head and threads, whereas it is almost impossible for a dentist to observe such damage and deterioration under the naked eye. The great advantage of performing this type of NDT in dental offices is that it will allow dentists to replace any severely damaged or deteriorated retaining screws with new ones. Failure to detect such deterioration and damage may lead in the future to a more complicated and expensive dental treatment for patients (Al Jabbari et. al, 2007b).

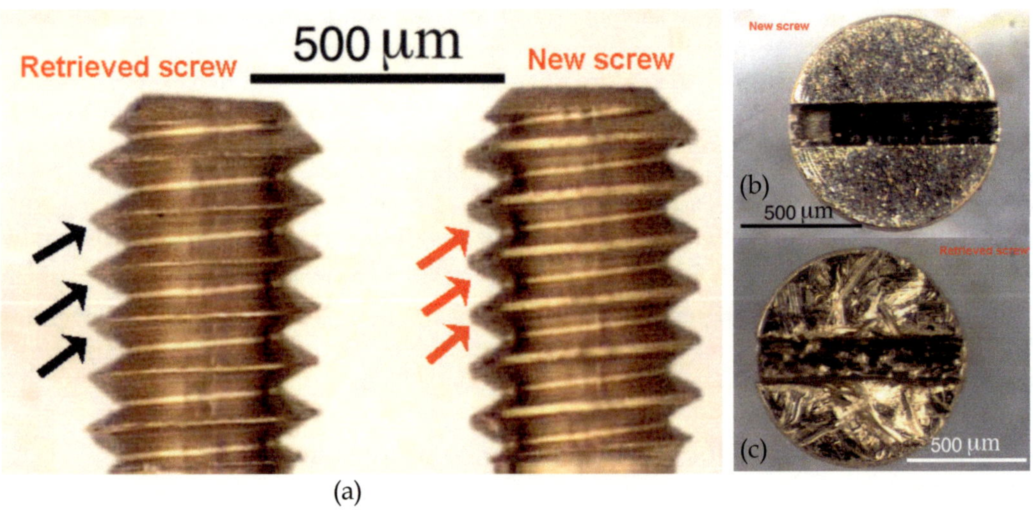

Fig. 2. Threaded segment (a) and slotted head segment (b) & (c) of new and retrieved implant retaining screws examined under stereomicroscope. Examination of these tiny dental implant devices reveals significant threads deterioration/thinning "black arrows" and screw head damage "(c)" when compared to the intact threads profile "red arrows" and normal slotted head (b).

As mentioned previously, dental technicians routinely evaluate the quality of their fabricated devices. For example, they use a stereomicroscope to examine devices made by casting. If they detect any external large casting porosity, they will correct and repair the problem by filling the pores with solder materials or they may need to remake the whole casted device before they pass it for clinical application. In addition, they regularly evaluate the passivity of the fit of their cast prosthesis superstructure utilizing a stereomicroscope. If they detect any misfit between the cast superstructure and the supporting teeth and/or implants, they will correct the problem using recasting, soldering, welding and electro-discharging machining techniques (Contreras et al. 2002; Ntasi et al. 2010; Romero et al. 2000; Zinelis 2007).

2.2 X-ray testing

Many dental devices (such as crowns, and fixed and removable partial dentures) are traditionally manufactured by casting dental alloys. Recently, the preparation of metal-free crowns and bridges has introduced the concept of casting ceramic materials. However, the mechanical stability and biocompatibility of dental devices depend on the properties of the materials and on the accuracy of the manufacturing process. Unfortunately, the dental casting procedure unavoidably leads to the development of pores in dental cast frameworks owing to gas entrapment or shrinkage, which may adversely affect the quality and efficacy of dental devices. The development of undesirable porosity is dependent on various factors and is a common complication for precious, semiprecious and base contemporary dental alloys (Elliopoulos et al. 2004; Li et al. 2010; Neto et al. 2003; Ucar et al. 2011; Zinelis 2000). This may negatively affect the long-term mechanical stability. For example, when there are cast external porosities, corrosion resistance decreases because of crevice formation and plaque accumulation in the oral cavity.

In industrial applications, internal voids can be readily investigated by employing X-ray examination, and the same technique has been adopted in dental practice as a non-destructive method for the same purposes. Pores can be easily distinguished as dark regions on radiographs, thereby providing significant information for the size, location and distribution of imperfections. The same methodology can be readily applied to identified internal voids in dental cast frameworks (Dharmar et al. 1993; Eisenburger et al. 2002; Eisenburger et al. 1998; Elarbi et al. 1985; Mattila 1964; Wictorin et al. 1979). However, the visibility of voids and the picture quality depend on the combination of the material to be tested and the analytical conditions applied for X-ray testing (Eisenburger & Addy 2002).

Going back to the principle of this technique, it must be noted that the attenuation of a narrow beam of monoenergetic photons with specific E and intensity I_o passing through a homogeneous material of thickness t is determined by Lambert's low:

$$I=I_o\, e^{-\mu(\rho,\, Z,\, E)*t} \tag{1}$$

where I is the photon intensity that goes out to the sample, and μ is the linear attenuation coefficient that depends on the material density, atomic number Z and beam energy. Therefore, the X-ray absorption depends on the atomic number, the density of the material and the energy of radiation. Accordingly, different dental alloys can be penetrated by X-rays to different extents during testing. As stated above, various precious, semiprecious and base metal alloys are used in the dental field that have different elemental, mechanical and physical properties (Roberts et al. 2009; Wataha 2002). Table 1 shows some representative types of dental alloys with densities ranging from 4.51 g/cm^3 for Ti to 19.3 g/cm^3 for pure Au. The vast majority of dental frameworks are produced by casting of precious and base metal alloys (Roberts et al. 2009) with the exception of pure gold for use in dentistry, which employs an electroforming technique (Vence 1997). As can be expected from equation (1) and the density values of the base metal alloys in Table 1, lower levels of X-ray absorption facilitate X-ray penetration. Low absorption coefficients and a high energy beam are needed to penetrate the thick metallic parts of cast dental frameworks. Figure 3 demonstrates the attenuation coefficients of some of the dental alloys presented in Table 1. As can be seen in Figure 3, the absorption coefficients decrease with increasing beam energy. This is not the case for pure Au and Au-based dental alloys where the absorption coefficient rises at the energy of 80 kV because the K absorption edge limits the penetration of Au and Au-based dental alloys to 0.6 mm (Eisenburger & Tschernitschek 1998). Non-precious and base dental alloys such as Ni-Cr, Co-Cr and commercially pure Ti (cpTi) can be penetrated by up to several millimeters depending on the accelerating voltage and exposure time (Eisenburger & Addy 2002; Eisenburger & Tschernitschek 1998).

Brand name	Mass content (%)	Type	Density (gr/cm^3)
Electroforming*	Au: 99.9	Pure Au	19.3
IPS d.SIGN 98	Au:85.9, Pt:12.1, Zn:1.5, In<1.0 Ir<1.0, Fe<1.0, Mn<1.0, Ta<1.0,	High Au	18.9
Degudent U94*	Au:76.0,Pt:9.6,Pd:8.9,Ag:1.2,Cu:0.3 Sn:0.8,In:1.5,Ir:0.1,Re:0.2,Ta:0.2.	High Au	17.9
Degulor M*	Au:70.0,Ag:13.5,Au:8.8,Pt:4., Pd:2.0, Zn:1.2, Ir: 0.1	High Au	15.5
Degupal G*	Pd:77.3, Ag:7.2, Au:4.5, Sn:4.0, Ru: 0.5; Ga: 6.0; Ge: 0.5	Pd based	11.7
4 all	Ni:61.5, Cr:25.7, Mo:11.0, Si:1.5, Mn<1.0, Al<1.0	Ni-Cr	8.4
IPS d.SIGN 30	Co:60.2,Cr:30.1,Ga:3.9,Nb:3.2 Si:0.9, Mo:0.6, Fe:0.5, B:0.3, Li<1.0	Co-Cr	7.8
Ti*	Ti:balance,O< 0.2,N< 0.06, C<0.08, H< 0.013, Fe < 0.25	cp Ti (gradeII)	4.5

Table 1. Brand names, elemental compositions, alloy types and densities of some representative dental alloys used for the production of metalloceramic crowns and bridges. The alloys are sorted in descending order of the density. The attenuation coefficients of the alloys indicated by asterisks are shown in Figure 3.

As examples, Figure 4 shows dental devices casted from grade II cpTi and analyzed by X-rays at 70-kV tube voltage, 8-mA beam current and 0.32-s exposure time. In Figure 3a, it can be seen that large pores are located at the connector areas of a cast framework for a fixed partial denture. Similar sized pores can be seen in cast rectangular specimens used for the determination of metalloceramic bonding strength values (Figure 4b). The spherical shape of these pores is a strong indication of gas entrapment, which is a typically reported problem when cpTi and Ti alloys are cast (Elliopoulos et al. 2004; Zinelis 2000). Figures 4c and 4d show spherical pore formation on a cast framework for a removable partial denture located critically at the shoulder of a circlet clasp, which is considered to be a common region for such pore formation.

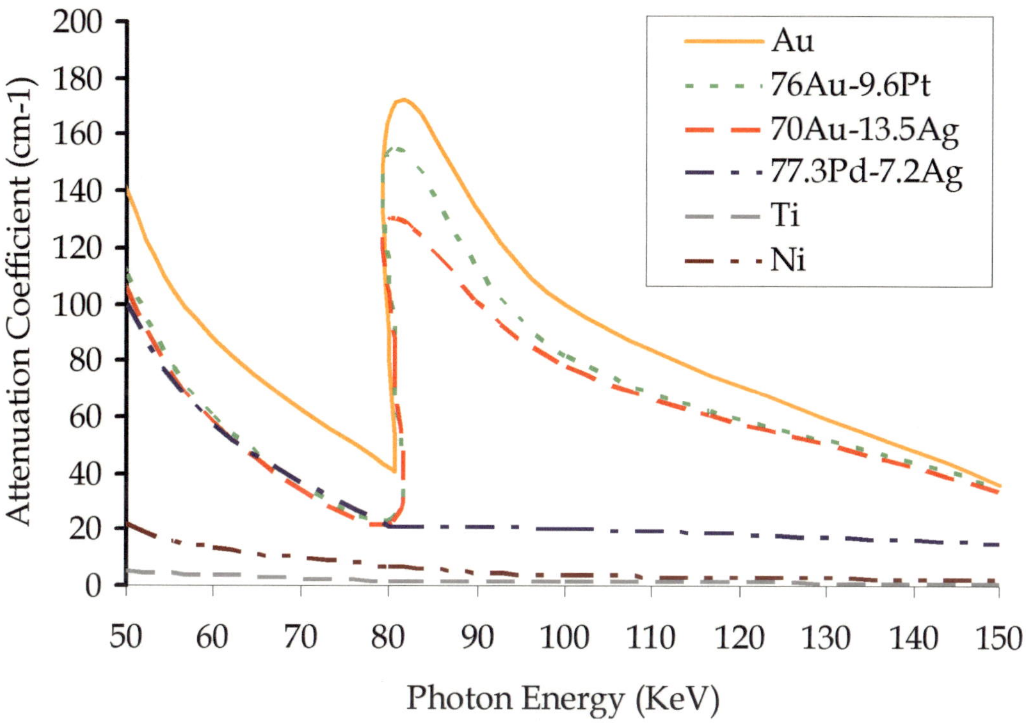

Fig. 3. Mass attenuation coefficient of three dental alloys (Eisenburger & Tschernitschek 1998). The same values for pure Au, Ni and Ti are presented for comparison purposes (Hubbell et al. 1996).

The main advantage of radiographic X-ray testing is the ability to provide quick valuable information about the quality of the metal framework casting by revealing the locations, numbers and sizes of internal defects that are not possible to detect by visual inspections. The occurrence of internal defects/porosity at critical regions such as the clasp shoulders of cast removable partial denture frameworks will lead to premature clinical failures. Therefore, early detection will allow proper correction of the problem prior to final delivery of a prosthetic device to a patient.

Previous studies have reported that fractures and premature failures of removable partial denture frameworks range from 16% to 19% owing to the occurrence of internal porosity in cast Co-Cr frameworks (Dharmar et al. 1993; Elarbi et al. 1985; Wictorin et al. 1979). In addition, they revealed that the frequencies of the common locations of these defects were 22% to 73% at major connectors, 5% to 43% at clasps/clasp shoulders and 6% to 8% at minor connectors (Dharmar et al. 1993; Elarbi et al. 1985). Therefore, and because of the high rates of occurrence of such defects, dentists and/or dental technicians are encouraged to perform radiographic X-ray inspections at early stages of removable partial denture fabrication.

Early detection of these casting defects and imperfections may allow easy repair or simple remaking of the metal framework. On the other hand, late detection may mandate remaking of the entire removable prosthesis, which is considered to be a costly choice to be performed by dentists, to avoid intraoral premature failures of a removable prosthesis.

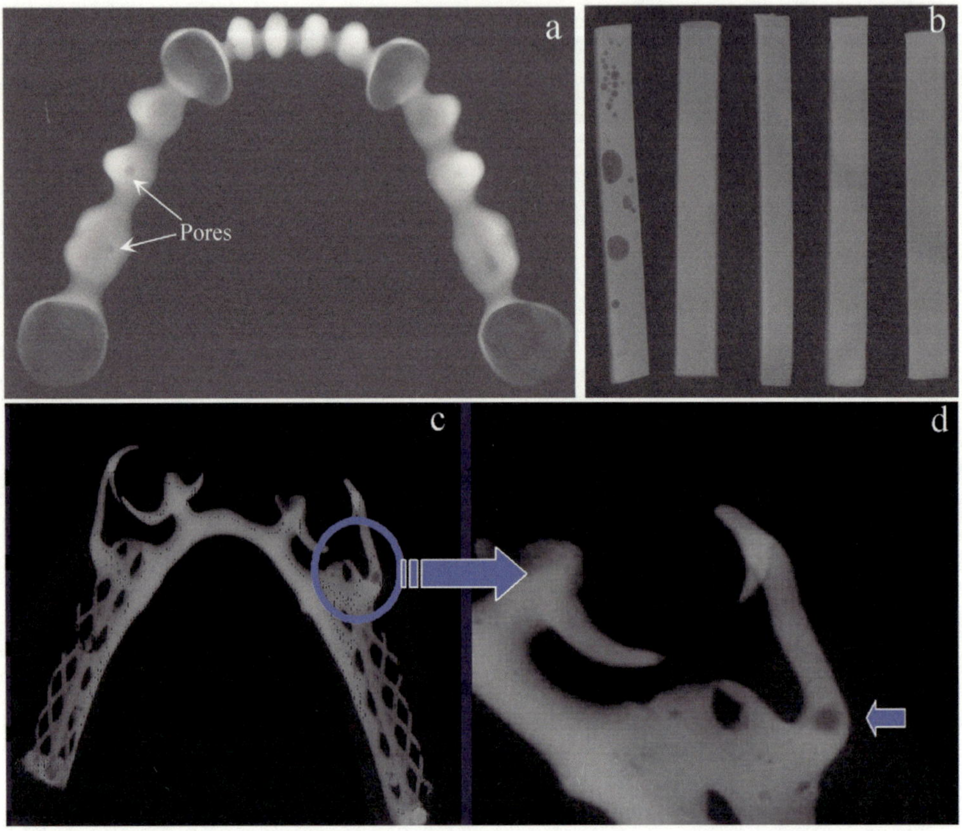

Fig. 4. X-ray images of dental casting. a) A dental cast framework for fixed partial denture, b) cast specimens for testing metalo-ceramic bond strength c) a cast framework for removable partial denture and d) High magnification of the highlighted region in (c) and it shows internal porosity occurrence (small arrow) at the clasp shoulder.

Another valuable dental application of radiographic nondestructive X-ray testing is in retrieval analysis studies. Conventional dental X-ray units can be readily used for X-ray analysis of dental alloys, except for pure Au and Au-based alloys where the penetration depth is limited to approximately 0.6 mm and thus increasing acceleration voltage to 120 kV is not possible with conventional dental X-ray machines. Microfocus X-ray systems (CRX 1000/CRX 2000; CR Technology Inc., Aliso Veijio, CA, USA) have been used to detect the occurrence of internal defects in prosthetic retaining implant screws made of gold-based alloys (Al Jabbari et al, 2007a). For NDT, utilization of a microfocus X-ray machine is possible and very valuable for evaluating various tiny dental devices made from precious, semiprecious and non-precious alloys. However, this type of machine is known to be mainly useful for dental biomaterials research purposes.

3. Applications of NDT for research purposes in the dental biomaterials field

Besides the valuable applications of NDT in quality assurance of dental devices, NDT makes significant contributions and has important applications in the dental biomaterials research field. Normally, dental biomaterials in the oral cavity are exposed to various aggressive conditions. When biomaterials are placed intraorally in the form of filling materials or prosthetic appliances, they go through a variety of degradation mechanisms such as fatigue, wear, corrosion and discoloration. Therefore, studies of the degradation mechanisms of biomaterials are important toward the design of new biomaterials with increased efficacy and longevity.

Laboratory or *in vitro* testing is widely applied to dental biomaterials to determine the properties of the materials. However, *in vitro* testing cannot provide any reliable information that can predict the *in vivo* behavior of biomaterials because the conditions of the oral environment cannot be simulated experimentally in research laboratories. Accordingly, NDT can be effectively used over a long period of time to monitor the occurrence of changes in a specific dental biomaterial or device resulting from intraoral aging. Generally, NDT of a retrieved specific dental biomaterial or device that has been placed in a patient's mouth for a reasonable time and comparison with the properties of a new (unused) biomaterial/device will provide more useful and significant information about the degradation mechanism resulting from long-term use *in vivo*. Under this general concept, a variety of NDT methods and techniques have been developed such as micro-XCT, VPSEM-EDS, optical profilometry and X-ray diffraction (XRD).

The main two limiting factors in conducting such research protocols are ethical and cost reasons. For example, a permanently placed filling dental biomaterial cannot be retrieved from the mouth for only research purposes. Retrieval of a biomaterial and replacement with a new one will subject the tooth to additional unnecessary clinical procedures. Multiple clinical procedures will cause repeated insults to a specific tooth that may lead to pulpal irritation and/or necrosis. Needless to say, that is ethically unacceptable. Similarly, a successful dental appliance or device cannot be retrieved from the mouth solely to conduct NDT before it fails, since making a new one is costly for the patient and there is no guarantee that the newly fabricated device will be as successful as the first one.

There are two additional obstacles that may also make the performance of NDT in retrieved dental biomaterials infeasible. First, it requires frequent patient follow-ups before the biomaterials and/or devices are retrieved. Unfortunately, not all patients will accept many follow-ups for research purposes only. Second, many retrieved dental biomaterials and/or devices are unsuitable for the conduction of certain types of NDT. For example, XRD analysis requires flat surfaces of a few square millimeters in dimension, a requirement that is hard to fulfill in dental devices.

3.1 Micro-XCT

Currently, computed tomography is extensively used in the medical field for diagnostic/treatment purposes. During the last two decades, new bench-top models have been introduced for use in the characterization of materials that employ similar principles to medical computed tomography. The only difference is that the bench-top models have an isotopic resolution capable of reaching a few tens of nanometers. Contrary to medical computed tomography machines, the specimens tested with bench-top models can be rotated while the detector and X-ray source are fixed within the machine. The micro-XCT scanning produces hundreds of horizontal slices for a tested specimen, which are then used to reconstruct the entire specimen. Reconstruction of a specimen is accomplished by two reconstruction algorithms commonly known as iterative and filtered back projection methods. In addition, computer software can be utilized in the development of three-dimensional models, pseudocoloring and quantitative determination of geometrical features of an irregular dental biomaterial device.

Figure 5 shows a good example of NDT utilizing a micro-XCT analysis of a fixed partial denture (FPD). It reveals the importance of this tool for nondestructively analyzing the internal structure of the whole ceramic FPD. The FPD was analyzed prior to permanent cementation of the prosthesis in the patient's mouth. The analysis revealed that the joining procedure of the three different parts of the alumina core was not done properly because of the entrapment of large voids at and within the bulk of the cementing material (Figure 5d). Unfortunately, after final insertion of the FPD in the patient's mouth, it did fail at the connector area after being in service for a short period of time. Therefore, it can be said that micro-XCT is a powerful tool for evaluating the quality of industrially and/or custom-made dental devices and for failure analysis of dental biomaterials.

3.2 SEM-VPSEM-EPMA

Scanning electron microscopy (SEM) combined with electron probe microanalysis (EPMA) is considered to be a powerful analytical tool for providing morphological and elemental information about tested samples at low (6×) and high (150,000×) magnifications. SEM is able to bridge the gap between optical stereomicroscopy and transmission electron microscopy. Recent advancements in SEM manufacturing technology can provide imaging of non-conductive specimens (low-vacuum SEM) and samples at 99% relative humidity (environmental SEM). These new operating modes are also known as VPSEM and confer tremendous capabilities to SEM. Additional information regarding the operation principles and applications of these new SEM models can be found in relevant previous reports (Bergmans et al. 2005; Danilatos; Danilatos 1991; Danilatos 1993; Danilatos 1994; Danilatos et al.; Kodaka et al. 1992).

In dentistry, brazing is the main joining technique for making metallic orthodontic appliances. A space maintainer is an example of the most commonly used orthodontic appliances. This appliance is made of two stainless steel tooth bands joined by a stainless steel orthodontic wire. The orthodontic wire and the two bands are joined by brazing utilizing low fusing silver brazing alloys (Figure 6). Figure 7 shows a high magnification SEM photomicrograph of an orthodontic space maintainer appliance, revealing that the soldered area joins the stainless steel bands to the orthodontic wire. NDT utilizing X-ray EDS analysis was performed at that area at two different times (before and after dental treatment). The purpose of the NDT and the analysis was to determine the effects of long-term use *in vivo* on the Ag-based brazing alloy. Small porosities (indicated by arrows in Figure 7) were used to identify the area for X-ray EDS analysis. The two spectra obtained at the two different times are shown in Figure 8, and reveal significant decreases in the Cu and Zn composition after intraoral aging.

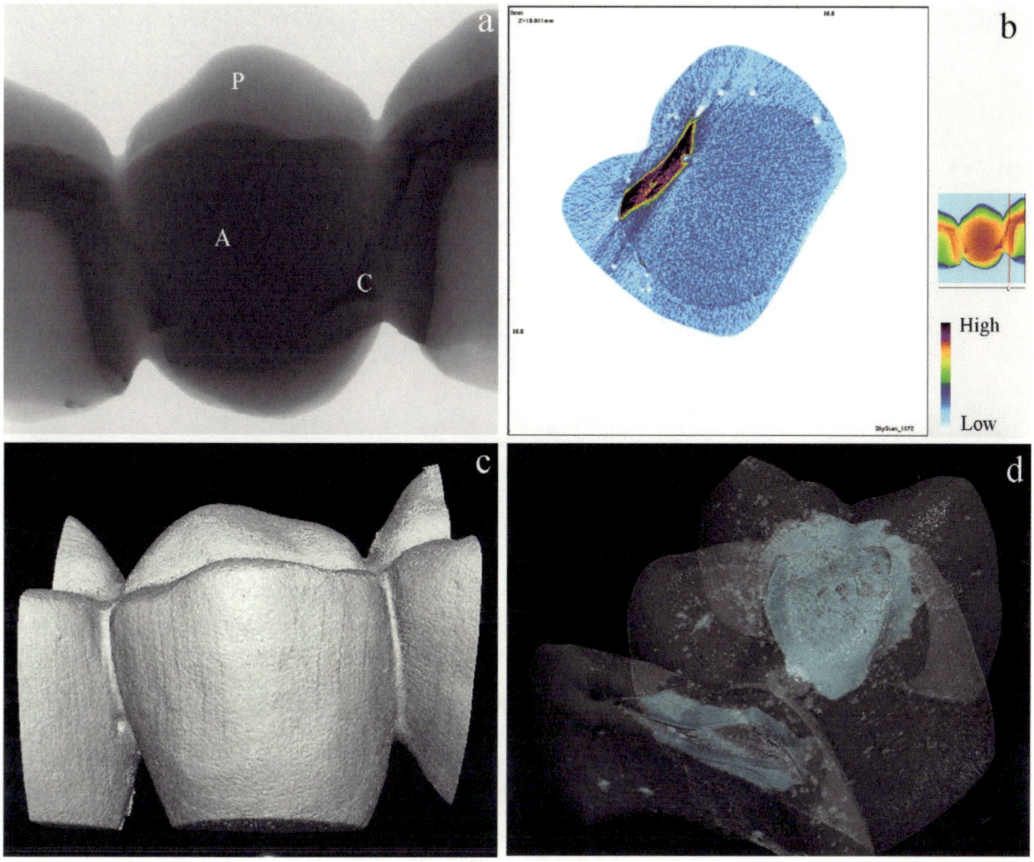

Fig. 5. Micro XCT analysis of a fixed partial denture (FPD) that was fabricated from all ceramic materials. The core of the FPD was made from alumina that was veneered with dental porcelain. **a)** FPD X-ray image before the FPD was cemented in the patient's mouth showing (A) the core alumina, (B) the veneering porcelain and (C) the connector area joining the three unites of the FPD. It is easy to distinguish between the three parts because the differences in X-ray absorption. **b)** A pseudocoloring reconstruction of a perpendicular cross section for the FPD showing the alumina core, the veneering porcelain and the material used for cementation at the connector area. Pores are easily identified as white circle areas in the regions of porcelain layer and at the core porcelain interface. Red line located on the small attached image indicates the cross sectional plane while the bar indicates the absorption scale. **c)** 3D image of the reconstructed structure after digital processing helpful in total volume determination. **d)** Alternative 3D image showing the distribution of internal pores in dental porcelain and the occurrence of big voids near and within the cement layer.

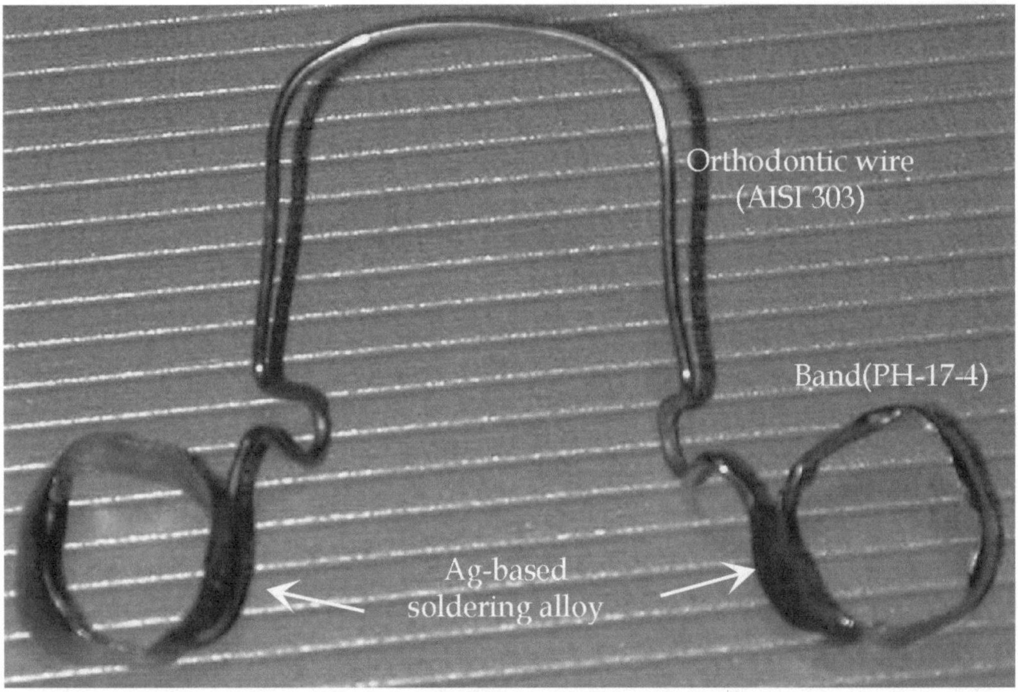

Fig. 6. An orthodontic device known as space maintainer made of two bands and a wire with two soldered joints (arrows). Bands and wires were manufactured from stainless steel whereas the soldering alloy is Ag-based alloy containing Cu, Zn and Sn.

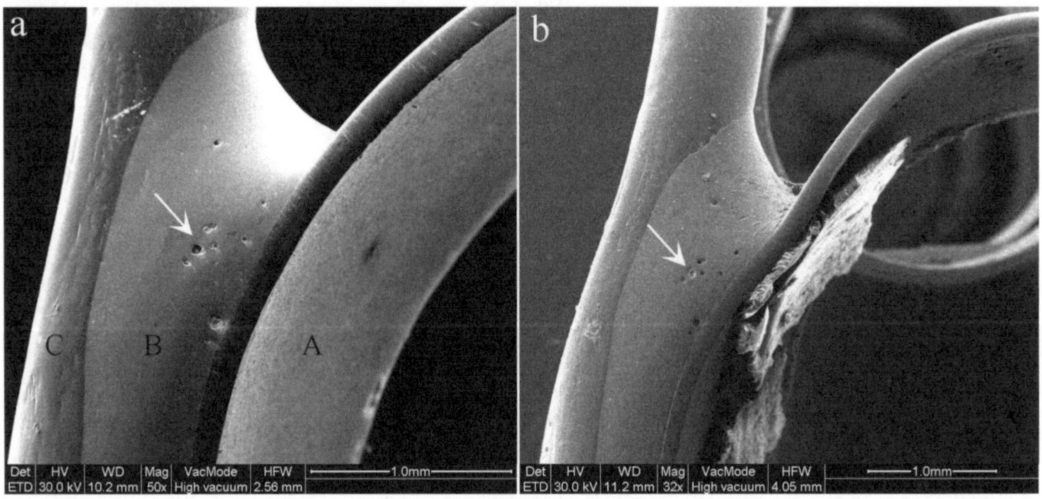

Fig. 7. Secondary Electron Images (SEI) of a joint area between the stainless steel band (A) and the orthodontic wire (C) soldered with Ag-based soldering alloy (B). (a) As-received appliance from the dental laboratory and before it was placed in the patient's mouth. (b) The retrieved appliance from the patient's mouth after it serviced for approximately 14 months of treatment period. External surface porosities (arrows) were used as a reference for locating exact area used for x-ray EDS analysis.

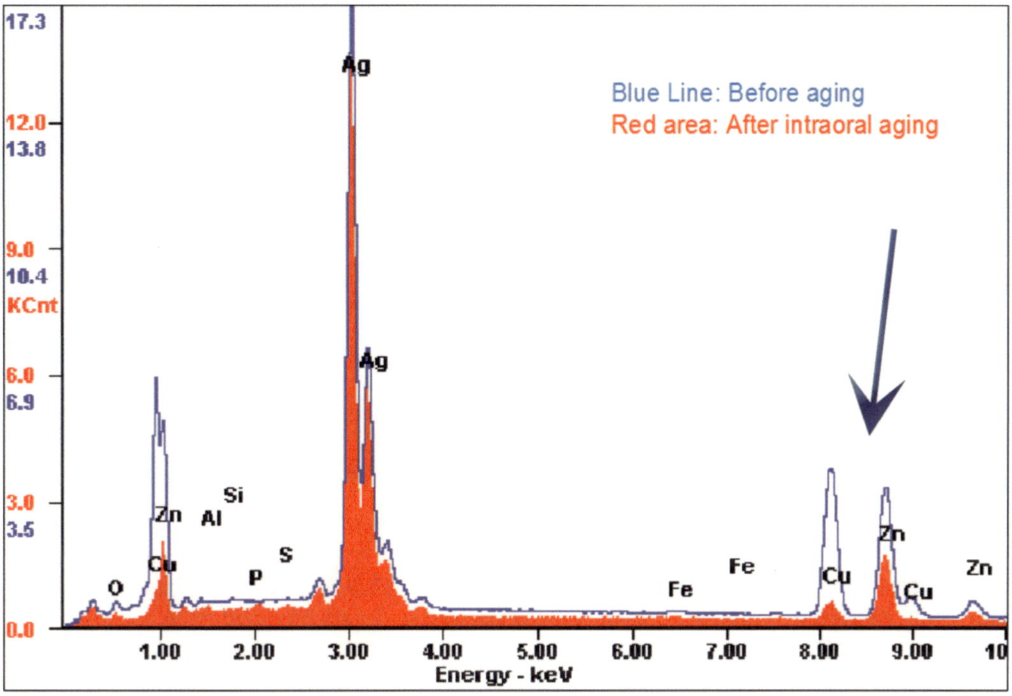

Fig. 8. EDS x-ray spectrum for the Ag-based alloy solder before and after intraoral aging. Note the decrease in Cu, Zn after long-term use *in vivo*.

Figures 8 and 9 provide additional information regarding the surface structure of the Ag-based soldering alloy. Both figures confirm that the significant decreases in Cu and Zn are appended to the dissolution of the low atomic contrast second phase, which is enriched in Cu and Zn. A possible explanation is that dental biomaterials with multiple phase structures might be prone to galvanic corrosion. Of course, the presence of the stainless steel bands and wire might have an additional effect on this phenomenon. However, this is only an assumption and the verification of galvanic corrosion requires further extended research. The important contribution of this NDT and X-ray EDS analysis is significant because it confirms other previous findings that Ag-based alloys are prone to corrosion and ionic release (Grimsdottir et al. 1992; Locci et al. 2000a; Mockers et al. 2002; Staffolani et al. 1999). These findings might be a reason for mucosal irritation, which was reported in a previous study (Bishara 1995). The significant release and dissolution of Cu and Zn during intraoral aging must be taken seriously because Cu ions have toxic effects on the human body (Locci et al. 2000b; Vannet et al. 2007; Wataha et al. 2002).

4. Dental applications of NDT as non-invasive diagnostic methods

Radiographic X-rays are used routinely in dental offices to non-invasively diagnose hard dental tissue diseases or to detect dental caries (tooth decay). However, in recent years, new technologies have been developed and introduced into the dental field for use as non-invasive diagnostic tools. The best two examples are RFA (Meredith et al. 1997) and fluorescence measurements (Jablonski-Momeni et al. 2011; Lussi et al. 2003; Rodrigues et al. 2010).

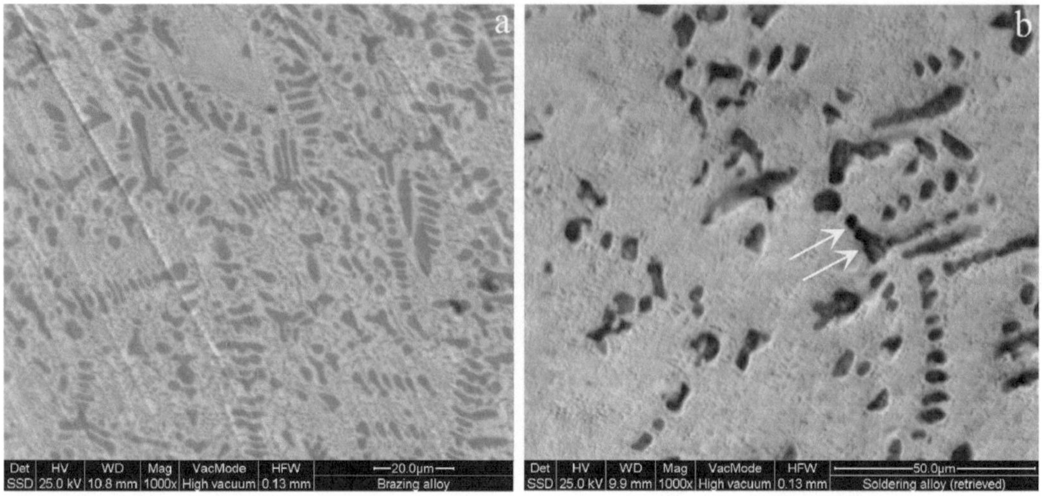

Fig. 9. Backscattered Electron Images after etching of Ag-based alloy solder before (a) and after (b) intraoral aging. The Ag based alloy has two different microstructure phases with dispersion of lower atomic contrast phase in the matrix. This second phase has been completely diluted leaving behind a random distribution of surface craters. (Original magnification: 1000X).

4.1 RFA

The main dental application of RFA as a diagnostic method is to quantify dental implant stability after surgical implant placement in the human jaw bone. Normally, when dental implants are placed in healthy human jaws, they will form and establish strong stable bond with the surrounding bone tissues after a period of several months and this phenomenon is known as *Osseointegration* (Albrektsson et al. 1986). It has been suggested in several studies that the stiffness of the bone-implant interface can be assessed by RFA (Valderrama et al. 2007; Sakoh J et al. 2006; Alsaadi G et al. 2007). Therefore, clinicians have been advised to utilize RFA to determine the strength and adequacy of the established bond (Osseointegration) before they restore implants with dental prostheses. A commonly used device for this purpose is the Osstell Mentor device (Integration Diagnostics, Goteborg, Sweden) (Figure 10a). As shown in Figure 10, part of the Osstell Mentor device comprises L-shaped transducers. These transducers will record all the information as an implant stability quotient (ISQ), which is a function of the bone-implant stiffness (N/µm) and the marginal bone height. The ISQ is a dimensionless quantity, and larger values indicate greater levels of interfacial bone-implant stiffness (meaning a higher established osseointegration with greater stability).

Besides the aforementioned beneficial diagnostic applications of RFA, it has been utilized extensively in research studies. Traditionally, research studies have evaluated the established bond between an implant and bone by histomorphometric tests. However, the main disadvantage of these tests is that they are destructive (Meredith N. 1998). Therefore, it has been suggested that RFA can be used periodically to evaluate the established bond (osseointegration occurrence) between an implant and bone without sacrificing the object in an *in vivo* study (Figure 10b) (Meredith N. 1998; Huang HM. et al. 2003).

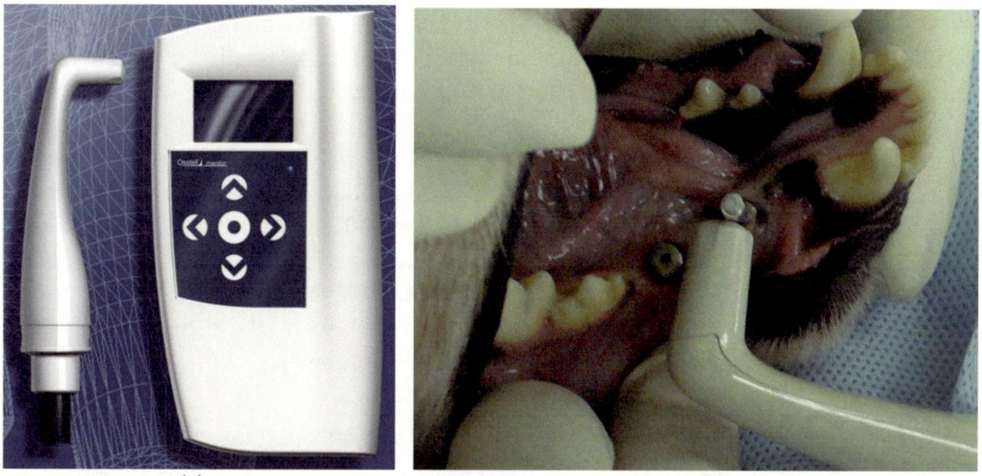

(a) (b)

Fig. 10. (a) The Osstell device utilized normally for RFA. (b) Illustrating conduction of RFA in an animal study. It is important that the transducer be placed in same position each time a measurement is taken for a specific dental implant. Different ISQ values could be obtained by positioning the transducer in different directions.

4.2 Fluorescence measurements

Dental caries are also known as tooth decay and result from demineralization of inorganic components of the outer layer (enamel) of the tooth structure. Released bacterial lactic acid will normally lead to tooth enamel demineralization. Detection of tooth caries at early stages is crucial because it only requires a non-costly simple treatment. However, the early stages of dental caries may not be easily detected by the naked eye during routine dental examinations. Therefore, fluorescence measurements have been recommended for early detection and diagnosis of enamel demineralization (Jablonski-Momeni et al. 2011; Lussi et al. 2003; Rodrigues et al. 2010). The structure of healthy and sound tooth enamel is characterized by a low baseline fluorescence level, while demineralized and infected enamel will have an increased fluorescence level. In addition, the fluorescence level increases as the caries process advances (Lussi et al. 2001). A recently developed device for detecting dental caries at its early stages based on fluorescence measurements is the DIAGNOdent device (KaVo, Biberach, Germany). The DIAGNOdent device emits red light at 655 nm and detects bacterial metabolites in the demineralized tooth structure (Lussi et al. 2003; Lussi et al. 2006a; Lussi et al. 2006b). The DIAGNOdent device then classifies the tested regions according to the calibrated fluorescence intensity as follows: scores 0–13 = no caries, scores 14–20 = early (incipient) enamel caries and scores above 20 = advanced dentine caries (Figure 10a) (Lussi & Hellwig 2006b).

Another recently developed fluorescence camera, VistaProof (Dürr Dental, Bietigheim-Bissingen, Germany), is used for early diagnosis of dental caries. The VistaProof emits blue light at 405 nm (Jablonski-Momeni et al. 2011) and records fluorescence from the probed tooth surfaces in the form of digital images (Rodrigues et al. 2008). Intact tooth structures show green fluorescent images, while infected and demineralized tooth structures show blue-violet fluorescent images. Infected areas with increased numbers of bacteria and bacterial byproducts show red fluorescent images (Figure 11b). The digital software utilized by the VistaProof quantifies the color components and provides scoring outcomes ranging from 0–4 that indicate the penetration depth of the dental caries within the tooth structure (Figure 11b). The scoring outcomes are good diagnostic values for the presence or absence and the severity of dental caries. The values are used as follows: 0–0.9 = sound and healthy tooth structure; 0.9–1.5 = initial (incipient) enamel caries; 1.5–2.0 = deep enamel caries; 2.0–2.5 = dentine caries; and above 2.5 = deep dentine caries. It is important to mention that, despite the reported reliable applications of the DIAGOdent and VistaProof for non-invasive diagnosis of dental caries, dentists routinely verify their findings by taking radiographic X-rays to diagnose the occurrence of dental caries.

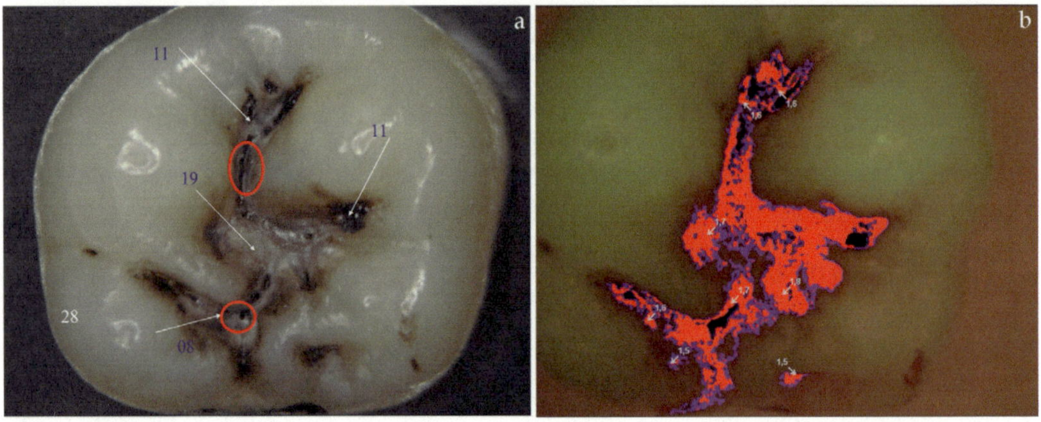

Fig. 11. Occlusal surface of a molar tooth with dental caries. The infected areas were diagnosed by DIAGNOdent (a) and VistaProof (b). Numbers in both images provide valuable information regarding the extent and penetration depth of dental caries within the tooth structure.

5. Conclusions

NDT plays very important roles in dentistry, and in the dental biomaterials research field in particular. However, the application of NDT on a daily basis for dental diagnosis purposes and for assuring adequate therapeutic quality is limited, mainly because of the significant increases in relevant time and cost. Luckily, NDT along with its various applications in the dental biomaterials research field is performed routinely by scientific researchers. Consequently, this NDT has led to noticeable enhancements of the quality, performance and biocompatibility of dental biomaterials that are placed daily into patients' mouths.

6. Acknowledgments

The author s would like to thank Prof. George Eliades, Dr. Triantaffilos Papadopoulos, Dr. Abdulaziz Al-Rasheed, Dr. Anil Sukumaran, Dr. Maysa Al-Marshood, Peter Tsakiridis, Dr. Alerxis Tagmatarxis and Maria Michalaki for their help in providing some of the pictures for this manuscript.

7. References

Al Jabbari YS., Fournelle R., Ziebert G., Toth J., Iacopino AM. (2007a). Mechanical behavior and Failure analysis of prosthetic retaining screws after long-term use in vivo. Part I: Charactrization of adhesive wear and structure of retaining screws. *Journal of Prosthodontics*, Vol.17, No.3, pp.168-180, ISNN 1532-849X

Al Jabbari YS., Fournelle R., Ziebert G., Toth J., Iacopino AM. (2007b). Mechanical behavior and Failure analysis of prosthetic retaining screws after long-term use in vivo. Part III: Preload and fracture tensile load testing. *Journal of Prosthodontics*, Vol.17, No.3, pp.192-200, ISNN 1532-849X

Alsaadi G., Quirynen M., Michiels K., Jacobs R., van Steenberghe D. (2007). A biomechanical assessment of the relation between the oral implant stability at insertion and subjective bone quality assessment. *Journal of Clinical Periodontology*, Vol.34, No.4, pp. 359-366, ISNN 0303-6979

Bergmans L., Moisiadis P., Van Meerbeek B., Quirynen M., Lambrechts P. (2005). Microscopic observation of bacteria: review highlighting the use of environmental SEM. *International Endodontic Journal*, Vol.38, No.11, pp. 775-88, ISNN 0143-2885

Bishara S. E. (1995). Oral Lesions Caused by an Orthodontic Retainer - a Case-Report. *American Journal of Orthodontics and Dentofacial Orthopedics*, Vol.108, No.2, pp. 115-117, ISNN 0889-5406

Contreras E. F. R., Henriques G. E. P., Giolo S. R., Nobilo M. A. A. (2002). Fit of cast commercially pure titanium and Ti-6Al-4V alloy crowns before and after marginal refinement by electrical discharge machining. *Journal of Prosthetic Dentistry*, Vol.88, No.5, pp. 467-472, ISNN 0022-3913

Danilatos G. D. (1991). Review and Outline of Environmental Sem at Present. *Journal of Microscopy-Oxford*, Vol.162, pp. 391-402, ISNN 0022-2720

Danilatos G. D. (1993). Introduction to the Esem Instrument. *Microscopy Research and Technique*, Vol.25, No.5-6, pp. 354-361, ISNN 1059-910X

Danilatos G. D. (1994). Environmental Scanning Electron-Microscopy and Microanalysis. *Mikrochimica Acta*, Vol.114, pp. 143-155, ISNN 0026-3672

Dharmar S., Rathnasamy R. J., Swaminathan T. N. (1993). Radiographic and metallographic evaluation of porosity defects and grain structure of cast chromium cobalt removable partial dentures. *Journal of Prosthetic Dentistry*, Vol.69, No.4, pp. 369-73, ISNN 0022-3913

Eisenburger M., Addy M. (2002). Radiological examination of dental castings -- a review of the method and comparisons of the equipment. *Journal of Oral Rehabilitation*, Vol.29, No.7, pp. 609-14, ISNN 0305-182X

Eisenburger M., Tschernitschek H. (1998). Radiographic inspection of dental castings. *Clinical Oral Investigations*, Vol.2, No.1, pp. 11-4, ISNN 1432-6981

Elarbi E. A., Ismail Y. H., Azarbal M., Saini T. S. (1985). Radiographic detection of porosities in removable partial denture castings. *Journal of Prosthetic Dentistry*, Vol.54, No.5, pp. 674-7, ISNN 0022-3913

Elliopoulos D., Zinelis S., Papadopoulos T. (2004). Porosity of cpTi casting with four different casting machines. *Journal of Prosthetic Dentistry*, Vol.92, No.4, pp. 377-381, ISNN 0022-3913

Grimsdottir M. R., Gjerdet N. R., Henstenpettersen A. (1992). Composition and Invitro Corrosion of Orthodontic Appliances. *American Journal of Orthodontics and Dentofacial Orthopedics*, Vol.101, No.6, pp. 525-532, ISNN 0889-5406

Hubbell J., Seltzer S. (1996) NIST X-ray Attenuation Databases. In NIST ed. NIST Databases.

Huang HM., Chiu CL., Yeh CY., Lin CT., Lin LH., Lee SY (2003). Early detection of implant healing process using resonance frequency analysis. *Clinical Oral Implants Research,* Vol.14, No.4, pp. 437-443, ISNN 1600-0501.

Jablonski-Momeni A., Schipper H. M., Rosen S. M. et al. (2011). Performance of a fluorescence camera for detection of occlusal caries in vitro. *Odontology,* Vol.99, No.1, pp. 55-61, ISNN 1618-1247

Kodaka T., Toko T., Debari K., Hisamitsu H., Ohmori A., Kawata S. (1992). Application of the environmental SEM in human dentin bleached with hydrogen peroxide in vitro. *Journal of electron microscopy (Tokyo),* Vol.41, No.5, pp. 381-6, ISNN 0022-0744

Li D., Baba N., Brantley W. A., Alapati S. B., Heshmati R. H., Daehn G. S. (2010). Study of Pd-Ag dental alloys: examination of effect of casting porosity on fatigue behavior and microstructural analysis. *Journal of Materials Science-Materials in Medicine,* Vol.21, No.10, pp. 2723-2731, ISNN 0957-4530

Locci P., Lilli C., Marinucci L. et al. (2000a). In vitro cytotoxic effects of orthodontic appliances. *Journal of Biomedical Materials Research,* Vol.53, No.5, pp. 560-567, ISNN 0021-9304

Locci P., Marinucci L., Lilli C. et al. (2000b). Biocompatibility of alloys used in orthodontics evaluated by cell culture tests. *Journal of Biomedical Materials Research,* Vol.51, No.4, pp. 561-568, ISNN 0021-9304

Lussi A., Hack A., Hug I., Heckenberger H., Megert B., Stich H. (2006a). Detection of approximal caries with a new laser fluorescence device. *Caries Research,* Vol.40, No.2, pp. 97-103, ISNN 0008-6568

Lussi A., Hellwig E. (2006b). Performance of a new laser fluorescence device for the detection of occlusal caries in vitro. Journal of Dentistry, Vol.34, No.7, pp. 467-471, ISNN 0300-5712

Lussi A., Longbottom C., Gygax M., Braig F. (2003). The influence of professional cleaning and drying of occlusal surfaces on DIAGNOdent readings. *Journal of Dental Research,* Vol.82, pp. B64-B64, ISNN 0022-0345

Lussi A., Megert B., Longbottom C., Reich E., Francescut P. (2001). Clinical performance of a laser fluorescence device for detection of occlusal caries lesions. *European Journal of Oral Sciences,* Vol.109, No.1, pp. 14-19, ISNN 0909-8836

Mano J. F., Sousa R. A., Boesel L. F., Neves N. M., Reis R. L. (2004). Bloinert, biodegradable and injectable polymeric matrix composites for hard tissue replacement: state of the art and recent developments. Composites Science and Technology, Vol.64, No.6, pp. 789-817, ISNN 0266-3538

Mattila K. (1964). A Roentgenological Study of Internal Defects in Chrome-Cobalt Implants and Partial Dentures. *Acta odontologica Scandinavica,* Vol.22, pp. 215-28, ISNN 0001-6357

Meredith N. (1998). Assessment of implant stability as a prognostic determinant. *International Journal of Prosthodontics,* Vol.11, No.5, pp.491-501, ISNN 0893-2174

Meredith N., Book K., Friberg B., Jemt T., Sennerby L. (1997). Resonance frequency measurements of implant stability in vivo. A cross-sectional and longitudinal study of resonance frequency measurements on implants in the edentulous and partially dentate maxilla. *Clinical Oral Implants Research*, Vol.8, No.3, pp. 226-33, ISNN 0905-7161

Mockers O., Deroze D., Camps J. (2002). Cytotoxicity of orthodontic bands, brackets and archwires in vitro. *Dental Materials*, Vol.18, No.4, pp. 311-317, ISNN 0109-5641

Neto H. G., Candido M. S. M., Junior A. L. R., Garcia P. P. N. S. (2003). Analysis of depth of the microporosity in a nickel-chromium system alloy - effects of electrolytic, chemical and sandblasting etching. *Journal of Oral Rehabilitation*, Vol.30, No.5, pp. 556-558, ISNN 0305-182X

Ntasi A., Mueller W. D., Eliades G., Zinelis S. (2010). The effect of Electro Discharge Machining (EDM) on the corrosion resistance of dental alloys. *Dental Materials*, Vol.26, No.12, pp. E237-E245, ISNN 0109-5641

Roberts H. W., Berzins D. W., Moore B. K., Charlton D. G. (2009). Metal-ceramic alloys in dentistry: a review. *Journal of Prosthodontics*, Vol.18, No.2, pp. 188-94, ISNN 1532-849X

Rodrigues J. A., Hug I., Diniz M. B., Lussi A. (2008). Performance of fluorescence methods, radiographic examination and ICDAS II on occlusal surfaces in vitro. *Caries Research*, Vol.42, No.4, pp. 297-304, ISNN 0008-6568

Rodrigues J., Diniz M., Hug I., Cordeiro R., Lussi A. (2010). Relationship between DIAGNOdent values and sealant penetration depth on occlusal fissures. *Clinical Oral Investigations*, Vol.14, No.6, pp. 707-711, ISNN 1432-6981

Romero G. G., Engelmeier R., Powers J. M., Canterbury A. A. (2000). Accuracy of three corrective techniques for implant bar fabrication. *Journal of Prosthetic Dentistry*, Vol.84, No.6, pp. 602-607, ISNN 0022-3913

Sakoh J., Wahlmann U., Stender E., Nat R., Al-Nawas B., Wagner W (2006). Primary stability of a conical implant and a hybrid, cylindric screw-type implant in vitro. *International Journal of Oral and Maxillofacial Implants*, Vol.21, No.4, pp. 560-566, ISNN 0882-2786

Siegwart, R. (2001). Indirect Manipulation of a Sphere on a Flat Disk Using Force Information. *International Journal of Advanced Robotic Systems*, Vol.6, No.4, (December 2009), pp. 12-16, ISSN 1729-8806

Staffolani N., Damiani F., Lilli C. et al. (1999). Ion release from orthodontic appliances. *Journal of Dentistry*, Vol.27, No.6, pp. 449-454, ISNN 0300-5712

Ucar Y., Brantley W. A., Johnston W. M., Dasgupta T. (2011). Mechanical Properties, Fracture Surface Characterization, and Microstructural Analysis of Six Noble Dental Casting Alloys. *Journal of Prosthetic Dentistry*, Vol.105, No.6, pp. 394-402, ISNN 0022-3913

Valderrama P., Oates TW., Jones AA., Simpson J., Schoolfield JD., Cochran DL. (2007). Evaluation of two different resonance frequency devices to detect implant stability: A clinical trial. *Journal of Periodontology*, Vol.78, No.2, pp.262-272. ISNN 0022-3492

Vannet B. V., Hanssens J. L., Wehrbein H. (2007). The use of three-dimensional oral mucosa cell cultures to assess the toxicity of soldered and welded wires. *European Journal of Orthodontics*, Vol.29, No.1, pp. 60-66, ISNN 0141-5387

Vence B. S. (1997). Electroforming technology for galvanoceramic restorations. *Journal of Prosthetic Dentistry*, Vol.77, No.4, pp. 444-9, ISNN 0022-3913

Wataha J. C. (2002). Alloys for prosthodontic restorations. J Prosthet Dent, Vol.87, No.4, pp. 351-63, ISNN 0022-3913

Wataha J. C., Lockwood P. E., Schedle A., Noda M., Bouillaguet S. (2002). Ag, Cu, Hg and Ni ions alter the metabolism of human monocytes during extended low-dose exposures. *Journal of Oral Rehabilitation*, Vol.29, No.2, pp. 133-139, ISNN 0305-182X

Wictorin L., Julin P., Mollersten L. (1979). Roentgenological detection of casting defects in cobalt-chromium alloy frameworks. *Journal of Oral Rehabilitation*, Vol.6, No.2, pp. 137-46, ISNN 0305-182X

Zinelis S. (2000). Effect of pressure of helium, argon, krypton, and xenon on the porosity, microstructure, and mechanical properties of commercially pure titanium castings. *Journal of Prosthetic Dentistry*, Vol.84, No.5, pp. 575-582, ISNN 0022-3913

Zinelis S. (2007). Surface and elemental alterations of dental alloys induced by electro discharge machining (EDM). *Dental Materials*, Vol.23, No.5, pp. 601-607, ISNN 0109-5641

4

Neutron Radiography

Nares Chankow
Department of Nuclear Engineering, Faculty of Engineering
Chulalongkorn University
Thailand

1. Introduction

A few years after the discovery of neutron by James Chadwick in 1932, H. Kallman and E. Kuhn started their work on neutron radiography in Germany using neutrons from a small neutron generator. Due to the second World War, their first publication was delayed until 1947. However, the first report on neutron radiography was published by Peters in 1946, a year before Kallman and Kuhn's. After research reactors were available, in 1956 Thewlis and Derbyshire in UK demonstrated that much better neutron radiographic images could be obtained by using intense thermal neutron beam from the reactor. Specific applications of neutron radiography were then started and expanded rapidly particularly where research reactors were available.

The radiographic technique was originally based on metallic neutron converter screen/film assembly. Neutron converter screen and film were gradually improved until early 1990's when computer technology became powerful and was available at low cost. Non-film neutron radiography was then possible to be used for routine inspection of specimens. After 2005, imaging plate specially designed for neutron radiography was available and could provide image quality comparable to the best image quality obtained from the gadolinium foil/film assembly with relative speed approximately 40 times faster. Nevertheless, neutron radiography has not been widely employed for routine inspection of specimen in industry like x-ray and gamma-ray radiography due to two main reasons. Firstly, excellent image quality still needs neutrons from nuclear reactor. Secondly, neutron radiography is only well-known among the academic but not industrial people. It is actually excellent for inspecting parts containing light elements in materials even when they are covered or enveloped by heavy elements. Nowadays, neutrons from small neutron generator and californium-252 source can give neutron intensity sufficient for modern image recording system such as the neutron imaging plate and the light-emitting neutron converter screen/digital camera assembly.

2. Principle of neutron radiography

Neutrons are fundamental particles which are bound together with protons within the atomic nucleus. Neutron is electrically neutral and has mass of nearly the same as a proton i.e. about 1 u. Once a neutron is emitted from the nucleus it becomes free neutron which is not stable. It decays to a proton and an electron with a half-life of 12 minutes.

Neutron radiography requires parallel beam or divergent beam of low energy neutrons having intensity in the range of only 10^4 - 10^6 neutrons/cm^2-s to avoid formation of significant amount of long-lived radioactive isotope from neutron absorption within the specimen. The transmitted neutrons will then interact with neutron converter screen to generate particles or light photons which can be recorded by film or any other recording media. Free neutrons emitted from all sources are fast neutrons while neutron radiography prefers low energy neutrons. To reduce neutron energy, neutron sources are normally surrounded by large volume of hydrogeneous material such as water, polyethylene, transformer oil and paraffin. Neutron collimator is designed to bring low energy neutron beam to the test specimen. As illustrated in Figure 2, attenuation coefficient of gamma-ray increases with increasing of the atomic number of element while attenuation coefficients of neutron are high for light elements like hydrogen(H), lithium (Li) and boron(B) as well as some heavy elements such as gadolinium (Gd), cadmium(Cd) and dysprosium (Dy). In contrast, lead (Pb) has very high attenuation coefficient for gamma-ray but very low for neutron. Neutron radiography therefore can make parts containing light elements; such as polymer, plastic, rubber, chemical; visible even when they are covered or enveloped by heavy elements.

Neutrons may interact with matter in one or more of the following reactions.

i. Elastic scattering: (n, n) reaction

Neutron collides with the atomic nucleus, then loses its kinetic energy. It should be noted that neutron loses less kinetic energy when it collides with a heavy nucleus. In contrast, it loses more kinetic energy when collides with a light nucleus. Hydrogen(^{1}H) is therefore the most effective neutron moderator because it is the lightest nucleus having mass almost the same as neutron (~1u). Elastic scattering is most important in production of low energy or slow neutrons from fast neutrons emitted from the source for neutron radiography. Water, paraffin and polyethylene are common neutron moderators. In fact, hydrogen-2 (^{2}H, so called "deuterium") is the best neutron moderator due to its extremely low neutron absorption probability. Heavy water (D_2O) has neutron absorption cross section only about 1/500 that of light water (H_2O) but heavy water is very costly.

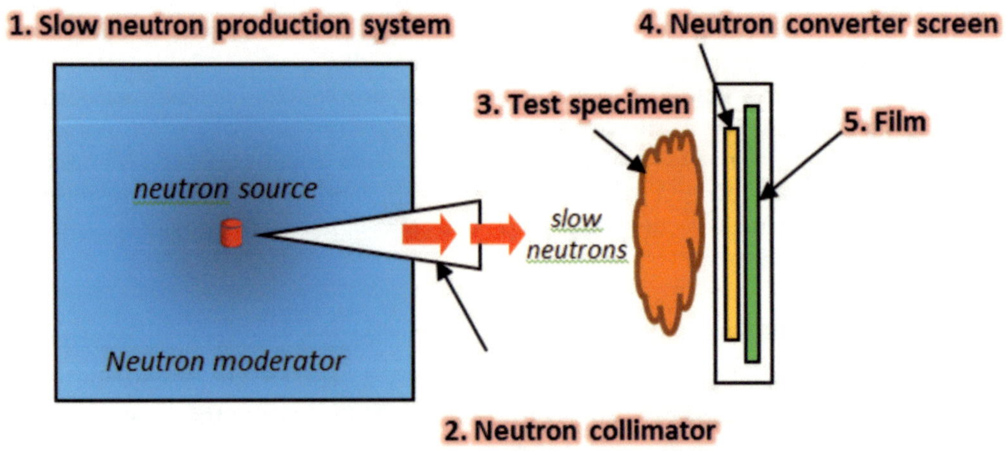

Fig. 1. Major components of typical neutron radiography system

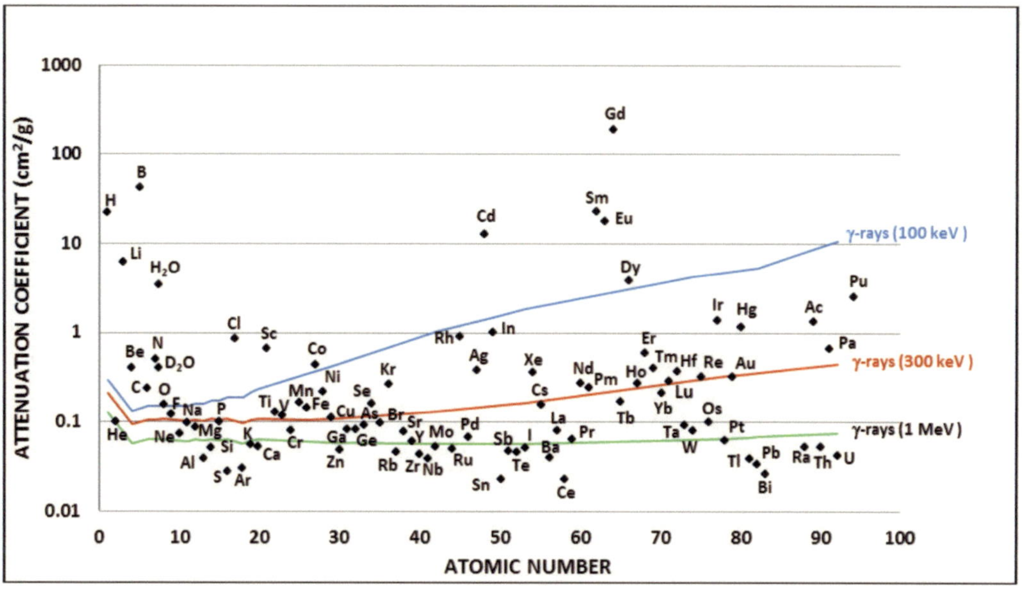

Fig. 2. Mass attenuation coefficients for thermal neutrons (♦) and gamma-rays as a function of atomic number of elements (reproduced from [3] with some modifications)

ii. Inelastic scattering: (n, n′) or (n, n′γ) reaction

Similar to elastic scattering but when a neutron collides with the atomic nucleus it has enough kinetic energy to raise the nucleus into its excited state. After collision, the nucleus will give off gamma-ray(s) in returning to its ground state. Even inelastic scattering also reduces energy of fast neutron but this is not preferable in neutron radiography because it increases in contamination of gamma-rays to the system.

iii. Neutron capture: (n, γ) reaction

A neutron can be absorbed by the atomic nucleus to form new nucleus with an additional neutron resulting in increasing of mass number by 1. For example, when cobalt-59 (^{59}Co) captures a neutron will become radioactive cobalt-60 (^{60}Co). The new nucleus mostly becomes radioactive and decays to beta-particle followed by emission of gamma-ray. Some of them are not radioactive such as ^{2}H, ^{114}Cd, ^{156}Gd, and ^{158}Gd. A few of them only decay by beta-particle without emission of gamma-ray such as ^{32}P. This reaction plays a vital role in neutron radiography when metallic foil screen is used to convert neutrons into beta-particles and gamma-rays.

iv. Charged particle emission: (n, p) and (n, α) reactions

Most of charged particle emission occurs by fast neutrons except for the two important (n, α) reactions of lithium-6 (^{6}Li) and boron-10 (^{10}B). The $^{6}Li(n, \alpha)^{3}H$ and $^{10}B(n, \alpha)^{7}Li$ reactions play important roles in neutron detection and shielding. In neutron radiography, these two reactions are mainly employed to convert neutrons to alpha particles or to light. The (n, p) reaction is not important in neutron radiography but it may be useful when solid state track detector is selected as the image recorder.

v. Neutron producing reaction: (n, 2n) and (n, 3n) reactions

These reactions occur only with fast neutrons which require a threshold energy to trigger. They may be useful in neutron radiography particularly when utilizing 14-MeV neutrons produced from a neutron generator. By inserting blocks of heavy metal like lead (Pb) or uranium (U) in neutron moderator, low energy neutron intensity can be increased by a factor of 2 - 3 or higher from (n, 2n) and (n, 3n) reactions.

vi. Fission: (n, f) reaction

Fission reaction is well-known for energy production in nuclear power plant and neutron production in nuclear research reactor. A heavy nucleus like uranium-235 (^{235}U), plutonium-239(^{239}Pu) undergoes fission after absorption of neutron. The nucleus splits into 2 nuclei of mass approximately one-half of the original nucleus with emission of 2 - 3 neutrons. When uranium (U) is used to increase neutron intensity by the above (n, 2n) and (n, 3n) reactions, fission reaction also contributes additional neutrons to the system. The degree of contribution depends on the ratio of uranium-235 to uranium-238 in uranium.

Attenuation of neutron by the specimen depends on thickness and the attenuation coefficient similar to gamma-ray as follows.

$$I_t = I_0 \exp(-\Sigma t) \qquad (1)$$

Where

I_t is the transmitted neutron intensity (n/cm^2-s)

I_0 is the incident neutron intensity (n/cm^2-s)

t is the specimen thickness (cm)

Σ is the macroscopic cross section (cm^{-1})

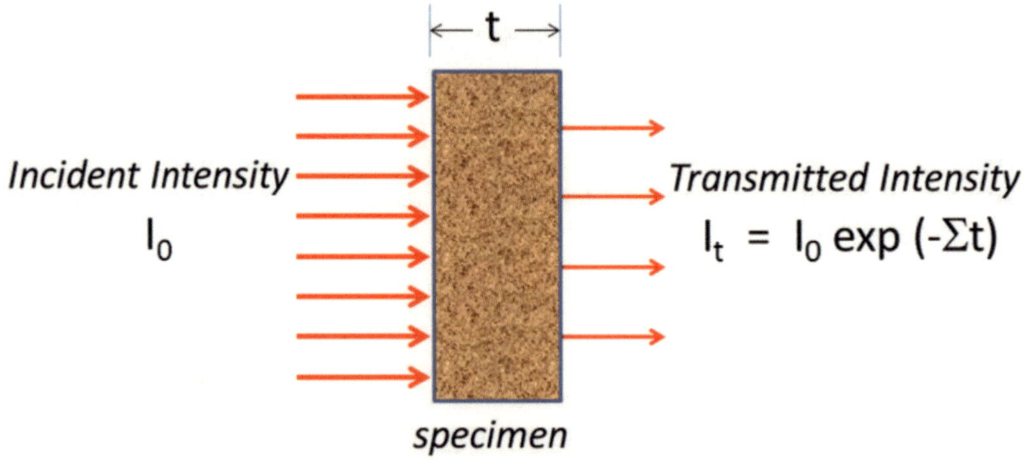

Fig. 3. Attenuation of neutrons by specimen

Σ is equivalent to the linear attenuation coefficient of gamma-ray (μ) and is the characteristic of elements in the specimen. Σ and μ are the product of atom density of elements contained in the specimen (in atoms/ cm^3) and their effective microscopic cross sections (σ) to the reactions of interest (in cm^2). σ is the effective cross section, not the actual physical cross section of the nucleus. It indicates probability of occurrence for each neutron interaction. For examples, $\sigma_{(n,\gamma)}$ indicates the probability of (n, γ) reaction and σ_s indicates the probability of scattering reaction which combines elastic (n, n) and inelastic (n, n′) scattering cross sections. Σ and σ of pure elements and common compounds or mixtures (such as water, heavy water and concrete) can be found in literatures. Figure 4 illustrates percentage of 0.0253 eV neutron transmission through different kinds of material having thickness of 1 cm.

Element or Molecular	Atomic number	Nominal Density (g/cm³)	σ_a, barns	Σ_a, cm^{-1}
Boron	5	2.3	759	97.23
Carbon(graphite)	6	1.60	3.4×10^{-3}	2.728×10^{-4}
Heavy water		1.105	1.33×10^{-3}	4.42×10^{-5}
Water		1.0	0.664	0.02220
Aluminum	13	2.699	0.230	0.01386
Iron	26	7.87	2.55	0.2164
Copper	29	8.96	3.79	0.3219
Silver	47	10.49	63.6	3.725
Gadolinium	64	7.95	49000	1492
Dysprosium	66	8.56	930	29.50
Gold	79	19.32	98.8	5.836
Lead	82	11.34	0.170	5.603×10^{-3}

Table 1. Microscopic and macroscopic cross sections of some elements and compounds [4]

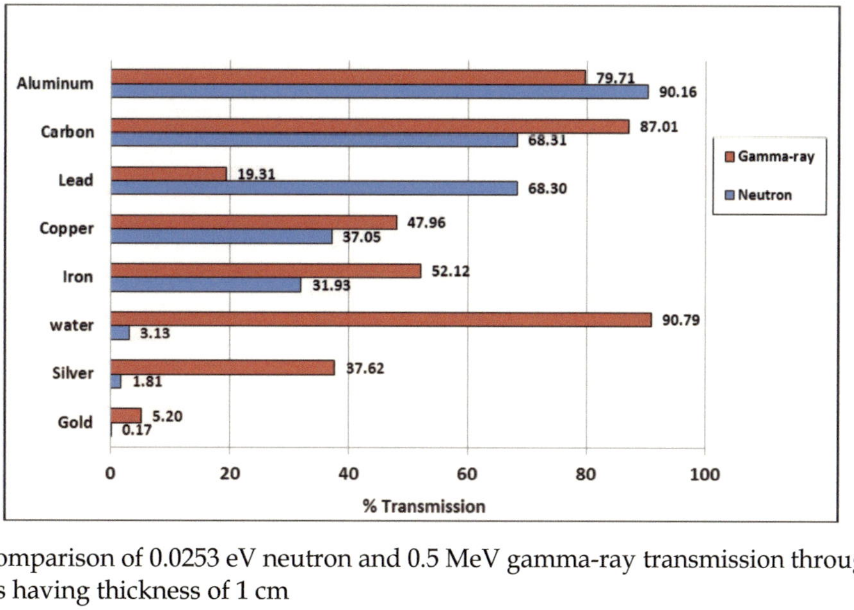

Fig. 4. Comparison of 0.0253 eV neutron and 0.5 MeV gamma-ray transmission through materials having thickness of 1 cm

3. Neutron sources

Neutron sources for neutron radiography can be divided into 3 groups. These are radioisotope source, electronic source and nuclear reactor.

i. Radioisotope neutron source

Nowadays, two radioisotope sources are appropriate and available for neutron radiography i.e. americium-241/beryllium (^{241}Am/Be) and californium-252 (^{252}Cf). ^{241}Am/Be produces neutron from (α, n) reaction by bombardment of beryllium (Be) nucleus with alpha particles from ^{241}Am. The average neutron energy and the neutron emission rate are approximately 4.5 MeV and 2.2 x 10^6 neutrons/second per 1 curie (Ci) of ^{241}Am with a half-life of 432 years. ^{241}Am/Be can be available up to several tens curies of ^{241}Am. ^{252}Cf emits neutrons from spontaneous fission with average neutron energy of 2 MeV and the emission rate of 4.3 x 10^9 neutrons/second per curie or 2.3 x 10^6 neutrons/second per microgram of ^{252}Cf. ^{252}Cf has a half-life of 2.6 years and is the best radioisotope source for neutron radiography due to its extremely high neutron output, low average emitted neutron energy and small size.

ii. Electronic neutron source

Particle accelerator and neutron generator are neutron emitting sources produced by nuclear reactions. Particles are accelerated to a sufficient energy and brought to hit target nuclei to produce neutrons. Compact neutron generators are now available for field use with neutron emission rate of 10^9 to 10^{12} neutrons per second. The reactions below are commonly used to produce neutrons.

$${}_1D^2 + {}_1D^2 \rightarrow {}_0n^1 + {}_2He^3 + 3.28\ \text{MeV}, \quad \textit{so called "DD Reaction"}$$

$${}_1T^3 + {}_1D^2 \rightarrow {}_0n^1 + {}_2He^4 + 17.6\ \text{MeV}, \quad \textit{so called "DT Reaction"}$$

$${}_1D^2 + {}_4Be^9 \rightarrow {}_0n^1 + {}_5B^{10} + 4.35\ \text{MeV}$$

$${}_1H^1 + {}_4Be^9 \rightarrow {}_0n^1 + {}_5B^9 - 1.85\ \text{MeV}$$

Energy of fast neurons produced from the above reactions is monoenergetic and depends on the incoming particle i.e. ^{1}H and ^{2}D. The DT reaction is a well-known fusion reaction for generating 14 MeV neutrons.

iii. Nuclear reactor

Nuclear reactor generally produces neutrons from fission reaction of uranium-235 (^{235}U). The fission neutron energy is in the range of 0 – 10 MeV with the most probable and the average energy of 0.7 and 2 MeV respectively. Fission reactions take place in the nuclear reactor fuel rods which are surrounded by neutron moderator. The moderator reduces the neutron energy to thermal energy or slow neutron. Due to its high slow neutron intensity in the reactor core of about 10^{12} – 10^{14} neutrons/cm^2 per second, good collimation of neutron beam can be easily obtained to give excellent image quality for neutron radiography.

Maximum neutron flux in moderator is a function of neutron emission rate from neutron source and neutron energy. Thermalization factor, as shown in Table 3, is the ratio of the neutron emission rate (in neutrons per second, s^{-1}) to the maximum neutron flux (in neutrons per second per square centimeter, cm^{-2} s^{-1}) in moderator. Neutron flux in water moderator per a neutron emitted from the neutron source at any distances can be obtained from Figures 6 and 7. For example, the thermalization factor of ^{252}Cf obtained from Table 3 is 100. The neutron emission rate of ^{252}Cf is 2.3 x 10^6 neutrons per microgram. If a 500 mg ^{252}Cf is used, the maximum flux in water can be calculated from 500 x 2.3 x 10^6/100 = 1.15 x 10^7 cm^{-2} s^{-1}. From a graph in Figure 6 for ^{252}Cf, the maximum flux is at 1 cm distance from the source which indicates the neutron flux of about 1 x 10^{-2} per a neutron emission from ^{252}Cf. Thus, the maximum neutron flux can be calculated from (500 x 2.3 x 10^6) x (1 x 10^{-2}) = 1.15 x 10^7 cm^{-2} s^{-1}. Neutron flux at other distances can also be obtained from Figure 6. It should be noted that the thermalization factor increases with increasing emitted neutron energy from the source.

As mentioned earlier, neutron radiography requires low energy neutrons. The lower neutron energy gives better image contrast. Fast neutron or high energy neutrons emitted from the source are slowed down by moderator such as water to produce slow or low energy neutrons. The slow neutron energy in moderator is dependent of moderator temperature and the energy distribution follows Maxwellian's for gas molecules and particles. The slow neutron is therefore called "thermal neutron". "Cold neutrons" can be produced by cooling the moderator/collimator down, such as with liquid helium, to obtain better image contrast. The cadmium ratio of cold neutrons is indicated in Table 2 where infinity (∞) means there is no epicadmium neutron (energy > 0.5 eV) in the beam.

Source	Comments
Radioisotope	constant neutron output, low cost, maintenance free, no operating cost, low neutron flux/long exposure time, acceptable image quality, mobile unit is possible
Accelerator	moderate cost, moderate operating and maintenance cost, medium neutron flux/medium exposure time, good image quality, mobile unit is possible for small neutron generator
Nuclear Reactor	constant neutron output, high cost, high maintenance and operating cost, high neutron flux/short exposure time, excellent image quality, mobile unit is impossible

Table 2. Neutron sources and their general characteristics

Source	Emitted neutron energy (MeV)		Normal thermal neutron flux (cm^{-2} s)		Therma-lization factor	L/D ratio	Cd ratio
	Range	Mean	in moderator	at specimen			
Radioisotope - ^{241}Am/Be (50 Ci)	0 - 10	4.5	10^8	10^3 - 10^4	400	30-50	5-20
- ^{252}Cf (1 mg)	0 - 10	2.0	10^9	10^3 – 10^5	100	50-100	5-20
Accelerator - ^{2}D(d, n)^{3}He	2.7	2.7	10^8 – 10^9	10^4 – 10^5	200	20-30	5-20
- ^{3}T(d, n)^{4}He	14.1	14.1	10^8 – 10^9	10^4 – 10^5	600	20-30	5-20
- ^{9}Be(p, n)^{9}B	1.15	1.15	10^8 – 10^9	10^4 – 10^5	50	20-30	5-20
- ^{9}Be(d, n)^{10}B	3.96	3.96	10^8 – 10^9	10^4 – 10^5	100	20-30	5-20
Nuclear Reactor	0-10	2.0	10^{11} - 10^{14}	10^6 – 10^8	100	100-300	100-∞

Table 3. Neutron sources and their technical characteristics [2, 6]

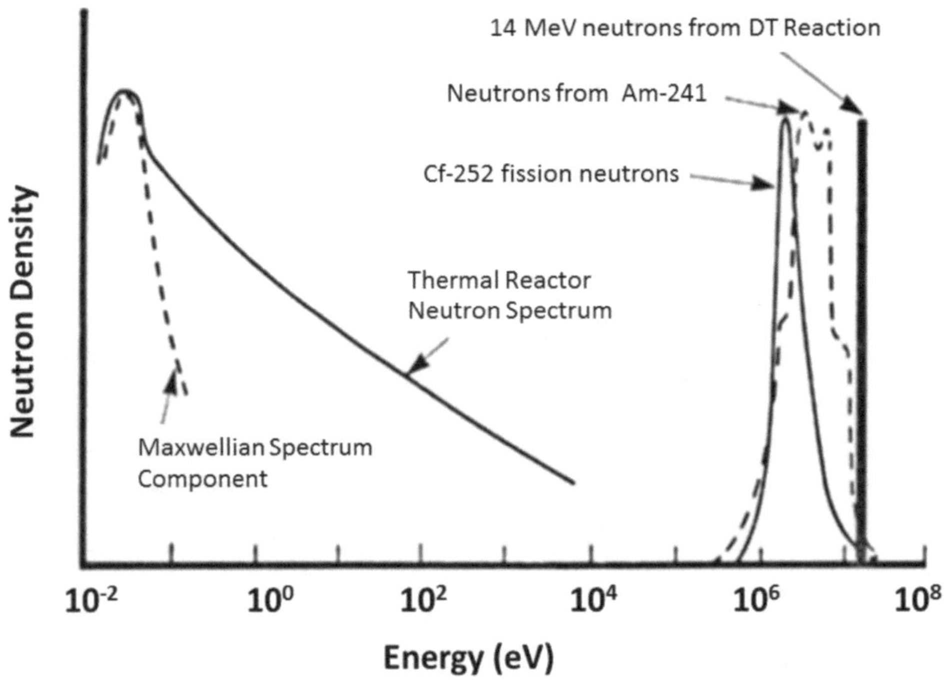

Fig. 5. Energy spectra of neutrons from neutron sources (reproduced from [5])

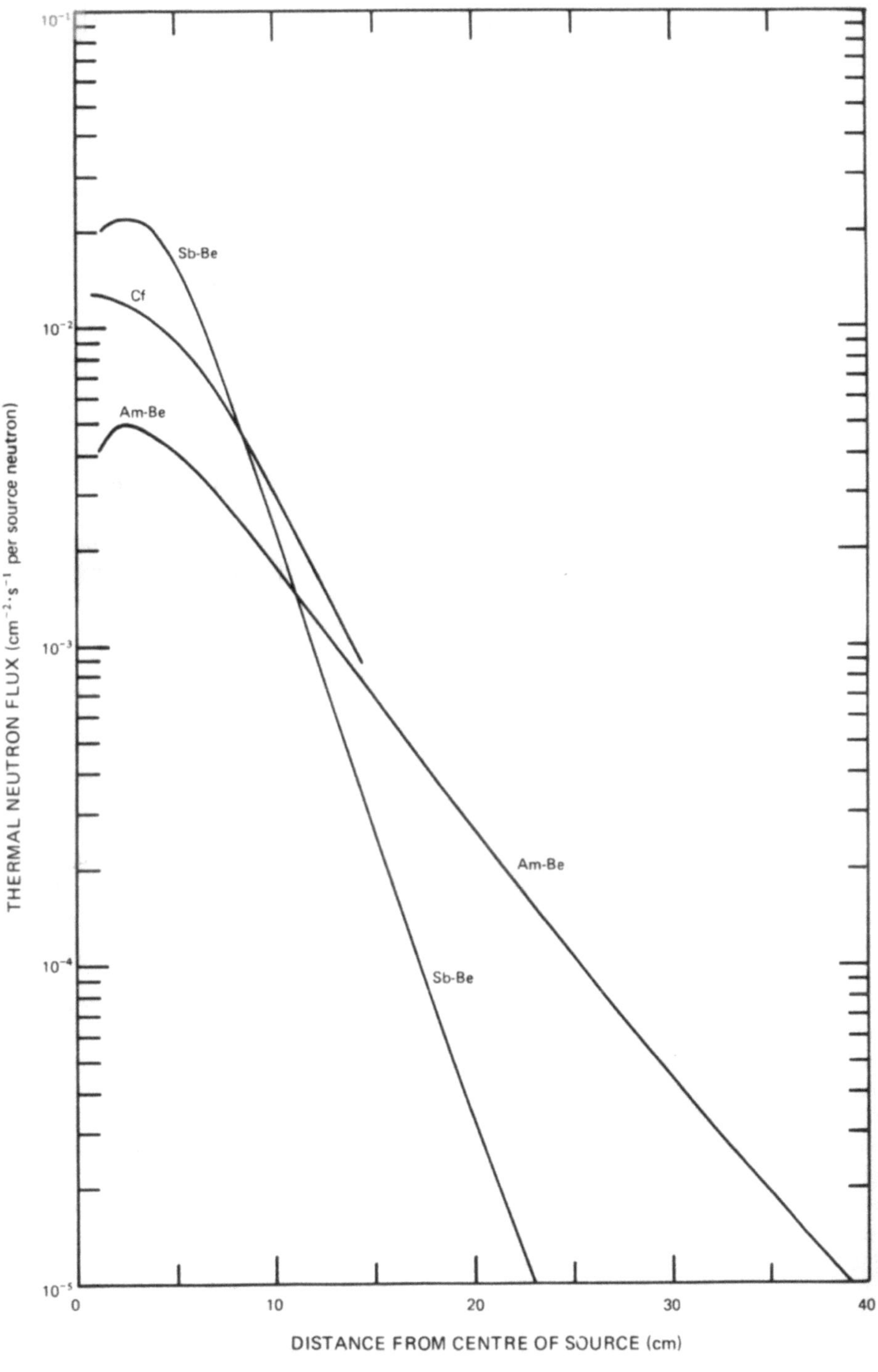

Fig. 6. Neutron flux in water per source neutron emitted from radioisotope neutron sources [6]

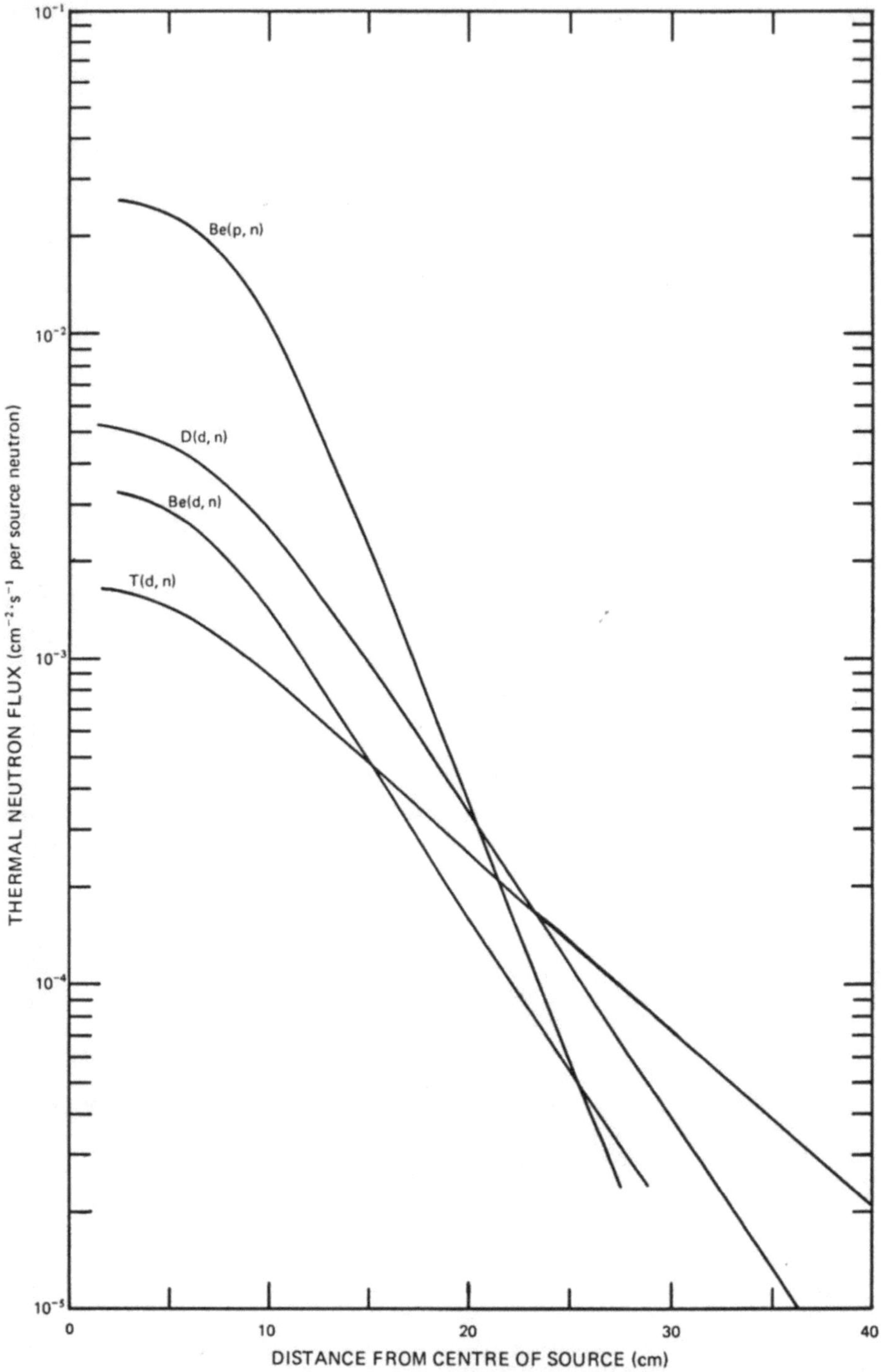

Fig. 7. Neutron flux in water per source neutron emitted from neutron producing accelerators [6]

4. Neutron collimators

Neutrons in moderator are scattered in all directions which are not suitable for radiography. Neutron collimator is a structure designed to extract slow neutron beam from the moderator to the specimen. Ideally, parallel neutron beam is preferred because it gives best image sharpness. If this is the case, Soller or multitube collimator is used. However, divergent collimator is easier to construct and gives good image sharpness depending on the geometrical parameters as will be discussed later.

i. *Soller or multitube collimator*: This collimator is constructed with neutron absorbing material; such as boron, cadmium and gadolinium; as illustrated in Figure 8 so as to bring parallel neutron beam to the test specimen. Neutrons can only get into the collimator from one end which is in the moderator then get out to the other end. Neutrons those are not travel in parallel with the collimator axis will hit the side of the tube or plate and are then absorbed allowing only neutrons travelling in parallel with the tube axis to reach the test specimen. This type of collimator is applicable to nuclear reactor where input neutron intensity to the collimator is high. The drawbacks are that the pattern of parallel plates or tubes may be seen on the image and it is more costly to construct in comparison to the divergent collimator.
ii. *Divergent collimator*: Divergent collimator is designed in the way that neutrons are allowed to get into the collimator only through a small hole from one end then diverge at the other end. The collimator is lined with neutron absorber to absorb unwanted scattered neutrons. It is easy to construct and can be used with non-reactor neutron source like radioisotope and accelerator where slow neutron input is low. The drawback is that image sharpness may not be as good as the Soller collimator. For low neutron intensity as in radioisotope system, neutron output at the specimen position can still be increased by making part of the collimator on the input or source side free from neutron absorber as shown in Figure 10. Neutrons can thus enter the collimator through this part resulting in increasing of neutron intensity. From experience with ^{241}Am/Be and ^{252}Cf sources, neutron intensity can be increased approximately by 10 - 60 % and the cadmium ratio can also be increased from about 5 to 20. In doing so, the image contrast is significantly improved while the image sharpness is a little poorer.

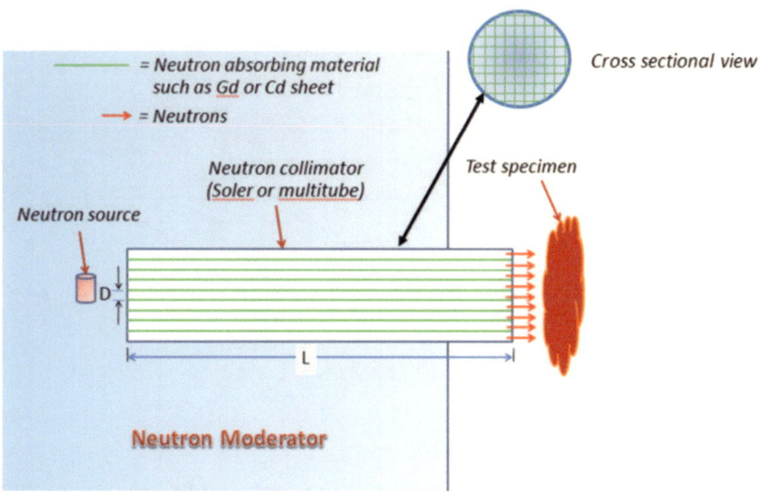

Fig. 8. Multitube neutron collimator

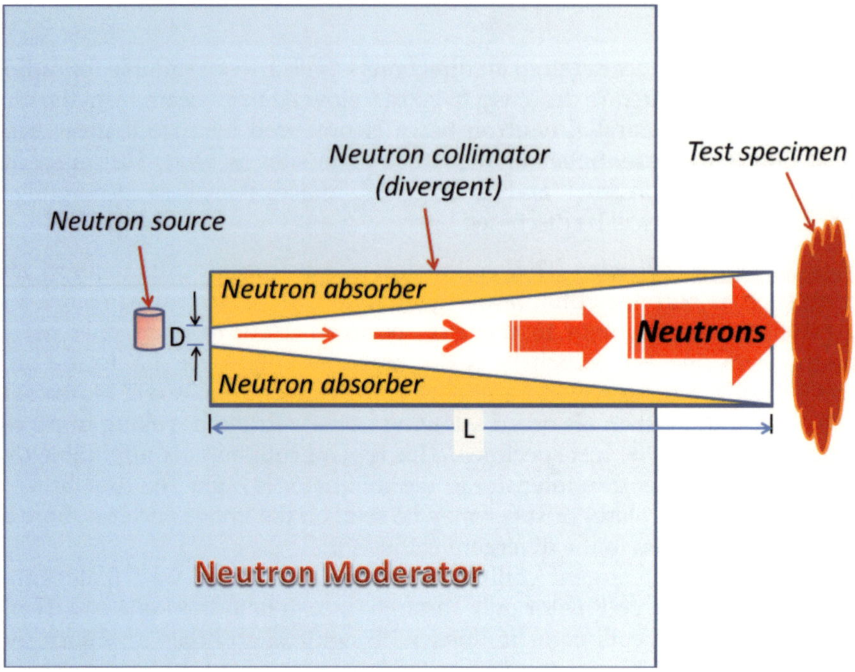

Fig. 9. Divergent neutron collimator allowing neutrons to get into the collimator only through the hole of diameter "D"

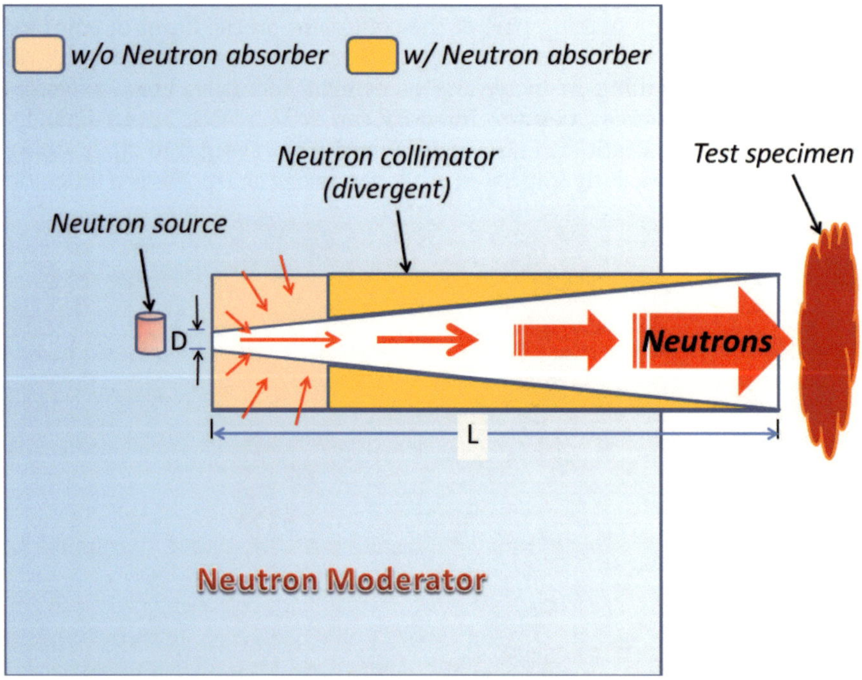

Fig. 10. Divergent neutron collimator with part of the source side contains no neutron absorber allowing more neutrons to get into the collimator

5. Neutron radiographic techniques

After neutrons pass through the specimen they interact with the converter screen to produce radioisotope, alpha particle or light which can be recorded by film, imaging plate, optical camera or video camera. Image recording medium must be selected to match with the particles or light emitted from the neutron converter screen so as to obtain the maximum efficiency. The neutron converter screen/image recording device assemblies commonly used in neutron radiography are described below.

i. *Metallic foil screen/film*: Metallic foil with high neutron cross section is employed to convert slow neutrons to beta-particles, gamma-rays and/or conversion electron while industrial x-ray film is normally used as the image recorder. Gadolinium (Gd) foil is the best metallic screen for neutron radiography in terms of having extremely high neutron absorption cross section, giving the best image resolution and not becoming radioisotope after neutron absorption. ^{155}Gd and ^{157}Gd are found 14.9 and 15.7 percent of natural Gd isotopes with neutron absorption cross sections of 61,000 and 254,000 barns respectively. ^{155}Gd and ^{157}Gd absorb neutrons then become ^{156}Gd and ^{158}Gd correspondingly which are not radioactive. Prompt captured gamma-rays emitted during neutron absorption can cause film blackening. More importantly, prompt gamma-rays may hit atomic electrons resulting in ejection of electrons from the atoms (so called "conversion electron") which are more effective to cause film blackening. It should be noted that less than a few percentage of gamma-ray photons cause film blackening. Electrons and beta-particles are preferred because they interact with film much more than gamma-rays.

Film may be replaced by imaging plate (IP) which has more than 10 times faster speed than the x-ray film. Gd foil/x-ray film requires relatively high neutron exposure thus it is not possible to carry out neutron radiography with low neutron flux system using radioisotope. About 5 years ago, Fuji started to produce neutron imaging plate by adding Gd into the imaging plate which can give the image quality comparable to that from the Gd foil/x-ray film assembly with approximately 50 times reduction of neutron exposure. It is therefore possible to be used with low neutron flux system.

Other metallic foil screens can also be used (as listed in Table 4) but the image quality is not as good as that obtained from Gd. This is mainly because low energy electrons emitted from Gd have very short ranges resulting in much better image sharpness. In case of having large gamma-ray contamination in the neutron beam and specimen containing gamma-ray emitting radioisotopes, dysprosium (Dy) is often used. To avoid gamma-ray exposure to x-ray film, the transfer method must be applied by exposing only the Dy screen with transmitted neutrons from the specimen. During exposure, radioisotopes ^{165m}Dy and ^{165}Dy are formed with half-lives of 1.26 minutes and 2.3 hours respectively. The Dy foil is then removed from the neutron beam and placed in close contact with an x-ray film to produce a latent image. The film density or film darkness is corresponding to the activity of Dy radioisotopes formed in each part of the Dy foil.

Formation of radioisotope from neutron irradiation follows the equation below.

$$A = n\sigma\phi\,(1 - e^{-\lambda T}) \qquad (2)$$

Where A is the radioactivity of radioisotope formed in disintegration per second (dps) after completion of neutron irradiation. n is the number of original stable isotope atoms. σ is the neutron absorption cross section of the original stable isotope in cm^2. ϕ is the neutron flux in $cm^{-2}\ s^{-1}$. λ is the decay constant of the radioisotope formed in s^{-1} and T is the irradiation time in second (s).

The decay constant (λ) can be obtained from :

$$\lambda = 0.693/T_{1/2} \tag{3}$$

where $T_{1/2}$ is half-life of the radioisotope. Form equation (1), more than 96 % of the maximum radioactivity can be obtained if the irradiation time is greater than 5 times of the half-life.

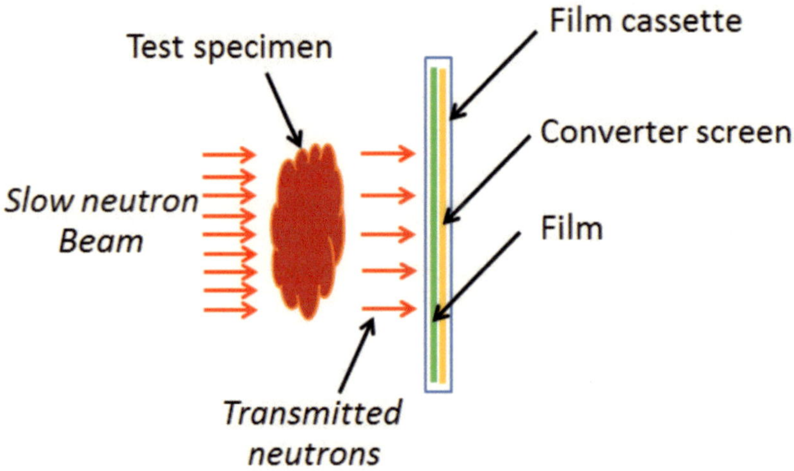

Fig. 11. Illustration of direct exposure method

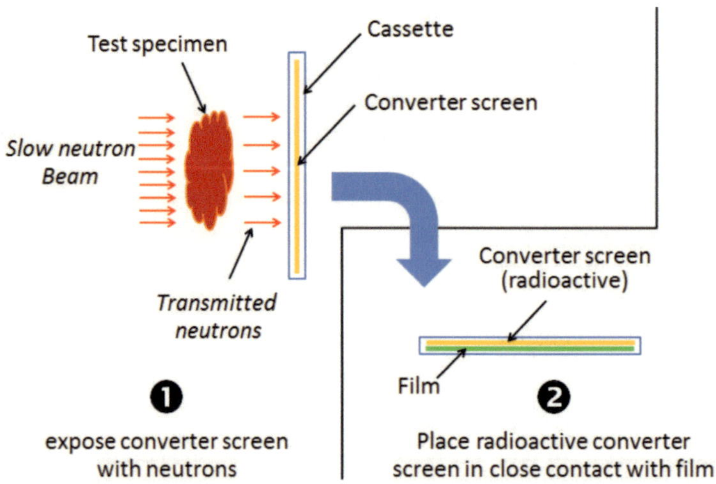

Fig. 12. Illustration of two steps of indirect or transfer exposure method

Material	Mode of Production of active Isotope	Cross-section (barns)	Half-life	Major particle emitted	Direct or Transfer technique
Lithium	$Li^6(n, \alpha)H^3$	935	Stable	α	Direct
Boron	$B^{10}(n, \alpha)Li^7$	3,837	Stable	α	Direct
Rhodium	$Rh^{103}(n, \gamma)Rh^{104}$	144	43 s	β^-	Direct
	$Rh^{103}(n, n)Rh^{103m}$		57 min	x-ray	
	$Rh^{103}(n, \gamma)Rh^{104m}$	11	4.4 min	β^-	
Cadmium	$Cd^{113}(n, \gamma)Cd^{114}$	20,000	Stable	γ	Direct
Indium	$In^{115}(n, \gamma)In^{116}$	45	14 s	β^-	Transfer
	$In^{115}(n, \gamma)In^{116m}$	154	54 min	β^-	
Samarium	$Sm^{149}(n, \gamma)Sm^{150}$	41,500	Stable	γ	Direct
	$Sm^{152}(n, \gamma)Sm^{153}$	210	46.7 h	β^-	
Gadolinium	$Gd^{155}(n, \gamma)Gd^{156}$	58,000	Stable	e^-	Direct
	$Gd^{157}(n, \gamma)Gd^{158}$	240,000	Stable	e^-	
Dysprosium	$Dy^{164}(n, \gamma)Dy^{165}$	800	2.3 h	β^-	Transfer
	$Dy^{164}(n, \gamma)Dy^{165}$	2,000	1.26 min	β^-	

Table 4. Characteristics of Some Possible Neutron Radiography Converter Materials [2, 7]

Radioisotope decays exponentially according to its half-life. If A_0 is the radioactivity of the radioisotope after completion of neutron irradiation, the radioactivity at any time t can be calculated from

$$A_t = A_0\, e^{-\lambda t} \tag{3}$$

For Gd foil, no radioisotope is formed during neutron irradiation. Emission of prompt gamma-rays and conversion electrons follows neutron absorption by ^{155}Gd and ^{157}Gd at the rate of $n\sigma\phi$ per second. In case of Dy foil, ^{165m}Dy and ^{165}Dy are formed with the radioactivity following equation (2). After removal from the neutron facility, ^{165m}Dy and ^{165}Dy will decay with half-lives of 1.26 minutes and 2.3 hours respectively. Film is exposed to emitted radiation while placing in close contact with the radioactive foil. Build-up and decay of a radioisotope is illustrated graphically in Figure 13.

ii. *Light emitting screen/film*: Light-emitting screen is a mixture of scintillator or phosphor with lithium-6 (^6Li) and/or boron-10 (^{10}B). Neutrons interact with ^6Li or ^{10}B to produce alpha-particles via (n, α) reaction. Light is then emitted from energy loss of alpha-particles in scintillator or phosphor. Light sensitive film, digital camera or video camera can be used to record image. This makes real-time and near real-time radiography possible. The most common light-emitting screen is NE426 available from NE Technology which is composed of ZnS(Ag) scintillator and boron compound. Gadolinium oxysulfide (terbium) [Gd_2O_2S (Tb), GOS] and lithium loaded glass scintillator are also common in neutron radiography. GOS itself is a scintillator. Conversion electrons as well as low energy prompt gamma-rays emitted from interaction of neutrons with Gd cause light emission. Glass scintillator is sensitive to charged particles such as alpha- and beta-particles. Lithium is added into the glass scintillator so that alpha-particle will be emitted from $^{6}Li(n, \alpha)^{3}H$ reaction resulting in emission of light.

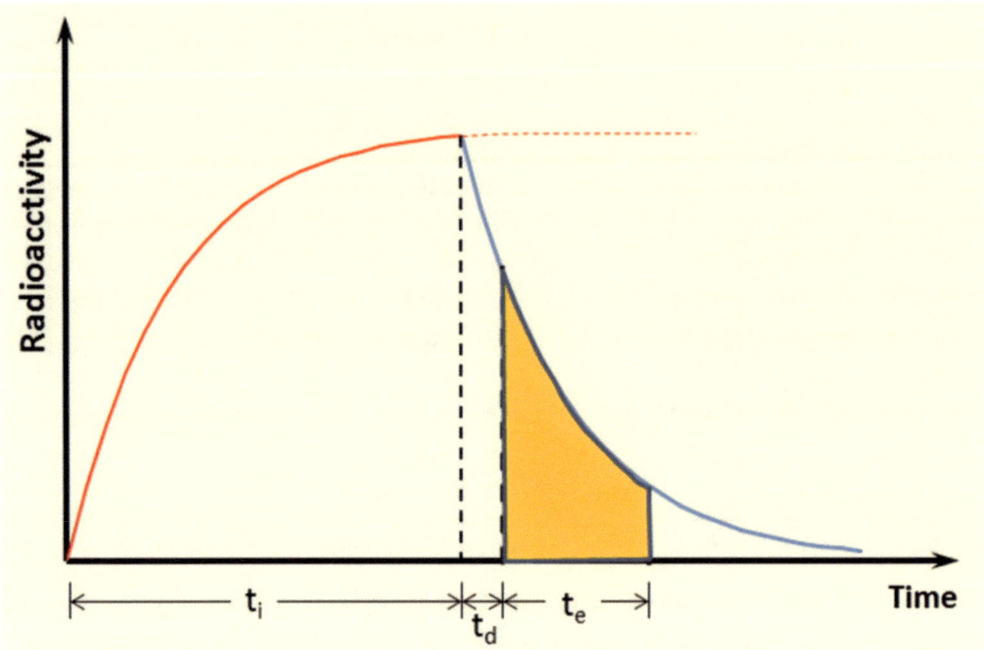

Fig. 13. Graphical illustration of build-up and decay of radioactivity of a radioisotopein neutron converter screen using indirect or transfer method

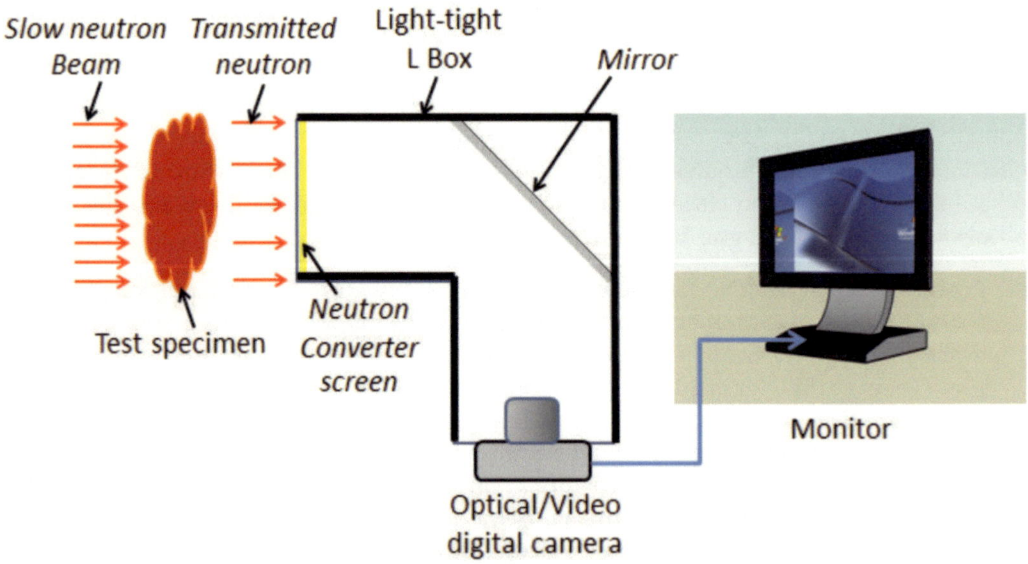

Fig. 14. Real-time or near real-time neutron imaging system using light-emitting neutron converter screen

Light-emitting screen offers highest speed but gives poorest image sharpness comparing to other screen/film assemblies. This is the only type of screen that can be used with low neutron flux system using radioisotope neutron source. From experience, photographic film is more suitable with the light-emitting screen than industrial x-ray film by the following two main reasons. Firstly, photographic film is less sensitive to gamma-ray. As a result, it gives better image contrast particularly when neutron beam is contaminated by large fraction of gamma-rays. Secondly, photographic film is cheaper and easily available.

iii. *Alpha-emitting screen/track-etch film*: Alpha-emitting screen is made of lithium and/or boron compound. Particles emitted from $^{6}Li(n, \alpha)^{3}H$ and $^{10}B(n, \alpha)^{7}Li$ reactions interact with track-etch film (or so called "solid state track detector, SSTD)") to produce damage tracks along their trajectories. The detector is later put into hot chemical solution to enlarge or "etch" the damage tracks. After etching, the damage tracks can be made visible under an optical microscope with a magnification of x 100 up. Radiation dose and/or neutron intensity can be evaluated by counting number of tracks per unit area. The area where track density is so large becomes translucent while the area with low track density is more transparent. The degree of translucence depends on track density resulting in formation of visible image on the film. However, contrast of the image is poor while sharpness is comparable to the Gd foil/x-ray film assembly. Methods for viewing the image is needed to improve image contrast such as reprinting the image on a high contrast film. It has been reported that the simplest method is to scan image on the track-etch film using a desktop scanner [8]. Track-etch film is not sensitive to light, beta-particle and gamma-ray. The alpha-emitting screen/track-etch film assembly can therefore be used to radiograph radioactive specimens by the direct method. No darkroom is needed for film processing. Kodak LR115 Type II, CA80-15 Type II and CN85 Type B have been widely used during the past two decades. They are cellulose nitrate films coated with lithium metaborate ($Li_2B_4O_7$). The optimum etching condition is 10 - 40 % sodium hydroxide (NaOH) at 60 ^{0}C for a duration of 30 – 40 minutes. Later, BE-10 screen of 93 % enriched boron-10 in boron carbide (B_4C) form manufactured by Kodak became available and has been widely used since then due to its highest neutron conversion efficiency. Kodak LR115, CA8015 and CN85 cellulose nitrate film without lithium metaborate are used with the BE-10 screen. CR39 plastic or poly(alyl diglycol) carbonate is also available and is used extensively for alpha detection due to its higher track registration efficiency. The CR39/BE-10 assembly will probably become the most common in track-etch neutron radiography.

6. Applications of neutron radiography: Facilities and sample images

Neutron radiography has been employed for non-destructive testing of specimens. Parts of test specimen containing light elements; such as rubber, plastic, chemicals; can be made visible even when they are covered or enveloped by heavy elements. Nuclear reactor gives the best thermal or cold neutron beam for neutron radiography as can be seen in Tables 3 and 6. The cadmium ratio is normally greater than 10 and can be as high as 300 or even infinity if required. The L/D ratio is always greater than 100 which indicate excellent image sharpness. Nuclear research reactor generally provides excellent beam ports for neutron experiments including for neutron radiography. All neutron converter screen/image recorder assemblies mentioned above can be employed for inspection of specimens but the exposure times vary considerably. Neutron exposure for some converter screen/film assemblies can be estimated by using the curves in Figure 15.

Converter Screen	Emitted Particles	Recording Medium	Comment
Metallic foil screen			
- Gd foil	electrons	Industrial x-ray film	Best image quality, needs high neutron exposure
- Dy foil	Beta-particles, Gamma-rays	Industrial x-ray film	Best for transfer method, good image quality, needs high exposure
Light-emitting screen			
- NE426 - GOS - ^{6}Li loaded glass scintillator	Light	Photographic film, Industrial x-ray film, Optical or video camera	Fastest speed, needs low exposure, acceptable image quality, allows real-time or near real-time imaging
Alpha-emitting screen			
- $Li_2B_4O_7$	Alpha-particles	Track-etch film e.g. CR39, cellulose nitrate	Good image quality, needs high exposure, low contrast, requires image viewing technique, does not require a darkroom for film loading and processing
- BE-10			

Table 5. Common neutron converter screens and image recording media

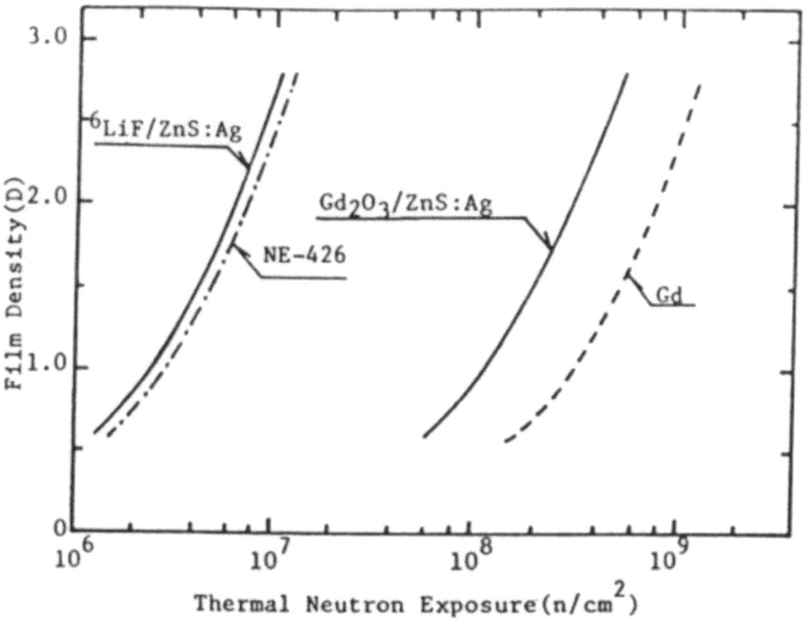

Fig. 15. Thermal neutron exposure required film for radiography [9]

For example, total thermal neutrons per square centimeter required for Gd metallic foil screen/film and NE-426 light emitting screen/film assemblies to make a density of 1.5 on film are approximately 5.5×10^8 and 5×10^6 respectively. If neutron flux at the specimen position is 10^6 cm^{-2} s^{-1}, the exposure time needed for the two screens are 550 and 5 seconds respectively.

When a 1 mg (1000 µg) Cf-252 is used as in Table 6, the maximum neutron flux in water will be 1000 µg x (2.3×10^6 s^{-1} µg^{-1})/100 = 2.3×10^7 cm^{-2} s^{-1}. The neutron flux at the specimen position for a circular cross-section, divergent collimator with an L/D of 12 (as in Table 6) can be calculated from:

$$\Phi_{exit} = \Phi_{source}\,(D/L)^2/16 \tag{4}$$

$$\Phi_{exit} = 2.3 \times 10^7\,(1/12)^2/16 = 9.98 \times 10^3 \approx 10^4\ \text{cm}^{-2}\text{s}^{-1}$$

Where Φ_{source} and Φ_{exit} are neutron fluxes at the source side and the specimen position respectively. The neutron flux obtained from calculation agrees with the value in Table 6. Thus, the exposure time required for the Gd metallic foil screen/film and NE-426 light emitting screen/film assemblies to make a density of 2.0 on film will be 5.5×10^4 and 500 seconds respectively. It is therefore impossible to use the Gd foil screen/film assembly with Cf-252 source. However, the neutron flux can still be increased by leaving part of the collimator on the source side without neutron absorber. The neutron flux will then be increased by a factor of (1 + 2a/L), where a is the length of the collimator without neutron absorber. For example, if the length of the total collimator (L) is 30 cm and the part without neutron absorber (a) is 10 cm, the neutron flux is increased by a factor of [1 + (2 x 10)/30] =

1.67 or 67 %. The exposure time will then be reduced from 500 seconds to 300 seconds. In doing so, the cadmium ratio is increased from about 5 to 15 -20 resulting in significant improvement in image contrast but the image sharpness is gradually reduced [10]. Use of the second neutron converter screen can also decrease the exposure time by a factor of up to 2.2 as shown in Table 7. During the past decades, the image recording devices have been rapidly improved in speed as well as graininess including film, imaging plate (IP), digital optical camera, digital video camera, CCD and CMOS chips. The new devices allow radiographers to perform non-film neutron radiography with neutron generator and Cf-252 neutron sources. The Fuji neutron imaging plate offers speed several ten times faster than that of the Gd/film assembly with comparable image quality [11-13]. The light-emitting screen coupled with a digital camera with light sensitivity from ISO 1600 and time integration mode makes non-film neutron radiography by Cf-252 possible. An image intensifier or a microchannel plate (MCP) is useful for real-time or near real-time neutron imaging in low flux system. Examples of neutron radiographic images taken from different neutron facilities and by different techniques are illustrated in Figures 18 to 23.

Source	Collimator		Typical beam characteristics		
	Position	Base flux ($cm^{-2} s^{-1}$)	Intensity ($cm^{-2} s^{-1}$)	L/D ratio	Cd ratio
Multi-purpose research reactor	Radial	10^{14}	10^{8}	250	2-5
	Tangential	10^{13}	10^{7}	250	10-50
	Cold source	2×10^{11}	10^{6}	100	∞
Radiography reactor	Radial	10^{12}	10^{6}	250	2-5
	Tangential	4×10^{11}	2×10^{6}	100	10-50
Be(d, n); 3 MeV, 400 μA	Radial	3×10^{9}	2×10^{5}	33	5-20
Be(γ, n); 5.5 MeV, 100 μA	Radial	4×10^{8}	8×10^{4}	18	5-20
T(d, n)+ U; 120 keV, 7 mA	Radial	10^{8}	2×10^{4}	18	5-20
^{252}Cf ; 5 mg + sub-critical Reactor assembly	Radial	3×10^{9}	2×10^{5}	18	5-20
^{252}Cf ; 1 mg	Radial	2×10^{7}	10^{4}	12	5-20

Table 6. Examples of common neutron radiography facilities [2, 6]

Converter Foil	Technique	Foil Thickness (μm)		Relative Speed
		Front	Back	
Rh/Gd	Direct	250	50	5.3
Rh/Rh	Direct	250	250	4.7
Gd/Gd	Direct	25	50	3.7
In/In	Direct	500	750	3.7
Dy/Dy	Direct	150	250	3.7
Cd/Cd	Direct	250	500	3.3
Ag/Ag	Direct	450	450	2.7
Dy	Direct	-	250	2.5
Gd	Direct	-	25	2.4
Cd	Direct	-	250	2.2
Rh	Direct	-	250	2.1
In	Direct	500	-	1.7
Dy	Transfer	250	-	16.4
In	Transfer	50	-	11.2

Remark : The relative speed for the direct and transfer methods are not comparable

Table 7. Relative Speed of Neutron Converter Screens [2]

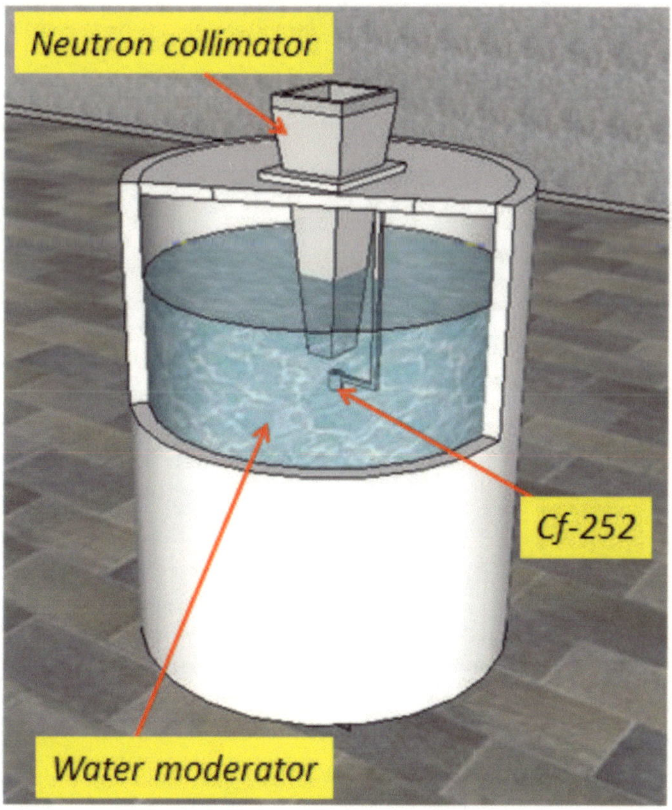

Fig. 16. An example of Cf-252 based neutron radiography system

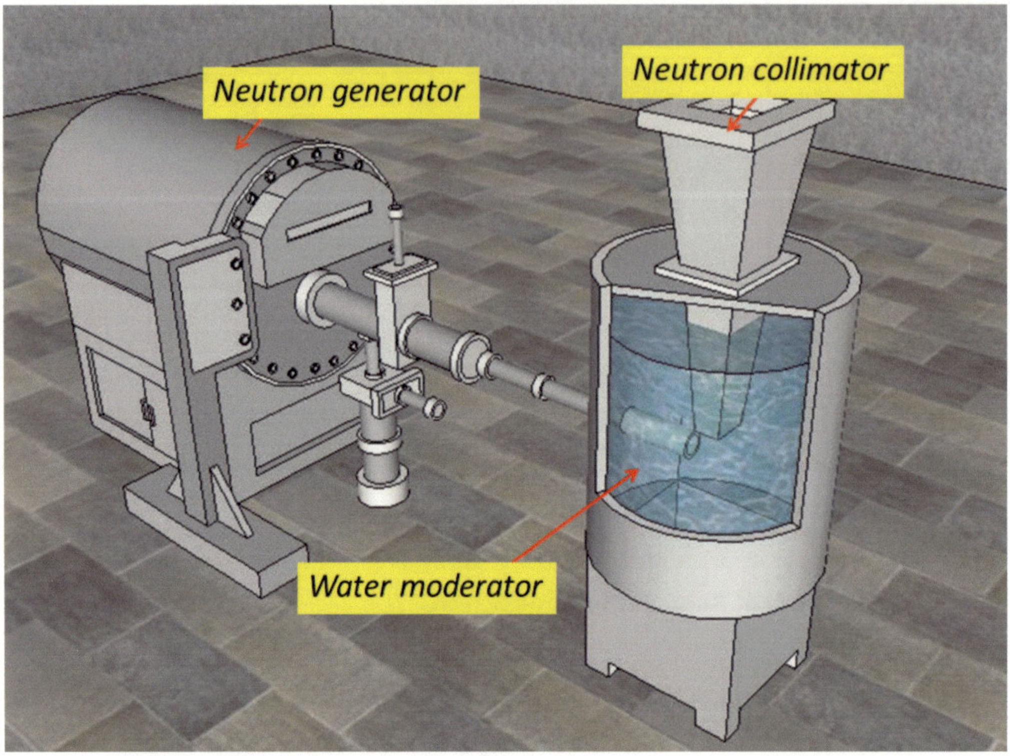

Fig. 17. An example of neutron generator based neutron radiography system

Fig. 18. A neutron radiograph of pistol bullets [14] (research reactor, Gd foil/film technique)

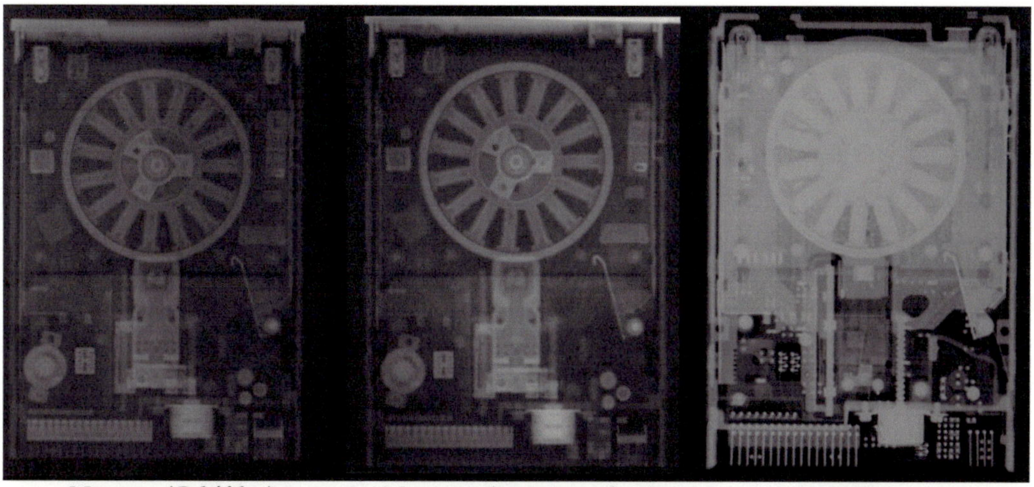

Neutron(Gd/film) Neutron (imaging plate) X-Ray

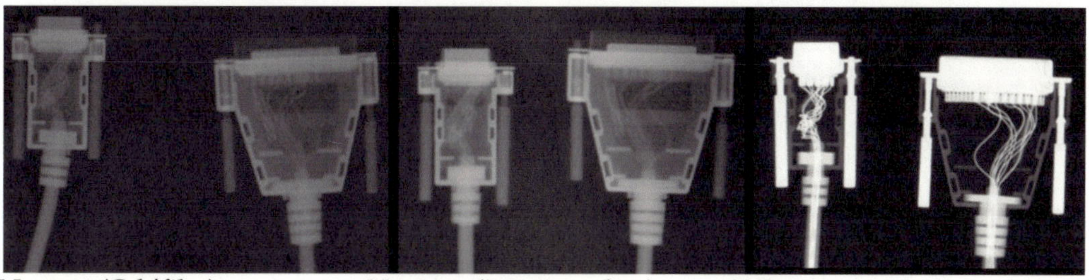

Neutron(Gd/film) Neutron (imaging plate) X-Ray

Fig. 19. Neutron radiographs with a floppy disk drive (above) and RS-232 connectors (below) using neutron beam from a TRIGA Mark III research reactor in comparison with x-ray radiograph [12]

Neutron (imaging plate) X-Ray

Fig. 20. Neutron radiograph of a Buddha statue using neutron beam from a TRIGA Mark III research reactor in comparison with x-ray radiograph [13] (Clay can be clearly seen in the middle part of the body in the neutron radiograph)

Neutron (imaging plate) X-Ray

Fig. 21. Neutron radiograph of an ancient lacquerware using neutron beam from a TRIGA Mark III research reactor in comparison of x-ray radiograph [13] (Pattern of embroidered bamboo thread can be seen clearer in the neutron radiograph)

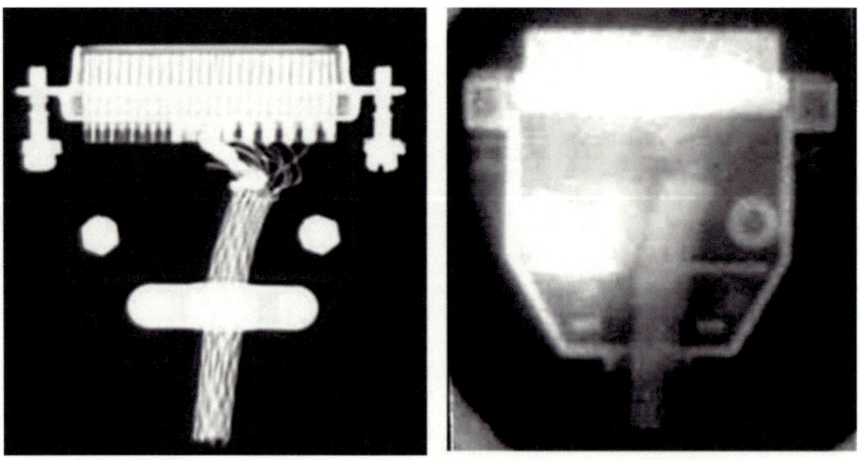

X-Ray Neutron

Fig. 22. Neutron radiograph of an RS-232 connector using low intensity neutrons from Cf-252 and NE426 light-emitting screen/photographic film [15]

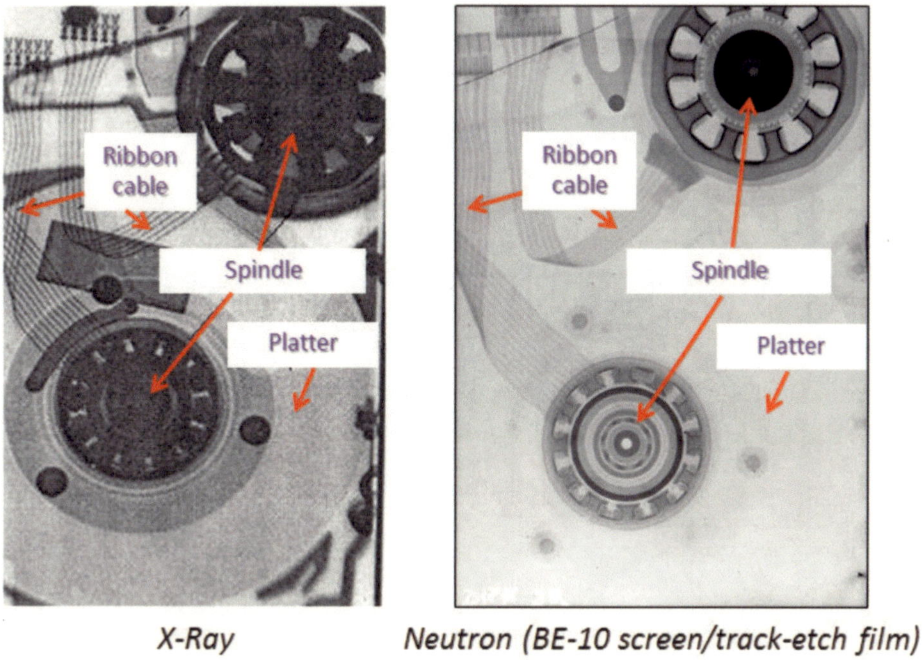

(a) (b)

Fig. 23. Neutron radiograph (b) of a hard disk drive using neutrons from a research reactor in comparison with x-ray radiograph (a) [16] (The neutron radiograph from the track-etch film was scanned by using a desktop scanner with a shiny polished metal sheet used as the light reflecting surface.)

7. Quality control of neutron radiographic image

Quality of neutron radiograph is affected by various factors not only the L/D ratio and the cadmium ratio as mentioned previously but also the gamma-ray content, the geometric unsharpness, type of converter screen, type of image recording medium and film processing. The image sharpness is improved with increasing of the L/D ratio while the image contrast is improved with increasing of the cadmium ratio. The gamma-ray content in neutron beam will deteriorate the image contrast. The other factors affect the neutron radiographs in the same way as in x-ray and gamma-ray radiography. The ASTM Beam Purity Indicator (BPI) and the ASTM Sensitivity Indicator (SI) are common neutron beam quality indicators as illustrated in Figures 24 and 25 respectively. Department of Nuclear Engineering of Chulalongkorn University also developed a neutron beam quality indicator, so called "CU-NIQI", as illustrated in Figure 26. The CU-NIQI is used with a 3 mm thick lead (Pb) plate to determine gamma-ray content. The quality indicators are based on the same principles. Teflon and polyethylene are hydrogeneous materials used for indicating proportion of slow neutrons to fast neutrons. Cadmium and boron are good neutron absorbers used for indicating proportion of neutrons of energy below 0.5 eV to beyond 0.5 eV. Lead strip or wire is the indicator for gamma-ray content. Film density readings at the positions corresponding to those materials can be used to evaluate quality of the neutron beam.

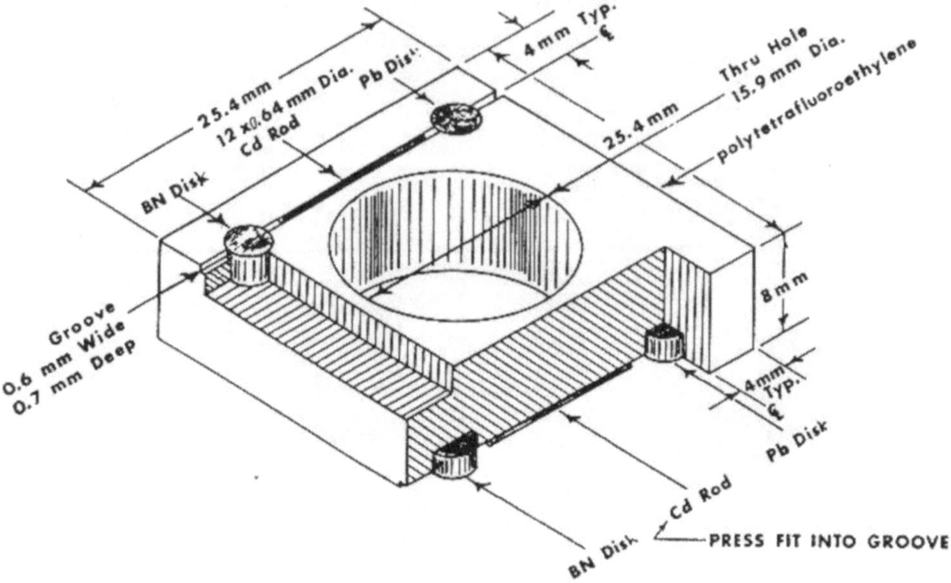

Fig. 24. The ASTM Beam Purity Indicator [2]

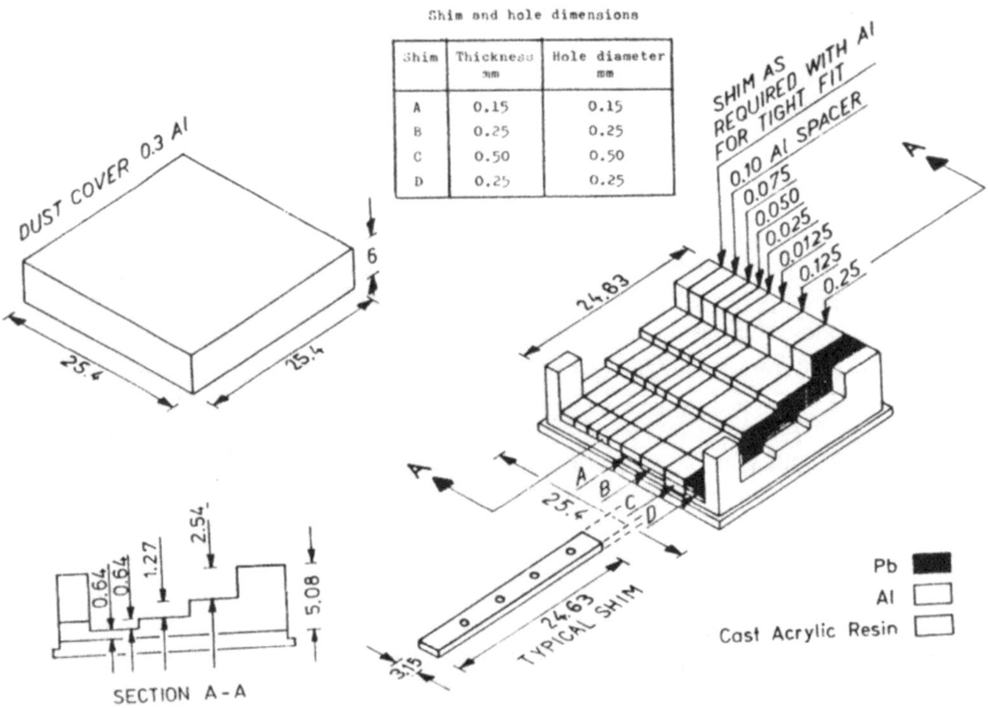
Shim and hole dimensions

Shim	Thickness mm	Hole diameter mm
A	0.15	0.15
B	0.25	0.25
C	0.50	0.50
D	0.25	0.25

Fig. 25. The ASTM Sensitivity Indicator [2]

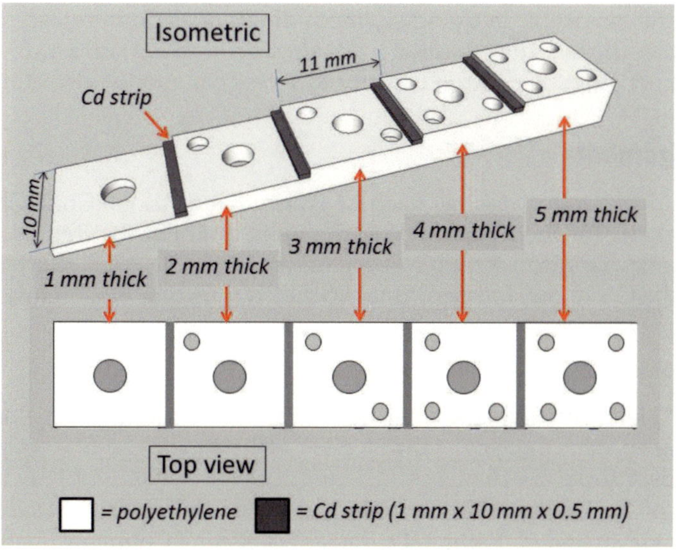

Diameter of the central through hole = 3 mm, diameter of hole no. 1 - 4 = 2 mm.
Depth of hole no. 1, 2, 3 and 4 are 1, 2, 3 and 4 mm respectively.

Fig. 26. The CU-NIQI Neutron Beam Purity Indicator developed by the Department of Nuclear Engineering of Chulalongkorn University

8. Methods for determining neutron exposure

In practice there is no standard procedure for constructing an exposure curve for neutron radiography as in x-ray and gamma-ray radiography. For specific application, the radiographer can first begin with trial and error to choose the best exposure. Without knowing the specimen composition and thickness to determine the neutron attenuation, the exposure cannot be obtained. The best method, so far, is to measure transmitted neutron intensity by using a small neutron detector at a few positions behind the specimen before placing the screen/film assembly. The transmitted neutron intensity is inversely

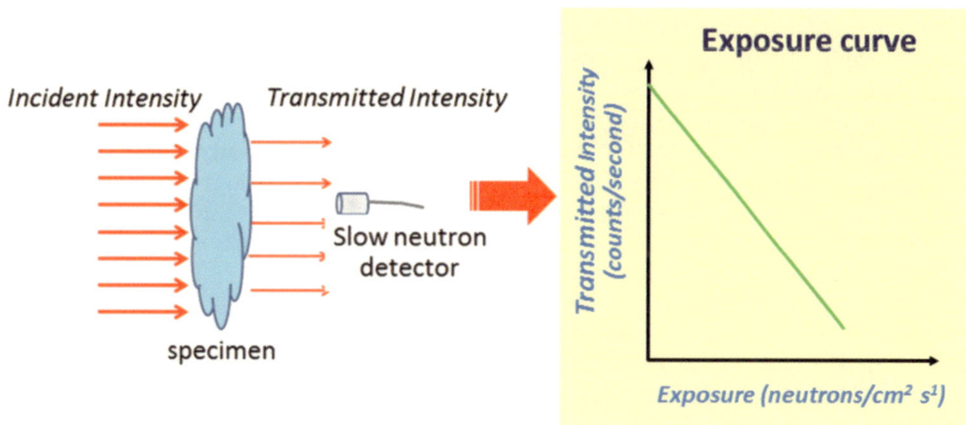

Fig. 27. Illustration of a method for determining the neutron exposure by measuring transmitted neutron intensity with a small neutron detector

proportional to the exposure. In doing so, the radiographer do not need to perform decay and distance corrections. This method can also be applied in x-ray and gamma-ray radiography by changing the neutron detector to x-ray/gamma-ray detector.

9. Acknowledgements

The author would like to express his deepest gratitude to Mr. JateChan Channuie and Mr. Kittiwin Iaemsumang for their assistance in the preparation of the figures and the tables in this chapter. Thanks are also extended to my eldest daughter, Miss Katriya Chankow, and my wife, Mrs. Julie Chankow, for their times spent in correcting the English translation.

10. References

[1] J.P. Barton. "Neutron Radiography - An Overview", *Practical Applications of Neutron Radiography and Gauging*. ASTM STP 586, American Society of Testing and Materials, 1976, p. 5 - 19.

[2] Commission of the European Communities. *Neutron Radiography Handbook*. Edited by P. von der Hardt and H. Röttger, D. Reidel Publishing Company, Dordrecht, 1981.

[3] D.A. Garrett and H. Berger. "The Technological development of Neutron Radiography", *Atomic Energy Review*. Vol. 15(2), 1977, p.123 - 142.

[4] J.R. Lamarsh. *Introduction to Nuclear Engineering*. 2nd Ed., Addison-Wesley, New York, 1983.

[5] A.A. Harms. "Physical Processes and Mathematical Methods in Neutron Radiography", *Atomic Energy Review*. Vol. 15(2), 1977, p.143 - 168.

[6] M.R. Hawkesworth. "Neutron Radiography : Equipment and Methods", *Atomic Energy Review*. Vol. 15(2), 1977, p.169 - 220.

[7] H. Berger. "Detection System for Neutron Radiography", *Practical Applications of Neutron Radiography and Gauging*. ASTM STP 586, American Society of Testing and Materials, 1976, p. 35 - 57.

[8] N. Chankow. "A Simple Method for Viewing Track-Etch Neutron Radiographic Images Using a Scanner". *Proceedings of the 2nd International Meeting on Neutron Radiography System Design and Characterization*, Yokosuka, Nov. 12-18, 1995.

[9] Y. Suzuki. et al. "Development of Imaging Converter", *Proceedings of the 3rd World Conference on Neutron Radiography*, Osaka, 1989.

[10] S. Jaiyen. , "Development of A Prototype for Low-Flux Thermal Neutron Radiography System Using Cf-252", Master Thesis, Department of Nuclear Technology, Chulalongkorn University, 2002.

[11] S. Wonglee. "Neutron radiography using Neutron Imaging Plate", Master Thesis, Department of Nuclear Technology, Chulalongkorn University, 2007.

[12] N. Chankow, S. Punnachaiya and S. Wonglee. "Neutron radiography using Neutron Imaging Plate", *Applied Radiation and Radioisotopes*. Vol.68 (4), 2010, p. 662-665.

[13] R. Funklin. "Application of Neutron radiography using Neutron Imaging Plate for Inspection of Ancient Objects", Master Thesis, Department of Nuclear Technology, Chulalongkorn University, 2009.

[14] P. Suksawang. "A Study of Neutron radiography", Master Thesis, Department of Nuclear Technology, Chulalongkorn University, 1981.

[15] P. Orachorn. "Cf-252 Based Neutron radiography using Neutron Imaging Plate", Master Thesis, Department of Nuclear Technology, Chulalongkorn University, 2009.

[16] W. Ratanathongchai. "Neutron radiography using BE-10 Neutron Converter Screen", Master Thesis, Department of Nuclear Technology, Chulalongkorn University, 1997.

5

Flaw Simulation in Product Radiographs

Qian Huang and Yuan Wu
South China University of Technology
China

1. Introduction

X-ray inspection machines have been widely used in modern industry to explore the interior of products, and to control product quality. With the advancement of automatic product inspection, we need lots of flaw radiographs to validate the system's sensitivity and to help tune the automatic inspection parameters. Flaw simulation in product radiographs has been receiving an increasing attention for over two decades.

Before implementing the program to inspect the work pieces, a large number of sample images are needed to tune the algorithm, examine its performance, and ensure its accuracy. Product radiographs with defects from the production line are the best, however, they are often not available in sufficient quantities or variety. Simulation of casting defects is an alternative approach to deal with this problem.

Recently, the CAD model method, the Monte Carlo method and the generative image method have become the three main approaches for flaw simulation.

1. The CAD model method enables simulations to be produced for complex three dimensional (3D) casting objects. Ray-tracing, together with calculation of the X-ray attenuation, is the basis of this model to produce 3D casting defect simulations.
2. Monte Carlo simulation is a method for iteratively evaluating a deterministic model using sets of random numbers as inputs. The creation of a proper physical model is the most critical task for the simulation.
3. Generative image model for flaw simulation of the product radiographs has been developed from the idea of the superimposition technique, and is based on defect analysis.

In this chapter, we will discuss flaw simulation in product radiographs. The chapter is organized as follows. Section 2 introduces the CAD models for simulation, the three main parts of CAD models will be discussed including the X-ray source, the geometric and material properties of the objects, and the imaging process. Section 3 presents the applications of the Monte Carlo method for X-ray image simulation, a brief introduction of the Monte Carlo method and how to combine it in the simulation application are discussed. Section 4 concludes with the generative image model for flaw simulation in radiographs, the generative models and simulation results are outlined, the authors themselves have produced significant results in this field. Section 5 concludes and offers suggestions for further research. Finally, a reference list is provided for further reading.

2. CAD Models for simulation

The CAD model simulates the entire X-ray imaging process. There are three main components of an X-ray penetration model, the X-ray generator, the interaction between the X-ray and the object and the imaging process (Tillack et al., 2000). In CAD simulation, the X-ray source, the geometric and material properties of objects as well as the imaging process are simulated independently. For product flaw simulation, the CAD model can be used in the two approaches, which are for a whole product with defects or flaws embedded, or for the flaws only but using post-processing for its applications.

2.1 CAD Models for radiographs

The X-ray beam comes from the X-ray source, penetrates the object, attenuates and finally projects into the image detector, see Fig. 1. So each pixel (u, v) in the simulated image can be projected back though the modelled object to the X-ray beam source. Since it is very difficult to make the X-ray source from the x-ray tube into a single point light source, this can be seen from Fig. 1. The "source", which refers to the area of the emitter aperture should be very small, to minimise the blurring of the edges of the work piece in a radiograph or on the film, the enlarged circle in Fig. 1.

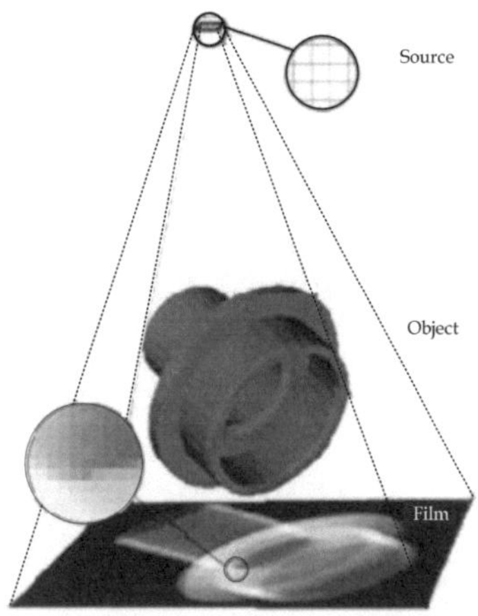

Fig. 1. The X-ray imaging system with the problems of the source and object edges in the radiograph

The X-ray attenuation law is the basis of the simulation process (Haken &Wolf, 1996):

$$\varphi = \varphi_0 exp(-\mu x) \tag{1}$$

where φ_0 is the incident radiation intensity, φ is the transmitted intensity, x is the thickness of the object, and μ is the energy attenuation associated with the material. The gray value of the simulated image can be computed as follows:

$$I = A\varphi_0(E)\Delta\Omega exp\left(-\sum_i \mu_i(E)x_i\right) + B \tag{2}$$

where $\varphi_0(E)$ is the incident radiation intensity with energy E, $\Delta\Omega$ is the solid angle that corresponds to the pixel observed from the source point, $\mu_i(E)$ represents the attenuation of material i at energy E, and x_i is the total path length through material i. Fig.2 is an example of a work-piece radiograph simulated from a CAD model and attenuation data.

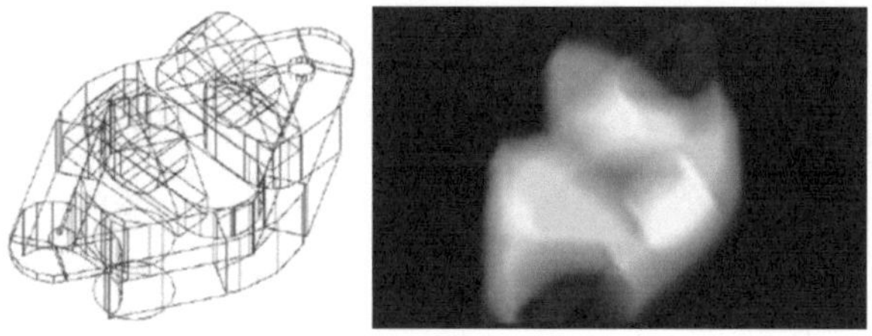

Fig. 2. A three-dimensional (3D) model of a workpiece and its generated radiograph

Some flaws, such as cavities, can be simulated as nil X-ray absorption and others, such as incrusted material, with a different attenuation coefficient μ_d. We rewrite the equation (1) as following:

$$\varphi = \varphi_0\, exp\big(-\mu(x-d)\big)\, exp(-\mu_d d) = \varphi(x) exp\big(d(\mu-\mu_d)\big) \quad (3)$$

Where d is the thickness of the flaw. If the flaw is a cavity with zero absorption, $\mu_d = 0$ (Mackenzie & Totten, 2006). Fig. 3 shows the small cavities in a workpiece on the left and on the right the simulated X-ray image compared with its real radiograph in the middle.

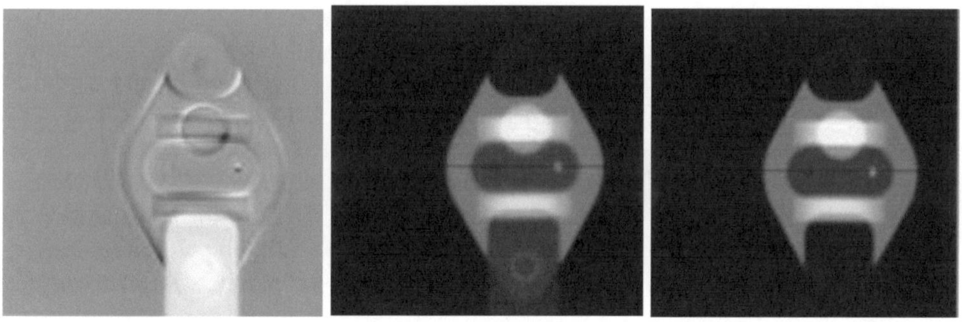

Fig. 3. The 3D image of a work-piece at left volume and its radiograph in the middle and the simulated image at right

In this technique, the entire object, including its defects, will be built up by independent CAD software. It offers excellent flexibility for setting objects and flaws, but it is not convenient and user friendly. The object itself may be complicated. It is time consuming to build up such a complicated model, moreover, in which every point should be calculated according to the attenuation law to form a simulated radiograph. What we need is to show flaws where they may appear on a real product. It is possibly easier to build up the flaw model only and superimpose the simulated flaws onto a real X-ray radiograph.

2.2 CAD models for flaw only

There is a serious disadvantage in the first approach. Building up an entire CAD model of the object and computing each pixel of the simulated image make the enormous computation. The second approach can be regarded as an improvement of the first one. In this technique, a 3D modelled flaw is projected and superimposed onto a real X-ray image.

The gray value displayed from the captured CCD image can be expressed as a linear function of the transmitted radiation:

$$I(x) = A\varphi(x) + B \tag{4}$$

where x is the thickness of the object. If there is a cavity with thickness d, according to equation (3), the equation will be changed into:

$$\mathrm{I(x\text{-}d) = A\varphi(x)exp(\mu d) + B} \tag{5}$$

According to equation (4), $A\varphi(x)$ can be substituted by $I(x) - B$:

$$I(x-d) = (I(x) - B)\exp(\mu d) + B \tag{6}$$

In an X-ray image, the brightest point I_{max} is obtained when the thickness is zero and the darkest point I_{min} is obtained when the thickness is a maximum.

$$I_{max} = A\varphi_0 + B \tag{7}$$

$$I_{min} = A\varphi_0\exp(-\mu x_{max}) + B \tag{8}$$

With equation (6) and (7), B can be estimated as following:

$$B = I_{max} - \frac{I_{max}-I_{min}}{1-exp(-\mu x_{max})} \tag{9}$$

In this technique, a 3D flaw model, the material absorption coefficient μ, and a real X-ray image are needed. Then, parameter B is estimated according to equation (9). The 3D flaw model with thickness d is projected and superimposed onto the real X-ray image using equation (6). The new gray value of a pixel, including the projected 3D flaw, comes from equation (6). All we need are the absorption coefficient and the flaw thickness (Mery, 2001). The original gray value provides us part of the information needed. The work required in this approach is much less than in the first approach. But in this approach, we don't consider the solid angle $\Delta\Omega$ of the X-ray beam that is projected onto the pixel, which generates an error.

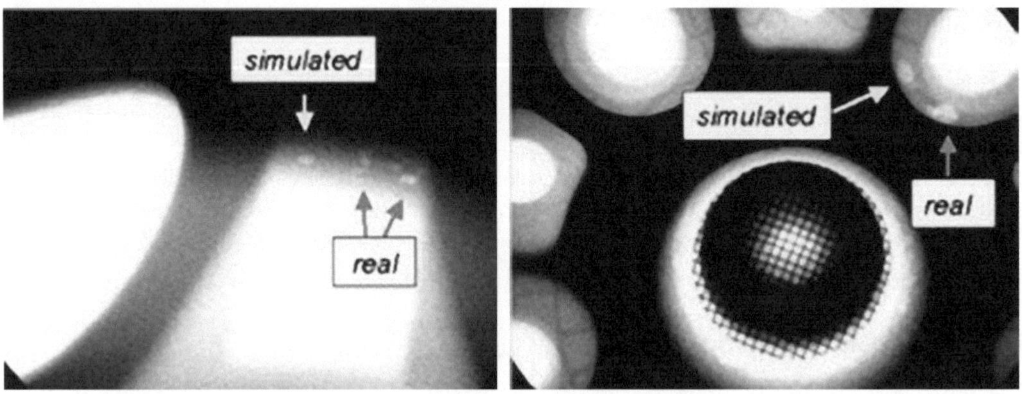

Fig. 4. Very simple simulated casting defects compared with the real ones

2.3 Coordinate system

The CAD model and the X-ray source use virtual 3D world coordinates, but the image uses 2D digital image coordinates. We need to transform between these two systems. 3D coordinates can be denoted as:

$$M = [X\ Y\ Z]^T \tag{10}$$

We assume that the digital image has the horizontal and vertical scale factors K_u and K_v. The focus distance is f. To maintain the projected system, the projected plane must be placed within the focus distance f. If the digital image has size (x_{max}, y_{max}) pixels, the projected plane must have the physical size of $(x_{max}/K_u, y_{max}/K_v)$millimetres. It means that the projected plane samples every $1/K_u$ millimetres horizontally and $1/K_v$ millimetres vertically during the simulation process. A pixel $m=(x, y)$ in the simulated image can be converted into

$$M = \left[\frac{U_0+x}{K_u}\ \frac{V_0-y}{K_v}\ f\right]^T \tag{11}$$

in the 3D coordinate system while (U_0, V_0) is the origin of the projected plane.

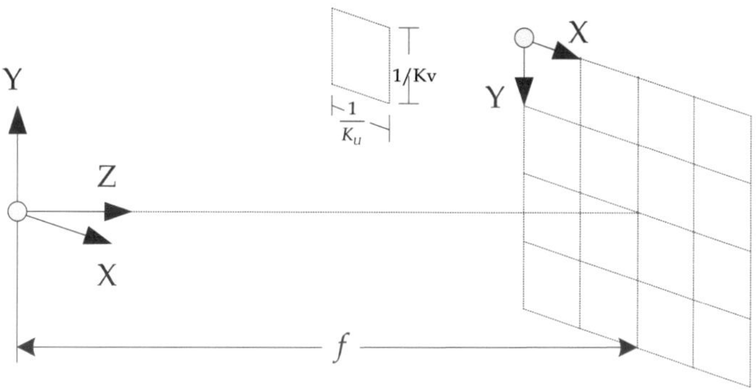

Fig. 5. The coordinate system.

To simulate the flaw there are 4 steps needed. First, capture a real X-ray image, and take the parameters f, K_u and K_v, (x_{max}, y_{max}) ; Second, build a 3D flaw model as a 3D mesh; Third, position the 3D flaw in the virtual coordinate system and simulate the X-ray capture process, obtaining a depth map matrix W(x,y) which stores the length travelled within the component of each ray that penetrated the 3D flaw. Finally, compute the new gray value with the depth matrix using equation (6). A single ray is emitted from every pixel of the 2D image to the X-ray source. The distance travelled by the ray inside the flaw will be calculated and stored in the depth matrix (Hahn & Mery, 2003).

2.4 Discussion of CAD model simulation

First, each intersection alternates the state of the ray from outside to inside. The final depth is the sum of all internal traces and the attenuation calculation of the X ray will be difficult for irregular product shape.

Second, the computation is enormous. Every pixel of the simulated image needs to emit a ray to penetrate the 3D model to the source. For simplification, we assume that the X-ray comes out from a point source. But actually, the X-ray source is not an ideal point source and it has a scatter effect. Another cause that extends the computation is the shape of the projected plane. The projected plane is by definition a plane with the central pixel closer to the X-ray source than the corner pixels. The intensity of the X-rays is stronger in the middle. We know a perfect detector would be a spherical one which is far more complicated in coordinate computation.

Third, the CAD model is not user friendly. A three-dimensional CAD model is needed as seen from Fig.2, but building up such a CAD model is a time consuming job. Although the second approach doesn't need an entire object model, the 3D flaw model is needed. However there are no prepared parameters for flaws normally. Considering the stochastic shape of the most defects, generating 3D CAD flaw model, which have to be created before all other calculation and projection, is a key technique, because using a fixed CAD model lacks randomization of flaw models.

Finally, there is a lack of description language and criteria for man-made 3D flaws. It's hard to tell whether man-made flaws are like or unlike real flaws.

3. Monte Carlo methods and their application in X-ray image simulation

3.1 Monte Carlo methods

A brief introduction of Monte Carlo methods is presented to better understand the concept and the four steps that Monte Carlo methods usually follow. An example of estimating π is summarised and the applications are discussed.

3.1.1 Introduction

The Monte Carlo method was developed in the 1940s by John von Neumann, Stanislaw Ulam and Nicholas Metropolis, while they were working on nuclear weapon projects (Manhattan Project) in the Los Alamos National Laboratory. It was named in homage to the Monte Carlo Casino, a famous casino, where Ulam's uncle would often gamble away his money (Metropolis, 1987).

Monte Carlo methods are a class of computational algorithms that rely on repeated random sampling to compute their results. Monte Carlo methods are often used in simulating physical and mathematical systems. These methods are most suited to calculation by a computer and tend to be used when it is infeasible to compute an exact result with a deterministic algorithm.

Monte Carlo methods are especially useful for simulating systems with many coupled degrees of freedom, such as fluids, disordered materials, strongly coupled solids, and cellular structures. They are used to model phenomena with significant uncertainty in inputs, such as the calculation of risk in business. They are widely used in mathematics, for example to evaluate multidimensional definite integrals with complicated boundary conditions. When Monte Carlo simulations have been applied in space exploration and oil exploration, their predictions of failures, cost overruns and schedule overruns are routinely better than human intuition or alternative "soft" methods.

3.1.2 Steps

Monte Carlo methods vary in different situations, but tend to follow these particular and simple steps:

1. Define a domain of possible inputs.
2. Generate inputs randomly from a probability distribution over the domain.
3. Perform a deterministic computation on the inputs.
4. Aggregate the results.

For example, consider a circle inscribed in a unit square. Given that the circle and the square have a ratio of areas that is $\pi/4$, the value of π can be approximated using a Monte Carlo method (Kalos &Whitlock, 2008):

1. Draw a square on the ground, and then inscribe a circle within it.
2. Uniformly scatter some objects of uniform size (grains of rice or sand) over the square.
3. Count the number of objects inside the circle and the total number of objects.
4. The ratio of the two counts is an estimate of the ratio of the two areas, which is $\pi/4$. Multiply the result by 4 to estimate π.

In this procedure the domain of inputs is the square that circumscribes our circle. We generate random inputs by scattering grains over the square then perform a computation on each input (test whether it falls within the circle). Finally, we aggregate the results to obtain our final result, the approximation of π.

To get an accurate approximation for π this procedure should have two other common properties of Monte Carlo methods. First, the inputs should truly be random. If grains are purposefully dropped into only the centre of the circle, they will not be uniformly distributed, and so our approximation will be poor. Second, there should be a large number of inputs. The approximation will generally be poor if only a few grains are randomly dropped into the whole square. On average, the approximation improves as more grains are dropped.

3.1.3 Applications

Monte Carlo methods are especially useful for simulating phenomena with significant uncertainty in inputs and systems with a large number of coupled degrees of freedom. Areas of application include: physical sciences, engineering, computational biology, applied statistics, games, design and visuals, finance and business, telecommunications, etc.

3.2 Monte Carlo and computer simulation

Computer simulation had developed rapidly just after the invention of the computer and the first large-scale deployment is to model the process of nuclear detonation. In this simulation, the scientists used a Monte Carlo algorithm.

3.2.1 Monte Carlo and computer simulation

A computer simulation is a computer program that attempts to simulate an abstract model of a particular system. Computer simulations have become a useful part of mathematical modelling of many natural systems in physics (computational physics), astrophysics, chemistry and biology, human systems in economics, psychology, social science, and engineering. Simulations can be used to explore and gain new insights into new technology, and to estimate the performance of systems too complex for analytical solutions.

Computer simulation developed hand-in-hand with the rapid growth of the computer, following its first large-scale deployment during the Manhattan Project in World War II to model the process of nuclear detonation. It was a simulation of 12 hard spheres using a Monte Carlo algorithm. Computer simulation is often used as an adjunct to, or substitute for, modelling systems for which simple closed form analytic solutions are not possible. There are many types of computer simulations; the common feature they all share is the attempt to generate a sample of representative scenarios for a model in which a complete enumeration of all possible states of the model would be prohibitive or impossible.

3.2.2 Monte Carlo simulation software

Many software systems and computer platforms can perform Monte Carlo simulation tasks, Matlab and C++ are the most well-known. Some specially-designed software systems for Monte Carlo simulation are very popular with researchers. Two of them will be introduced below.

1. EGSnrc

The EGSnrc system (EGSnrc, URL) is a package for the Monte Carlo simulation of coupled electron-photon transport. Its current energy range of applicability is considered to be 1 keV to 10 GeV. The EGS acronym stands for Electron Gamma Shower, and EGSnrc is an extended and improved version of the EGS4 package originally developed at Stanford Linear Accelerator Center (SLAC). In particular, it incorporates significant improvements in the implementation of the condensed history technique for the simulation of charged particle transport and better low energy cross sections.

2. GoldSim

GoldSim is the premier Monte Carlo simulation software solution for dynamically modelling complex systems in business, engineering and science (GoldSim, URL). GoldSim supports decision and risk analysis by simulating future performance while quantitatively representing the uncertainty and risks inherent in all complex systems. Organizations worldwide use GoldSim simulation software to evaluate and compare alternative designs, plans and policies in order to minimize risks and make better decisions in an uncertain world.

3.3 X-ray images simulation by Monte Carlo methods

As the X-ray imaging system must be modelled by Monte Carlo simulation, the physical principles of an X-ray imaging system will be introduced firstly. Two publications with the corresponding models and the simulation results are discussed.

3.3.1 Principles of the X-ray imaging system

Typical X-ray imaging systems consist of a point X-ray source and an image recording device (detector). The object to be imaged is placed between the X-ray source and the detector. The images are produced by local reduction in the intensity of the primary X-rays produced by the X-ray source reaching the detector through the internal structure of an object in their path. These local intensity reductions correspond to the distribution of absorption (attenuation) in the object under study. The special amount of attenuation by the object depends, in turn, on its thickness, its chemical composition, and the X-ray energy. Besides the source and detector, other auxiliary devices exist in a real imaging system. The schematic diagram is illustrated in Fig. 6.

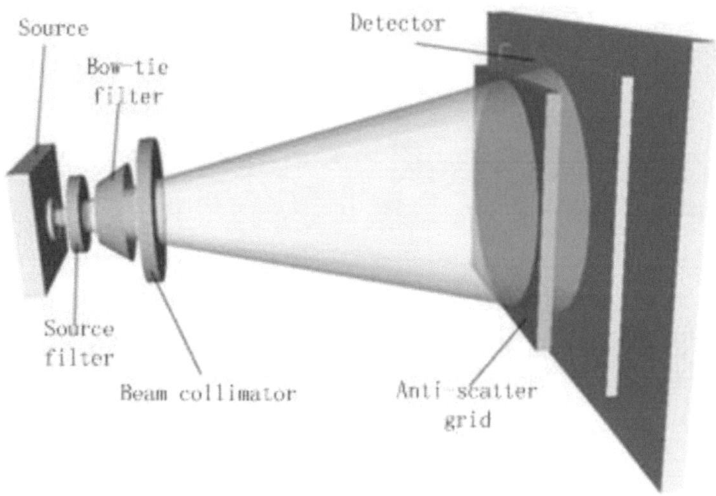

Fig. 6. The schematic depiction of an X-ray imaging system.

3.3.2 Direct imaging simulation model

U. Bottigli et al. developed a voxel-based Monte Carlo simulation model of X-ray imaging. They try to get the X-ray images from this simulation model directly. Two main steps can be summarised as below:

1. Modelling

The energy spectrum, polarization and profile of the incident beam can be defined so that X-ray imaging systems can be modelled. Photoelectric absorption, fluorescent emission, elastic and inelastic scattering are included in the model. The detailed model and the complicated mathematical descriptions can be found in the literature (Bottigli et al., 2004).

2. Simulation

The Monte Carlo method is employed in the course of simulation. The core of the simulation is a fast routine for the calculation of the path lengths of the photon trajectory intersections with the grid voxels. The selected simulation result can be seen in Fig. 7 below.

In this experiment, only the absorption is considered, i.e. the photons from scattering or fluorescence emission are discarded. This type of simulations is useful for testing the correctness and the speed of the procedure for numerical integration of the absorption along an arbitrary path inside a grid. The CPU time for the simulation of one projection image was about 3 seconds using a PC based on a PENTIUM IV 1800MHz processor.

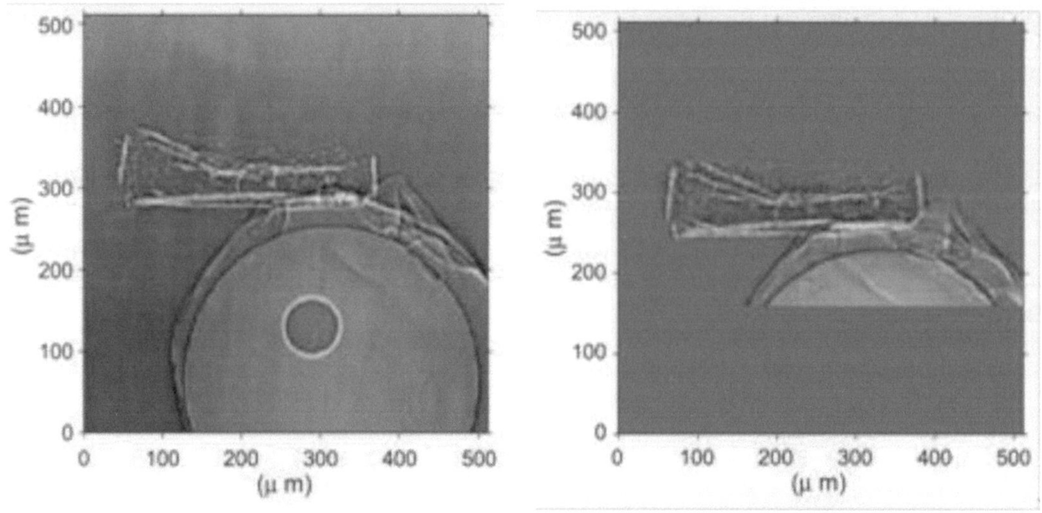

(a) Experimental image (b) Simulated image

Fig. 7. The simulation result of the voxel-based Monte Carlo simulation model. (a) is the experimental image, and (b) is the simulated image.

3.3.3 Scattered image simulation model

Different from the direct imaging model, the scattered image model focuses on simulating the scattered image. The final image can be obtained by subtracting the scattered image from the real image taken during the inspection. Frank Sukowski et al. provided a simple description of this model in (Sukowski & Uhlmann, 2011).

A lot of effects affect the image quality in radioscopy systems. By optimizing the throughput of the inspection system it is of essential interest to suppress all effects reducing the image quality. One effect e.g. is the scattered X-ray radiation from inside the specimen during inspection which hits the detector and reduces the contrast and sharpness of the projection. This effect leads to reduced possibility of detection of small defects. With the Monte Carlo simulation is it possible to simulate the scattering effects in the specimen and also the distribution of the scattered radiation on the detector. If we know the intensity distribution of the scattered radiation from the specimen on the detector, it can be subtracted from the real image taken during the inspection. With this operation it is possible to get images of the specimen with nearly zero intensity of scattered radiation resulting in better contrast and higher sharpness of the image. The author believes that simulation is the only way to get a realistic and not approximated intensity distribution of scattered radiation. In Fig. 8 the simulated projection of a step wedge and the simulated intensity distribution of the scattered radiation are shown.

3.4 Discussion about Monte Carlo simulation method

Monte Carlo simulation is a method for iteratively evaluating a deterministic model using sets of random numbers as inputs. This method is often used when the model is complex, nonlinear, or involves more than just a couple uncertain parameters. An X-ray based

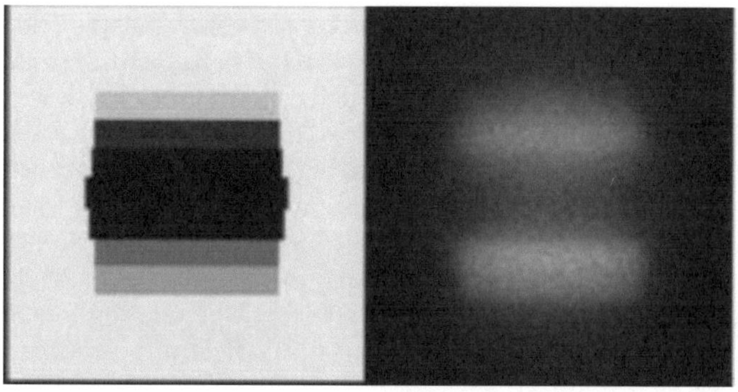

Fig. 8. The scattered image simulation. The left half is the simulated projection of a step wedge, and the right half is the simulated scattered image.

imaging system is a very complex system including lots of factors which can affect the final image. To simulate this kind of complex system with a Monte Carlo method, to build a proper model, is difficult and important. In this research area there is no readily available model that can be followed, even a tiny difference of hardware or device position can lead a very different model, and the objects under inspection can also affect this model. Another problem is the computing cost of simulation, which may be extremely high due to the model complexity. However, the Monte Carlo method is a very powerful tool to solve complex simulation problems, and its application in X-ray imaging systems is still being explored.

4. Generative image for flaw simulation in product radiographs

Defect superimposition is another technique that attempts to simulate casting defects. It differs from the CAD approach in the process of defect creation using 2D image technology, to superimpose the simulated casting defects onto real radioscopic images (Filbert et al., 1987). It needs neither complex 3D software packages nor a model of the casting specimen under test, furthermore, it offers a radioscopic image of a real product with a range of possible defects to test, validate and measure the accuracy of different radiograph analysis procedures, or for tutorial and training purposes.

4.1 Image generation technology

The image generation technology for flaw simulation of product radiographs is developed from the idea of the defect superimposition technique, and is based on defect analysis. The descriptions of defects or flaws defined by different standard organizations for specimens or products use a language and high level semantics which can be understood by engineers and workers in the field. Though many sample images are used as standards and are available for demonstration, they cannot include all because almost any defects in the products are distinguished greatly from others in their shapes, grey contrasts, salience of edges, size of areas, girth of counters etc. The defects are recognized from a product X-ray image normally depending on an understanding of the descriptions or definitions of the flaw but also rely on the calculation of their image parameters. We can generate flaw images according to the image semantic interpretation rather than depend heavily on the system parameters of the X-ray imaging or CAD models which provide the geometrical parameters of products.

The flaw generation process, as shown in Fig. 9, consists of five steps. The first step is defect image collecting. Though the definition for one kind of defect is uniform or unique, images have great variety, i.e. polymorphism in the defect formation. From a certain number of defect samples, the qualitative visual features are extracted and analysed. In the second step, which is a key step called "seed feature"or main high frequency feature creation process, the defect images are formed or grown from it. Simultaneously, the fuzzy image representations for each type of feature are expected to apply to the image generation after obtaining the qualitative features and statistical analysis, fuzzy membership is needed and is applied to these image features in the third step. The fourth step is generating low frequency features in the regions of the creative defect, to subject the defect appearances to their semantic definitions. The last step is image synthesizing or superimposing the created flaw image onto real product radiographies.

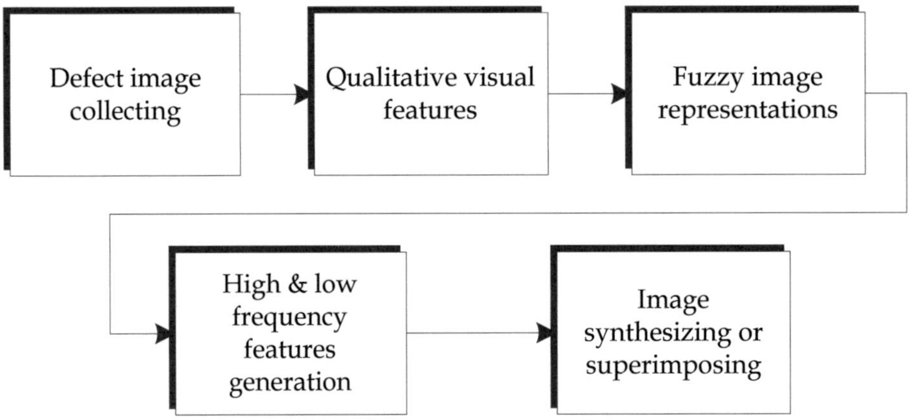

Fig. 9. Phases of flaw image generation for product radiographs

4.2 Defect image collecting

As X-ray radiation is passed through the material under test, and a detector senses the radiation intensity attenuated by the material, a defect in the material modifies the expected radiation received by the sensor. The contrast in the X-ray image between a flaw and a defect-free area of the specimen is distinctive. In an X-ray image we can see that the defects, such as voids, cracks, or bubbles, show up as bright areas compared to their background, i.e. the imaging of defect-free material.

Defects are defined according to their appearances, as many different kinds are described by the natural language along with some radiograph images, as we can see in Fig.10. The images shown in Fig. 10 are chosen from the reference radiographs presented by The American Society for Testing and Materials (ASTM) E155 (Hoff, 1999). These are standard radiographs of castings produced to help the radiographer make appropriate assessments of the defects in casting components.

As discussed in detail in the NDT Resource Center (Iowa State University, 1996), the characteristics of casting defects, the discontinuities produced by gas porosity or blow holes, sand and dross inclusions, various kinds of shrinkages, are summarized in 'radiography-NDT course material' (Iowa State University, 1996) are quoted here:

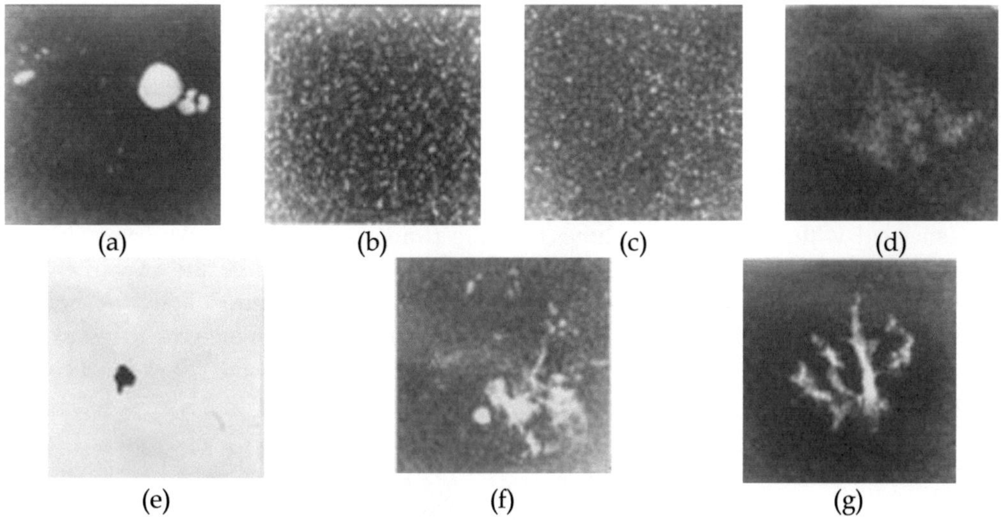

Fig. 10. The sample patterns of the American Society for Testing and Materials (ASTM). (a) gas porosity or blow holes (b) shrinkage of needle cavity(c) shrinkage of round cavity (d) sponge shrinkage (e) sand inclusions or dross (f) filamentary shrinkage (g) dendritic shrinkage

1. "Gas porosity or blow holes are usually smooth-walled rounded cavities of a spherical, elongated or flattened shape caused by accumulated gas or air which is trapped by the metal. The gas or air will be trapped as the molten metal begins to solidify while the sprue is not high enough to provide the necessary heat transfer needed to force the gas or air out of the mold.
2. Sand inclusions and dross are nonmetallic oxides, which appear on the radiograph as irregular, dark blotches. These come from chipped parts of the mold or core walls and/or from oxides which have not been skimmed off prior to the introduction of the metal into the mold gates.
3. Shrinkage has various forms of discontinuity occurring in all cases as the molten metal shrinks as it solidifies in all parts of the final casting. Shrinkage in its various forms can be recognized by a number of characteristics on the radiographs. There are at least four types of shrinkage: (a) cavity; (b) dendritic; (c) filamentary; and (d) sponge types. Some documents designate these types only by number to avoid misunderstanding. Their characteristics are summarized in the following:

(a) Cavity shrinkage appears as areas with distinct jagged boundaries. It may be produced when metal solidifies between two original streams of melt coming from opposite directions to join a common front. Cavity shrinkage usually occurs at a time when the melt has almost reached solidification temperature and there is no source of supplementary liquid to feed possible cavities.

(b) Dendritic shrinkage is a distribution of very fine lines or small elongated cavities that may vary in density and are usually unconnected.

(c) Filamentary shrinkage usually occurs as a continuous structure of connected lines or branches of variable length, width and density, or occasionally as a network.

(d) Sponge shrinkage shows itself as areas of lacy texture with diffuse outlines, generally toward the mid-thickness of heavier casting sections. Sponge shrinkage may be dendritic or filamentary. Filamentary sponge shrinkage appears more blurred as it is projected through the relatively thick coating between the discontinuities and the film surface."

The X-ray image is usually captured with a frame-grabber and stored in a matrix. The size of the image matrix corresponds to the resolution of the image. The gray value between 0 and 255 associated with the image brightness denote from the one hundred percent black to white in an image. Tuned by the voltage and current for the X-ray tube, the gray intensity of a product and its defects will vary significantly, that means the contrast between the defect and its background will change, the examples shown in Fig. 11, in which the tube current is 1.5mA but the tube voltages are varied from 110KV to 105KV and 75KV in each from left to right.

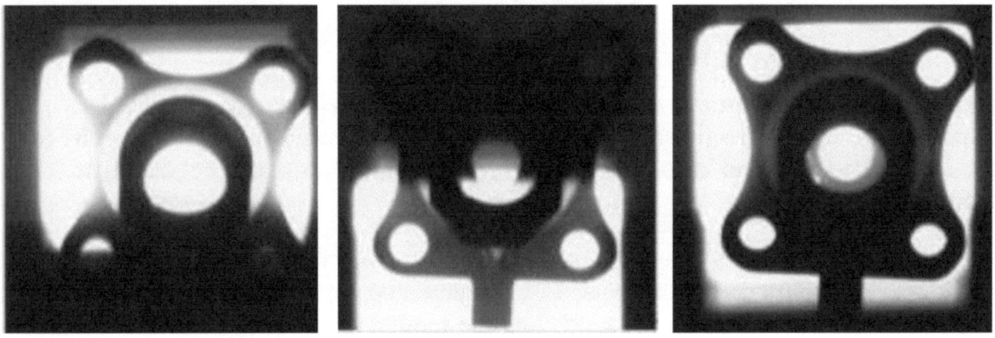

Fig. 11. The radiographs with different tube voltage for the same specimen, the tube voltage used from left to right are separately 110KV, 105KV and 75KV and all with tube current 1.5mA

From the pictures in Fig. 11 we can see that not only the contrast of the defects against their backgrounds vary in every image but the brightness of the defects also changes. Although using high voltage, the defect in the left image is the darkest compared with the defects in the other two images with relatively low voltages, and the contrast around the defect in the left image associated with its background is very low.

It is necessary to collect the sample defect radiographs and classify them into different semantic categories and if possible, get their radiographs with different voltages.

4.3 Visual feature analysis and nonparametric sampling

Deduced from example-based technique for image generation, the nonparametric sampling and modelling that relies on the qualitative visual feature analysis contributes to set up nonparametric models for both main construction of defects, their texture synthesis and other low frequency information. The introduction of nonparametric sampling and defect image reconstruction leads to an intricate feature analysis problem. Here we mention two different classical nonparametric sampling and modelling approaches, one is a stochastic defect counter based on a hierarchical template model and the other is a skeletonization-based model.

4.3.1 Counter based hierarchical template model

Most casting defects are located on the inner part of the work pieces, as seen in Fig. 12, where the two diagrams on the left are typical of actual radiographs with defects generated from the production line, the images sizes are 540×540 pixels, and the others are the diagrams of what is seen when the porosity defects are cut from the actual images and magnified to display with image sizes of 200×200 pixels. The intensities of casting defects in the radiographs are usually lighter than the parts around them. If the work pieces have different structures and shapes, the defects with the same size often appear as different grey values on different work pieces. This is a result of the different absorption of the X-rays by the metal due to the different thicknesses. The energies required for X-ray penetration of the thicker structures are significantly higher, and the grey values of the casting defects are strongly affected by the corresponding backgrounds. This is illustrated in Fig. 12, showing casting defects in casting products. Though it is not easy to classify them into type one or type two defects, all of them are single large and discontinuous. The aim of this paper is to generate simulated defects similar to these real ones.

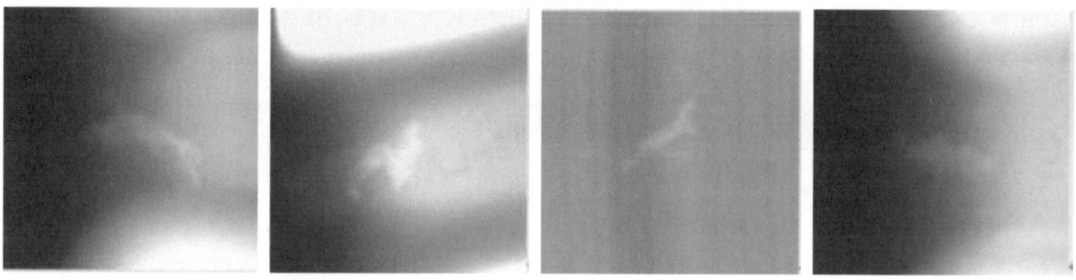

Fig. 12. The sample defects in real casting pieces

The cavities, as shown in Fig. 13 (a) and (b), are both single large casting defects, however, they are considered separately as blow hole and shrinkage defects, and they differ greatly in appearance and volumetric measurement from the porosity defect shown in Fig. 13 (c).

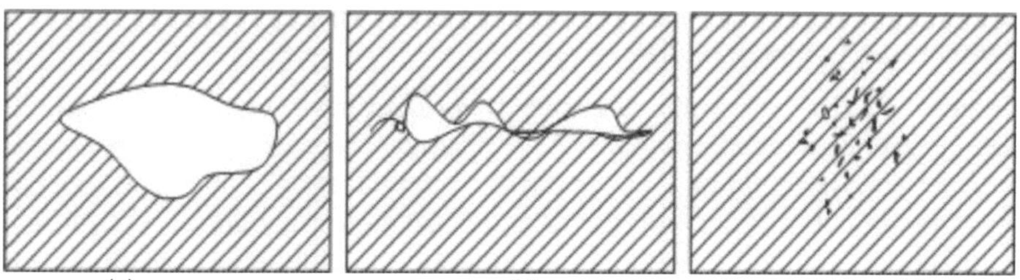

(a) (b) (c)

Fig. 13. Sample defects in real casting pieces

By careful inspection of the visual features of about 2000 images of casting defects, we have found that stochastic shapes with closed defect contours are the most prominent features. Based on the principle of Active Qualitative Vision (Aloimonos, 1990), an initial shape needs to be designed to provide a starting point followed by the adjustment of the following image parameters during the simulation: grey level value, shape, size, orientation, shading effects, and contrast with background. The random shape template is designed as three embedded nesting stencil-plates, as illustrated in Fig. 14, in which the shapes of each template can be given randomly but their sizes are limited so that every smaller one of three shapes fits entirely inside the larger one for the appropriate grey value adjustment. The analysis and the measurement of the profiles of the largest cavity defects, from all of our casting defect images, leads to a size for the profile of the largest stencil-plate of 78x46 pixels (1cm=31.889 pixels), which approximates to the size of the single defect profiles that appear normally in casting pieces.

From Fig. 14 and from the physics of the x-ray process it is easy to find that the middle of the cavities is normally brighter than their edges. For convenience the higher gray value is set to the middle of the entire shape; and also there is a general requirement for the template set that their areas vary, with the smaller being totally enclosed by the larger. With the proposed defect creation algorithm, the stencil-plates are not limited in shape, and changes in their formation will not significantly influence the outcome of the generated defects. Furthermore, the templates used are binary images. The white parts in the templates are the shapes used in the defect simulation. Though these white parts are not the exact final shape of the simulated defects, they approximate to the final shape having some of the features of real casting defects with smooth, continuous, and irregular contours.

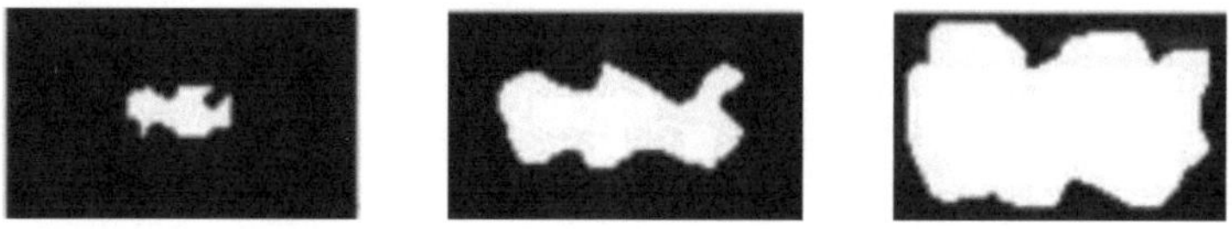

Fig. 14. The sample defects in real casting pieces

We use one of shapes in Fig.14 to design the superposing operation to locate the simulated defect onto a specified region of the original X-ray image; this superposing operation can be used by all stencil-plate superposition, the details can be found in reference paper (Huang, et al., 2009).

4.3.2 Skeletonization based model

The skeleton of some dendritic flaws such as filamentary shrinkage and dendritic shrinkage in castings or cracks in any materials can be the main feature or so called 'seed feature' to let simulated flaws grow from it. Skeletonization is a process for reducing foreground regions in a binary image to a skeletal remnant that largely preserves the extent and connectivity of the original region while throwing away most of the original foreground pixels(Gonzalez, 2008). It can be a key step when extracting the nonparametric features from a sample defect. To see how this works, imagine that the foreground regions in the input binary image are made of some uniform slow-burning material. Light fires simultaneously at all points along the boundary of this region and watch the fire move into the interior. At points where the fire travelling from two different boundaries meets itself, the fire will extinguish itself and the points at which this happens form the so called "quench line". This line is the skeleton. Under this definition it is clear that thinning produces a sort of skeleton.

The dash line in set A denotes its skeleton S(A), as shown in Fig. 15,the notion of a skeleton.

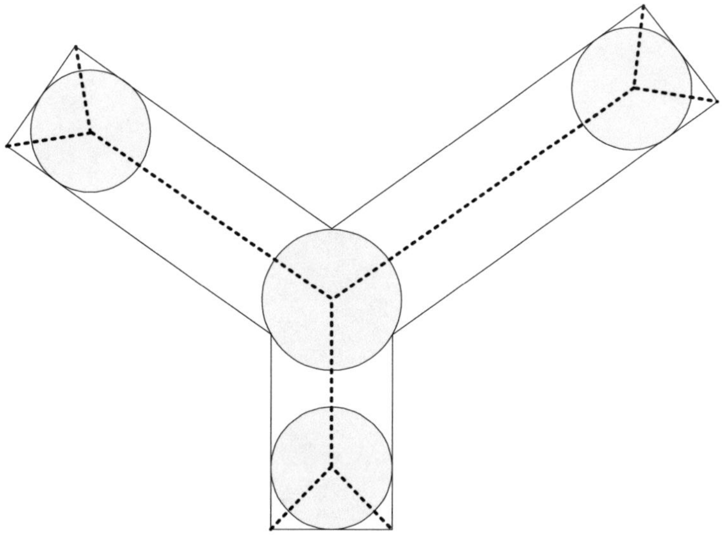

Fig. 15. Set A and its skeleton denoted by dashed lines

The skeleton of set A can be expressed in terms of erosions and openings as following:

$$S(A) = \bigcup_{k=0}^{k} S_k(A) \tag{12}$$

With

$$S_k(A) = (A \oplus kB) - (A \oplus kB) \circ B \tag{13}$$

Where B is a structuring element, and (A⊕kB) indicates k successive erosions of A K times,

$$(A \oplus kB) = (\cdots (A \oplus B) \oplus B) \oplus \cdots) \oplus B \tag{14}$$

and K is the last iterative step before A erodes to an empty set. In other words,

$$K = max\{k | (A \oplus kB) \neq \emptyset\} \tag{15}$$

The formulation states that S(A) can be obtained as the union of the skeleton subsets $S_k(A)$. Also, it can be shown that A can be reconstructed from these subsets by using the equation

$$A = \cup_{k=0}^{K}(S_k(A) \oplus kB) \tag{16}$$

Where $(S_k(A) \oplus kB)$ denotes k successive dilations of $S_k(A)$;that is

$$(S_k(A) \oplus kB) = ((\cdots (S_k(A) \oplus kB) \oplus B) \oplus \cdots) \oplus B \tag{17}$$

Another term medial axis transform (MAT) is often used interchangeably with skeletonization but they are not quite the same. The skeleton is simply a binary image showing the simple skeleton, while MAT is a graylevel image where each point on the skeleton has an intensity which represents its distance to a boundary in the original object, shown in the Fig. 16.

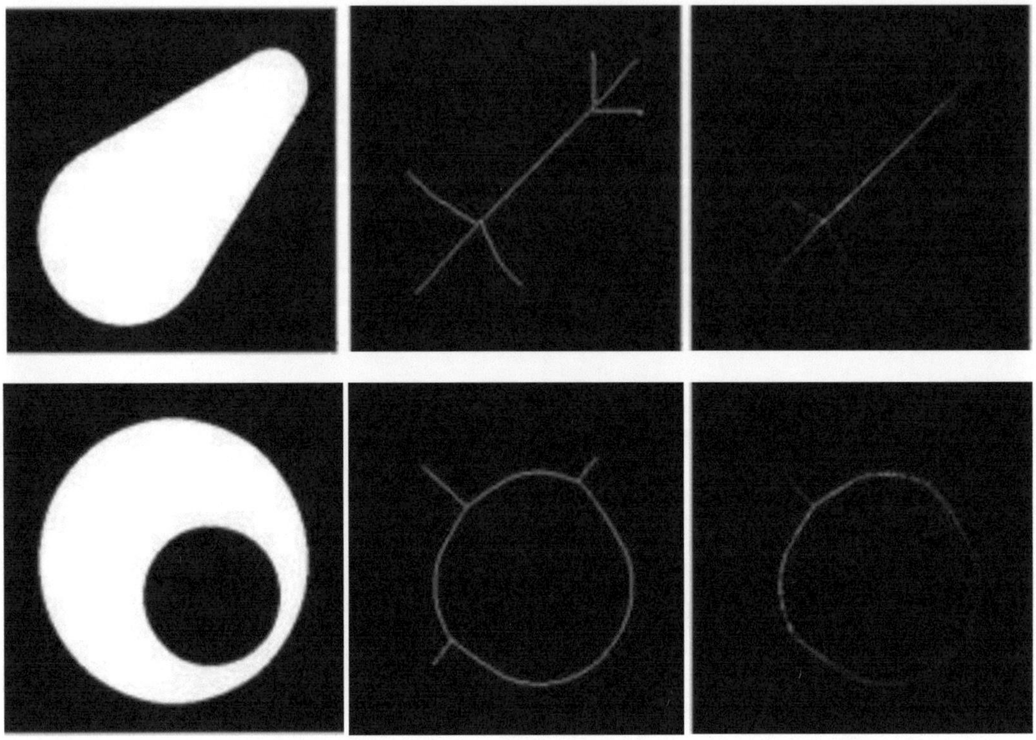

Fig. 16. Skeletonization of a binary image, from left to right: the original binary image, normal skeletons, the skeleton image formed by the medial axis transformation

The medial axis of an object is the set of all points having more than one closest point on the object's boundary. In mathematics the closure of the medial axis is known as the cut locus. In 2D, the medial axis of a plane curve S is the locus of the centres of circles that are tangent to curve S in two or more points, where all such circles are contained in S. (It follows that the medial axis itself is contained in S.) The medial axis of a simple polygon is a tree whose leaves are the vertices of the polygon, and whose edges are either straight segments or arcs of parabolas. The medial axis together with the associated radius function of the maximally inscribed discs is called the medial axis transform , which is a complete shape descriptor, meaning that it can be used to reconstruct the shape of the original domain.

According to American Society for Testing and Materials (ASTM), Filamentary shrinkage usually occurs as a continuous structure of connected lines or branches of variable length, width and density, or occasionally as a network. From Fig.17 we can see that, the most obvious feature is the continuous tree structure, wider in the trunk and thinner in the branches. In grayscale, the parts closer to the center of the structure are darker. The outside part is lighter and the branches are between the centre and the outside part. Because of its structure and features, the skeleton description is one of the best. In the application, several layers of skeletons are needed, as showing in Fig.17, the simple and the complicate skeletons for the left dendritic shrinkage are as Fig. 17(c) and (d), which will be used to represent different parts of the defects, some like trunks and some branches.

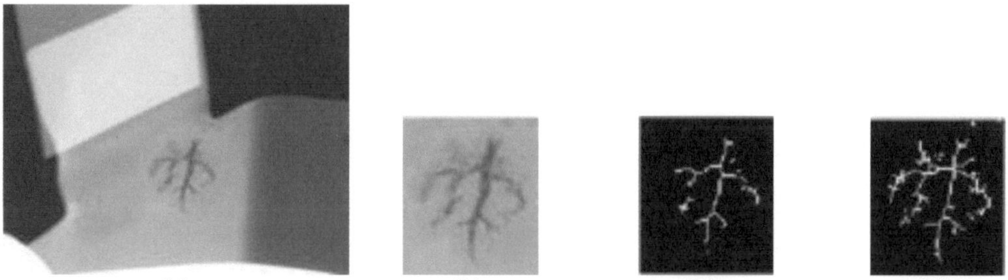

Fig. 17. Skeletonization of a real dendritic shrinking for flaw simulation: (a) a radiograph of a workpiece, (b) zoomed dendritic shrinking, (c), (d) the skeleton images based on binarized images of different thresholds

Fig. 18. Randomly created skeleton for crack generation and simulation

Apart from sample defects, skeletons can also be created randomly such as for cracks or filamentary shrinkages, and so on. Please refer Fig. 18.

4.4 Visual feature analysis and fuzzy representation

Since the qualitative visual features are fuzzy, so a fuzzy representation of them is appropriate. Set the image grayscale range *[1, Ng]*, μ is defined as the membership function of the gray value in a specified image area, it makes the gray value mapped to [0, 1]. On a view, like a factor image with piexls point to a single point of ambiguity in the original image, its corresponding values indicate the membership degree that the image point belongs to the object region. All the pixels in the image form a single dot fuzzy matrix expressed with μ *(I)*.

The following eight fuzzy geometric features impact on the accuracy and speed of operation with more good compromise effects. The image size is $M \times N$, the μ *(I)* is denoted by μ, the membership definitions of these fuzzy geometric features and the classification functions for the object structures are as follow:

1. Area:

$$a(\mu) = \sum \mu \tag{18}$$

Area is the weighted sum of the pixels in an object image range.

2. Circumference:

$$p(\mu) = \sum_{i=1}^{M} \sum_{j=1}^{N-1} |\mu(i,j) - \mu(i,j+1)| + \sum_{i=1}^{M} \sum_{j=1}^{N-1} |\mu(i,j) - \mu(i+1,j)| \tag{19}$$

Circumference pixel horizontal distance and vertical distance weighted and is used to measure the undulations of the image area.

3. Firmness

$$comp(\mu) = \frac{a(\mu)}{p^2(\mu)} \tag{20}$$

Firmness reflects the degree of fuzzy compact region

4. Height and width

Height

$$h(\mu) = \sum_{i=1}^{M} max_j\{\mu(i,j)\} \tag{21}$$

Width

$$w(\mu) = \sum_{j=1}^{N} max_j\{\mu(i,j)\} \tag{22}$$

Height is the total number of the largest membership of the rows of the image, and width isthe total number of the largest membership of the columns of the image. The height and width can be used, respectively, to reflect the image of certain areas in the overall vertical and horizontal directionfeatures.

5. Length and width

Length

$$l(\mu) = max_j\{\sum_i^M \mu(i,j)\} \tag{23}$$

Width

$$b(\mu) = max_i\{\sum_j^N \mu(i,j)\} \tag{24}$$

Length is the total number of the largest membership of the rows of the image, width is the total number of the largest membership of the columns of the image. The use of the X-ray image can be described in certain regions in the vertical and horizontal distribution of the extreme characteristics of the membership.

6. Area coverage

$$IOAC(\mu) = \frac{a(\mu)}{l(\mu)b(\mu)} \tag{25}$$

The characteristic quantity is the fuzzy area defined by the length and width of the moment shaped area ratio, length and width defined by the area of image certain areas of the largest possible area of fuzzy.

4.5 Generation of the low frequency information

No matter how the generative flaw image is rendered from a creative counter or from a simple skeleton, the counters or skeletons normally are the crucial characteristics of most inner defects of the materials and are used mainly to produce the similar perceivable structures with the original flaw. While some types of flaws are not obvious in their high frequency information, but present as apparent textural features such as shrinkages of needle cavity and small round cavity as well as sponge shrinkage showing in Fig. 19 (Hoff, 1999).

Fig. 19. Defect texture, from left to right: shrinkages of needle cavity and small round cavity as well as sponge shrinkage

4.5.1 Texture generation

Efros and Leung (1999) have proposed a nonparametric model to synthesize an output image from a given metric sample. Their approach works surprisingly well for a wide range of textures. Their algorithm consists of the following steps:

1. The output image is initialized with a random noise whose histogram is equalized with respect to the histogram of the input sample.

2. For each pixel in the output image, in scan-line order, the already-synthesized values in neighbourhood of current pixel *z* of specific size *N(z)* is considered and is compared with all possible neighbourhoods from the input sample. The value of the input pixel x with the most similar *N(x)* is then assigned to *z*.

Efros & Leung's algorithm is significantly useful for some kinds of defect simulation. After, initialize a synthesized texture with a 3x3 pixel "seed" from the source defect texture, for every unfilled pixel which borders some filled pixels, find a set of patches in the source image that most resemble the unfilled pixel's filled neighbors. Choose one of those patches at random and assign to the unfilled pixel a grey value from the center of the chosen patch, and do it repeatedly. The only crucial parameter is how large a neighbourhood you consider when searching for similar patches. As the "window size" increases, the resulting textures capture larger structures of the source image. The finer details of this approach can be found in the Efros' paper. The other problem which arises from radiograph simulation is to merge together the generated flaws with the workpiece or other parts of the background image. In a similar manner to the heat diffusion equation, the image diffusion technique is adopted, which is discussed below.

4.5.2 Imagediffusion technique

Embedding simulated flaws into the objects' radiographs, which is called augmented reality in the simulation research field, is not an avoidable step and is important both in simulation application and the technology of image superimposing. Take the X-ray penumbra into account, we know that the edges of the defects are blurred and sometimes fused with their background, furthermore, the grey values of the defect vary when the background changes, constrained by the law of X-ray attenuation and simultaneity, the grey values of the work-piece and its defect vary enormously in the large range of tube potential.

Similar to the effects of the heat diffusion equation, image diffusion is suitable for edge preserving blur and can be used widely in a flaw superimposing into a real radiograph, the apparent visual characteristics of blurred edges by diffusion can be seen from Fig. 20.

Fig. 20. The original panda image with the diffused ones on their right

To an initial image *u(x,y,0)* change to be image *u(x,y,t)* after a time *t*, the anisotropic diffusion first proposed by Perona and Malik (1990) is denoted as:

$$\frac{\partial(x,y,t)}{\partial t} = div[g(x,y,t)\nabla u(x,y,t)] \tag{26}$$

where $\nabla u(x,y,t)$ is the gradient of the image and $g(x,y,t)$ is diffusion coefficient, which can be isotropic or anisotropic and depends on different applications. Teboul et al.(1998) further discussed the conditions of the coefficient $g(x,y,t)$ and three constraints are given as follows:

1. $lim_{|\nabla I|\to 0}\, g(x) = Z$，$0 < Z < \infty$;
2. $lim_{|\nabla I|\to \infty}\, g(x) = 0$;
3. $g(x)$ is a strictly decreasing function of $|\nabla I(x)|$.

There are many research contributions to the diffusion coefficient function, such as Gauss Smoothing Diffusion(Catte, et al. 1992), Morphological Diffusion(Segall& Acton, 1997), Degeneration Diffusion Mode(You, et al. 1995)Statistical Diffusion Mode(Black, et al. 1998), Curvature Diffusion Mode(Alvarez, et al. 1992), but the anisotropic diffusion models are ill-posed, it is difficult to determine the convergence condition. In the flaw simulation, we still rely on the qualitative visual feature analysis of real radiographs.

4.6 Synthesizing and generating radiographs

The flaws generated in Fig. 21, follow a similar procedure though their 'seed features' differ from greatly from each other. After analysis of sample images and grouping them into different defect categories, extract their qualitative visual features; choosing the 'seed feature' and decide the flaw rendering approach such as the template method or the skeletonization method or texture synthesizing. Based on statistical analysis, calculate thefuzzy image representations for each type of feature, which are expected to be relevant for the image generation, fuzzy membership as needed is applied to these image features during the image generation process. As the seed feature grows, simultaneously, the low frequency features in the regions of the creative defect should be taken out to ensure that the defect appearances reflect their semantic definitions. Finally, image superimposing onto real product radiographs is also an important step that contributes to the success of the simulation application. The super impositions rely on the location as well as the associated gray values, which generally are implanted in the simulated flaws according to the contrast of the defect against its background.

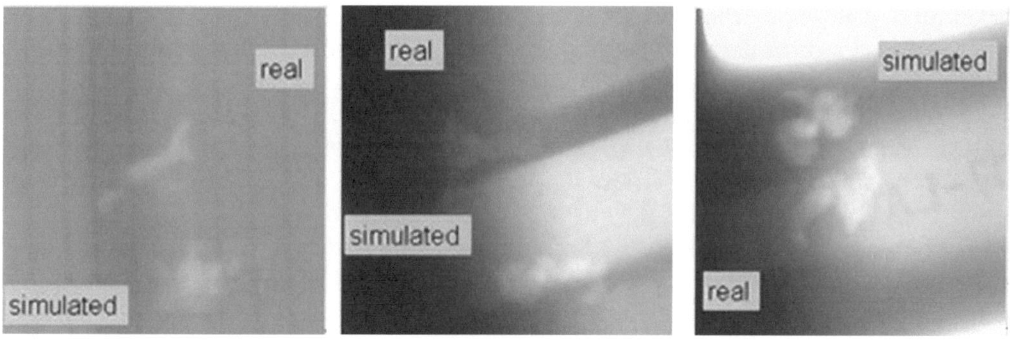

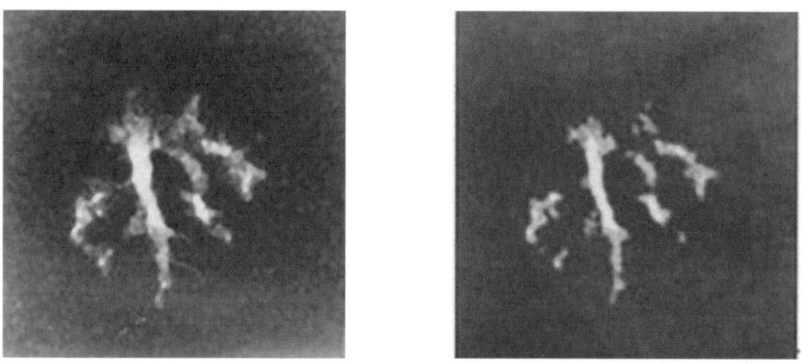

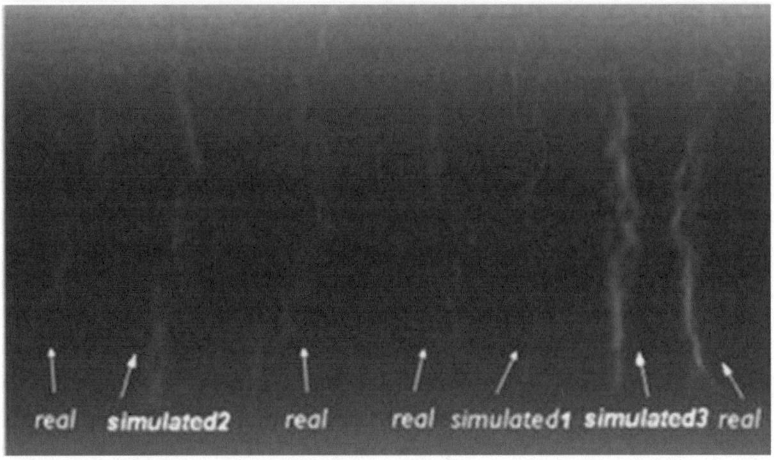

Fig. 21. Comparison simulated flaws with real defects on different products, at the third row the original is on the left and the simulated is on the right

5. Conclusion

X-ray defect simulation is associated with many uncertainties in terms of shape, inconsistancy of grey values and the blurring of simulated defect edges against the background. The approaches discussed in this section have shown how to simulate and create various kinds of defects including stochastic large discontinuity defects and regular shape defects such as circular and elliptical spots, which can be used to build a sample library for testing and tuning automatic inspection programs. The new defect superposition methods using nesting stencil-plates, texture generation as well as skeletonization have been shown to be able to simulate the normal range of casting defects. These new methods are not only capable of simulating the casting defects, but also able to be used to simulate irregular objects in different kinds of X-ray images. Apart from the crucial high frequency image features that were adopted to render the structures of the flaws, the anisotropic diffusion method has received more attention as nearly all the objects appearing in radiographs have shading effects, which could be visually represented by diffusion technology.

Compared with the CAD model and the Monte Carlo method, the advantages and characteristics of image generation method, that is the augmented reality simulation method, can be summarized as: (1) It has more randomness and flexibility and provides a better match for the shape variability of the defects: (2) The 'seed features' are not limited in shape or size, and changes in their form will not significantly influence the simulated defects. (3) Many and varied defects can be created by the user and added to a real casting piece image. The program is easy to use and, by choosing and tuning the parameters, it can be used for the majority of casting pieces without any advanced measurements on the work piece. (4) The program works directly on the radiographs, and the location for the simulated defect can be chosen randomly by the user in any X-ray image. (5) Since no part of the programme is based on a CAD platform, there is no need of a CAD software package or any other database, the defect simulation function can be organized as a program module and integrated into the casting automatic inspection system to share the stored data.

6. References

Aloimonos, J. (1990). Purposive and Qualitative Active Vision, *Proceedings of 10th International Conference on Pattern Recognition (ICPR)*, pp. 346-361

Alvarez, L. et al. (1992). Image selective smoothing and edge detection by nonlinear diffusion II. SIAM *Journal of Numerical Analysis*, (1992), Vol.29, No.3, pp. 845-866

Black, M. et al. (1998). Robust anisotropic diffusion, *IEEE Transactions on Image Processing*, (1998), Vol. 7, No.3, pp. 421-432, ISSN: 1057-7149

Bottigli et al. (2004). Voxel-based Monte Carlo Simulation of X-ray Imaging and Spectroscopy Experiments, *Spectrochimica Acta Part B: Atomic Spectroscopy*, (2004), Vol.59, No.10-11, pp. 1747-1754

Catte, F. et al.(1992). Image selective smoothing and edge detection by nonlinear Diffusion, *SIAM Journal of Numerical Analysis*, et al. (1992), Vol.29, pp. 182-193

Efros, A. & Leung, K. (1999). Texture Synthesis by Non-parametric Sampling, Proceedings of *IEEE International Conference on Computer Vision*, pp. 1033-1038, Corfu, Greece, September 1999

EGSnrc (URL). Available from http://irs.inms.nrc.ca/software/egsnrc/

Filbert, D.; Klatte, R.; Heinrich, W. & Purschke,, M. (1987). Computer Aided Inspection of Castings. *Proceedings of the IEEE-IAS Annual Meeting*, pp. 1087-1095, Atlanta, USA

GoldSim(URL). Available from http://www.goldsim.com/Home/

Gonzalez, C. & Woods, E. (2008). *Digital Image Processing*, Prentice Hall, ISBN: 0201180758, 0130946508

Haken, H. &Wolf, H. (1996). *The Physics of Atoms and Quanta: Introduction to Experiments and Theory*. Berlin, Heidelberg: Springer, 5th edition, 1996

Hoff, A. (1999). Reference Radiographs, In: *The American Society for Nondestructive testing*, 03.08.2011, Available from http://www.asnt.org/publications/materialseval/basics/oct99basics/oct99basics.htm

Iowa State University. (1996), Radiograph Interpretation - Castings, In : *NDT resource center*, 04.08.2011, Available from http://www.ndt-ed.org/EducationResources/CommunityCollege/Radiography/TechCalibrations/RadiographInterp_Castings.htm

Kalos, H. & Whitlock, A. (2008). *Monte Carlo Methods*, Wiley-VCH, ISBN: 978-3527407606

Mackenzie, D.& Totten, G. (2006). *Anyalytical Characterization of Aluminum, Steel, And Superalloys*, CRC Press, Taylor & Francis Group, (2006). pp. 718-721, ISBN: 9780824758431

Mery, D. (2001). Flaw Simulation in Castings inspection by radioscopy, INSIGHT, *Journal of the British Institute of Non-Destructive Testing*, (2001), Vol.43, No.10, pp.664-668

Metropolis, N. (1987). *The Beginning of the Monte Carlo Method*, Los Alamos Science (1987 Special Issue dedicated to Stanislaw Ulam), pp. 125–130

Perona, P. & Malik, J. (1990). Scale-space and edge detection using anisotropic diffusion. *IEEE Transactions Pattern Anal. Mach. Intel.*, (1990), Vol.12, No.7, pp. 629–639, ISSN: 0162-8828

Segall, A. & Acton, T. (1997). Morphological anisotropic diffusion, *Proceedings of IEEE international conference on image processing*, (1997), No.1, pp. 356-361

Sukowski, F. & Uhlmann, N. (2011). *Applications of Monte Carlo Method in Science and Engineering*, Shaul Mordechai (Ed.), InTech, ISBN: 978-953-307-691-1

Teboul, S. et al. (1998). M. Barlaud. Variational approach for edge-preserving regularization using coupled PDEs. *IEEE transactions on Image Processing,* (1998), Vol 7, No.3, pp. 387-397, ISSN: 1057-7149

Tillack, G. et al. (2000). X-ray modeling for industrial applications, NDT & E International, Vol.33, No.7, (2000), pp. 481-488

You,Y. et al. (1995). On ill-posed anisotropic diffusion models. *Proceedings of International Conference on Image Processing*, 1995, No.2, pp. 268-271

You,Y. et al.(1996). Behavioral analysis of anisotropic diffusion in image processing. *IEEE Transactions on Image Processing*. (1996), Vol.5, No.11, pp. 1539-1553, ISSN: 1057-7149

6

Study of Metallic Dislocations by Methods of Nondestructive Evaluation Using Eddy Currents

Bettaieb Laroussi, Kokabi Hamid and Poloujadoff Michel
Université Pierre et Marie Curie (UPMC), Laboratoire d'Electronique et Electromagnétisme (L2E), Paris
France

1. Introduction

This chapter summarizes a work which we have conducted over the period 2005-2010 [Bettaieb et al., 2008; Bettaieb, 2009]. The aim was to better understand the Non Destructive Evaluation (NDE) method based on the use of eddy currents. A very simple example, well adapted for beginners, was the evaluation of a homogenous rectangular aluminum plate of constant thickness, in case of a calibrated crack. Such a choice had a great advantage: it allowed the development of simple theoretical analysis, and therefore a very good understanding of the physics of the problem.

Naturally, theoretical results must be checked experimentally. To this end, we built a Helmholtz arrangement of two circular coils, to create a convenient uniform alternating field within the plate. Then, to measure the reaction field of the plate, we had to use a sensor. We used alternately a SQUID and a Hall sensor. The choice between those is not obvious, as will be shown later. Indeed, the SQUID is about 1000 times more sensitive than the Hall sensor, but its sensitive part is more remote than the Hall sensor, and this may reduce significantly its advantage. We found this comparison extremely interesting [Bettaieb et al., 2010].

Indeed, if we cut a rectangular plate, the cut being parallel to one side and normal to the excitation field, it seems natural that the plate is equivalent to a homogeneous one. Therefore, the cut cannot be detected. Just in case, we made the experiment and there was an unexpected signal!! We rapidly realized the reason of this result: the cut through the plate did not interrupt any current, but the cutting tool had damaged the microstructure of the aluminum over a small part of each half plate. Therefore, none of the two half parts had remained homogeneous. We realized that the microstructure of a small part of the half plates had been modified. In fact, the two halves of the original plate were equivalent to a homogeneous plate, except along the cut. Therefore, *we did not detect the cut, but the metallic dislocations around it.* Some more theoretical investigations allowed to evaluate the importance of the change in the electrical resistivity associated with the dislocations, and their extension. Then, we started to investigate the dislocations due to hammer shocks and to mechanical flexions.

This chapter is a report of all our efforts. And we hope that other studies, by other investigators, will extend these results.

2. Theoretical study of a defect free rectangular aluminium plate

We consider the case of a homogeneous rectangular aluminum plate (resistivity $\rho = 5.82 \times 10^{-8}\Omega.m$). Its dimensions are $a = 110mm$, $b = 100mm$, $c = 5mm$ in the x, y, z directions respectively (figure 1). This plate is immersed in an alternative uniform magnetic induction field $\vec{B}$ parallel to the Oy direction, varying at the $180Hz$ frequency, with an amplitude of $0.15 \times 10^{-4}T$ if we use a SQUID and $2.45 \times 10^{-4}T$ if we use a Hall probe sensor.

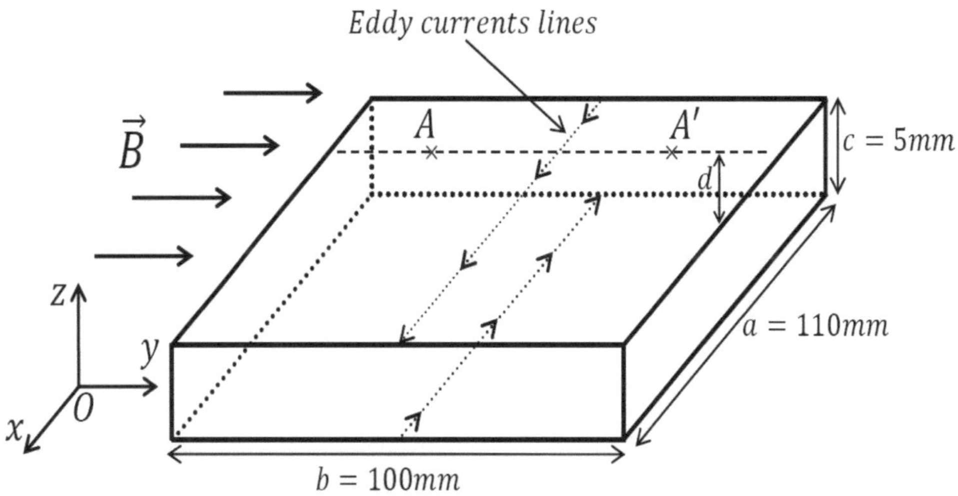

Fig. 1. Defect free rectangular aluminium plate under consideration

This excitation frequency has been chosen because it is one of those which are the less present in the general environment in and around our laboratory.

We shall study the induced current density using the more appropriate simplifying assumptions, in order to develop a good physical understanding of the phenomena.

In a first step, we assume that all the induced currents lines are straight lines parallel to the Ox direction from one side to the other (figure 1). If this is so, the external field is easy to evaluate by Biot and Savart law. An interesting remark is that the field of the induced currents, inside the plate, is quite much smaller than the excitation field.

In a second step, we improve the evaluation of the induced current lines in a plane *abcd* parallel to the plane xOz (figure 2).

In such planes, the value of $\vec{B}$ is given under the form of a double Fourier expansion

$$B(x, z) = \sum_{m,n} a_{mn} \cos\left(\frac{m\pi x}{a}\right) \cos\left(\frac{n\pi z}{c}\right) \quad (1)$$

and the values of the amplitude of the components j_x and j_z of the current density are

$$j_x = -\frac{\omega}{\rho}\frac{16ac^2}{\pi}B\sum_{m,n}\frac{1}{n}\frac{\sin(\frac{m\pi}{2})\sin(\frac{n\pi}{2})}{(m\pi a)^2+(n\pi c)^2}\sin\left(\frac{m\pi x}{a}\right)\cos(\frac{n\pi z}{c}) \quad (2)$$

$$j_z = \frac{\omega}{\rho}\frac{16ca^2}{\pi}B\sum_{m,n}\frac{1}{m}\frac{\sin(\frac{m\pi}{2})\sin(\frac{n\pi}{2})}{(m\pi a)^2+(n\pi c)^2}\cos\left(\frac{m\pi x}{a}\right)\sin(\frac{n\pi z}{c}) \quad (3)$$

respectively. It may be noted that j_x and j_z are the derivative of a "current function":

$$U(x,z) = \sum_{m,n}\frac{1}{\rho}(j\omega)(\frac{4ac}{\pi})^2\frac{B}{mn}\frac{\sin(\frac{m\pi}{2})\sin(\frac{n\pi}{2})}{(m\pi a)^2+(n\pi c)^2}\cos\left(\frac{m\pi x}{a}\right)\cos(\frac{n\pi z}{c}) \quad (4)$$

so that $j_x = -\frac{\partial U(x,z)}{\partial x}$ and $j_z = \frac{\partial U(x,z)}{\partial z}$.

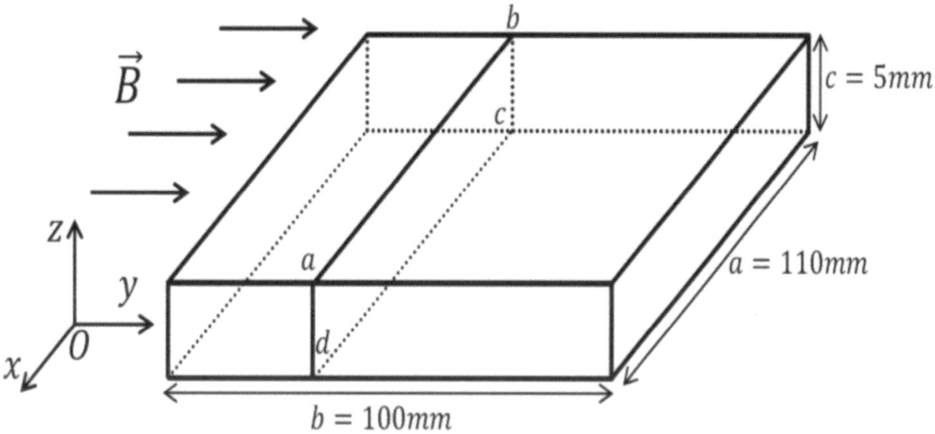

Fig. 2. Rectangular aluminium plate under consideration showing the cross section plane *abcd* for the evaluation of the induced currents lines

The important point is to understand that, in the case of figure 2 ($a = 110mm$, $c = 5mm$), the current density is essentially parallel to the Ox axis, and that this justifies the above first approximation: the current lines are parallel to Ox (figure 3).

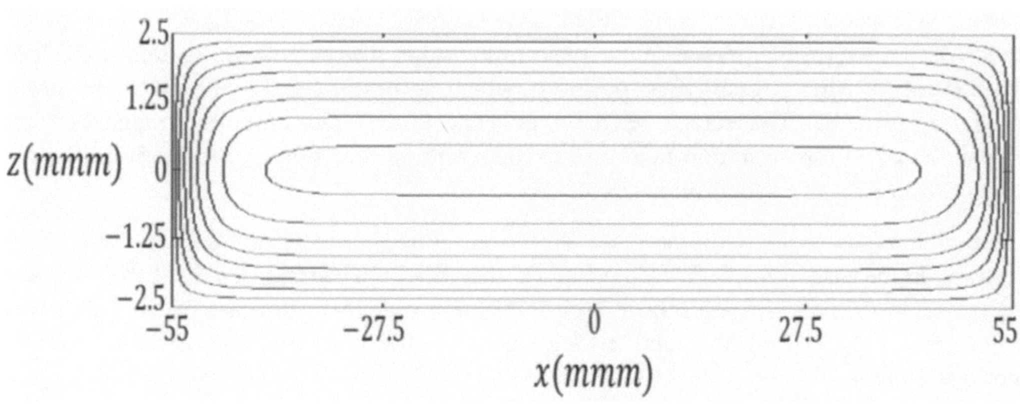

Fig. 3. Lines of induced currents for a defect free rectangular aluminium plate ($a = 110mm$, $c = 5mm$), the current density is essentially parallel to the Ox axis

3. Theoretical study of a rectangular aluminum plate with a calibrated crack

Consider now the above plate, where we have made a calibrated crack, as shown in figure 4.

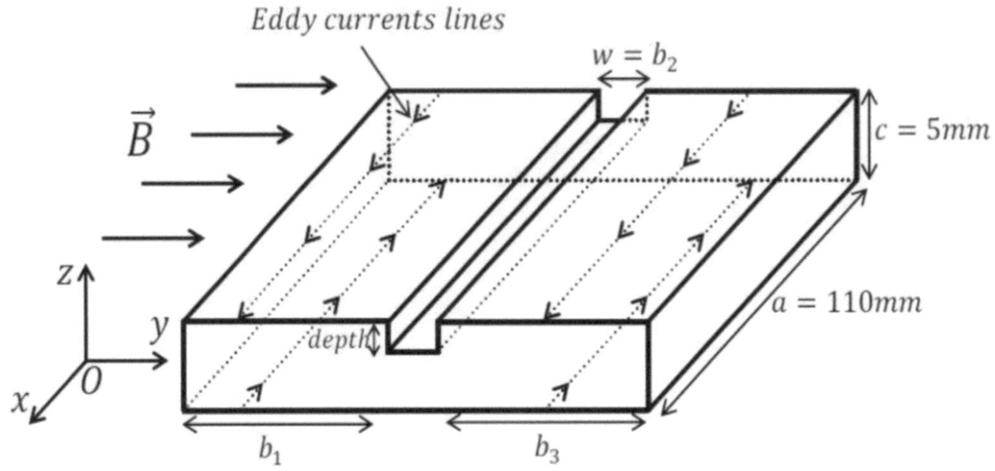

Fig. 4. Rectangular aluminium plate with a calibrated crack having a width ($w = b_2$) and a depth ($depth$)

We analyze the induced currents produced by the alternating excitation field $\vec{B}$, considering that the plate is equivalent to 3 sane plates of widths b_1, $b_2 = w$ and b_3 ($b_1 + b_2 + b_3 = 100mm$) and of thickness c, $c - depth$ and c respectively. The above analysis stands for each of those three plates; the numerical analysis is straight forward, and is given below (figures 8 and 9).

4. Experimental set-up

In our laboratory, we have established an experimental apparatus based on Helmholtz coils arrangement for excitation, and fixed magnetic sensors (SQUID and Hall probe) (figure 5). The structure has been made mainly of Plexiglas and other non magnetic materials (brass, copper, Al, wood…) in order to avoid any magnetic noise. The samples are moved underneath the sensor by means of an x-y scanning stage. The stage and the data acquisition are controlled with a LABVIEW program which is optimized to automate the entire measuring process. The sensor (SQUID or Hall probe) measures the magnetic field perpendicular to the excitation field and to the scanning direction. A lock-in amplifier has been used to achieve a synchronous detection.

The two versions of the equipment are shown in figure 6. The sensor is located in a fixed position. In the case of the SQUID, mechanical distance between upper side of the plate and the Dewar is $0.5mm$, and the sensor itself is $12.3mm$ higher. In the case of the Hall probe (right part of the figure), the total distance between the plate and the active part of the sensor is $1mm$.

The difference between the two sensors appears immediately. Indeed, the excitation field, when the SQUID is in use, is limited to $0.15 \times 10^{-4}T$ to avoid the saturation. When the Hall probe is in use, this limit reaches $2.45 \times 10^{-4}T$ with our generator; but it might be higher with better equipment. In both cases, the frequency is $180Hz$. The sensitivity of the SQUID measuring sensor is $4.25V/\mu T$, and the sensitivity of the Hall sensor is $5mV/\mu T$.

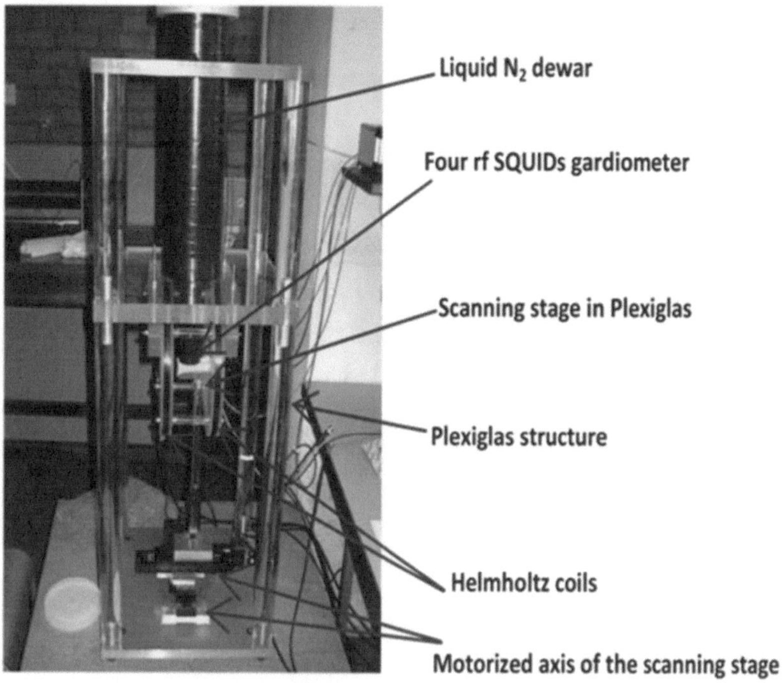

Fig. 5. High T_c SQUID NDE instrumentation set-up

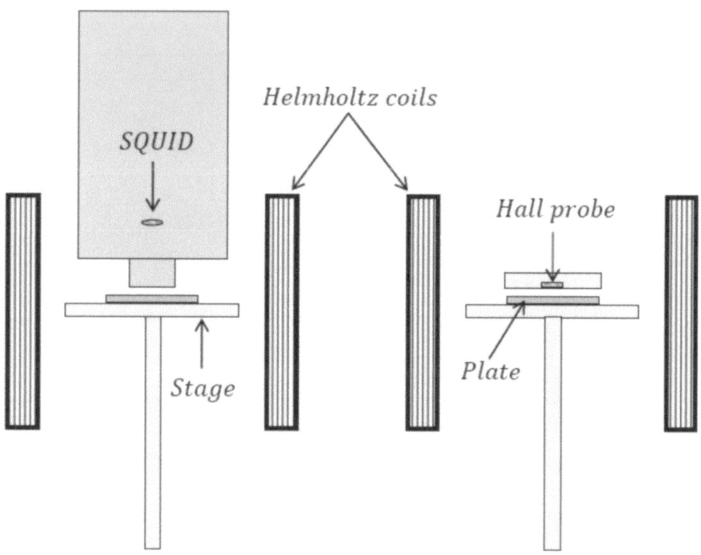

Fig. 6. The two versions of the experiment set-up: on the left is a SQUID sensor and on the right a Hall probe

5. Experimental verification

Figure 7 shows the plate with a calibrated crack immersed in the uniform excitation field described above ($180Hz, 0.15 \times 10^{-4}T$ if we use a SQUID and $2.45 \times 10^{-4}T$ if we use a Hall probe sensor).

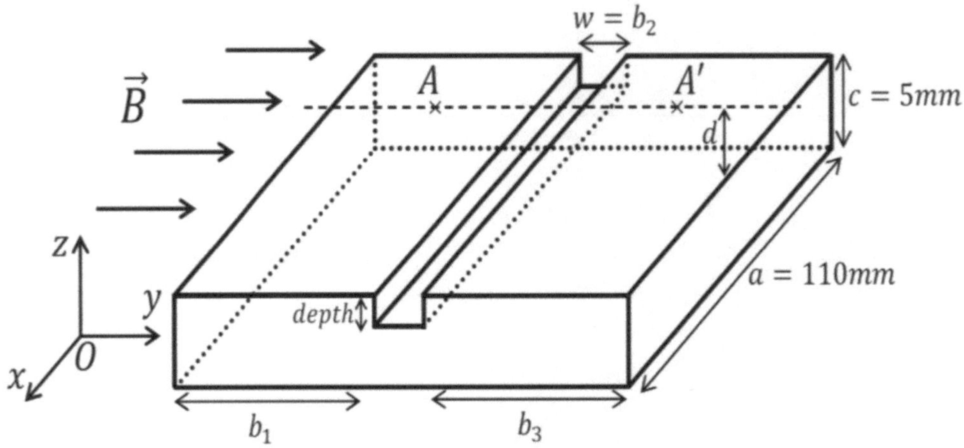

Fig. 7. Rectangular aluminium plate with a calibrated crack immersed in a uniform excitation field created by Helmholtz coils

The vertical component of the resulting induction field, measured between A and A' is given at a vertical distance d of $1mm$ (measured with the Hall probe), or at a vertical distance d of $12.8mm$ (measured with the SQUID). It is clear that the computed values and the measured values tally in two very different cases (figures 8).

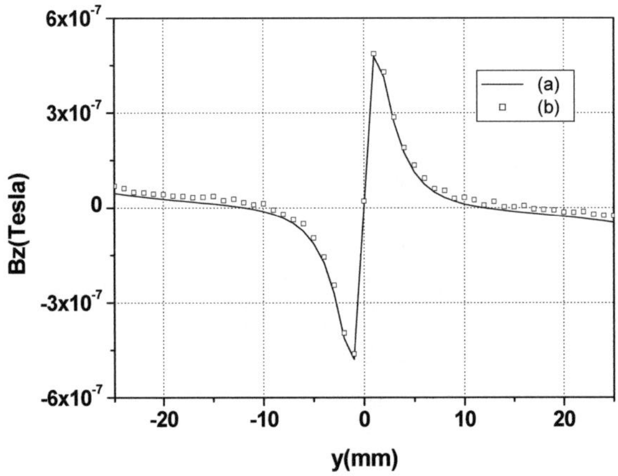

(A)

Fig. 8.A. Computed (a) and measured (b) values of B_z between A and A'in the case of a rectangular aluminium plate with a calibrated crack (width $w = 1mm$, $depth = 1mm$): measurement with a Hall probe at $d = 1mm$ above the plate

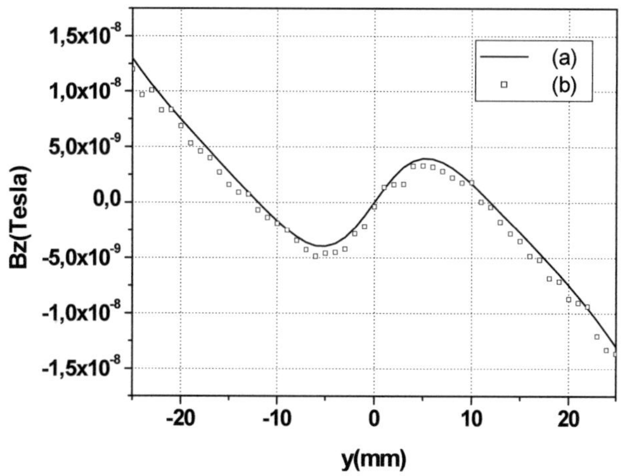

(B)

Fig. 8.B. Computed (a) and measured (b) values of B_z between A and A'in the case of a rectangular aluminium plate with a calibrated crack (width $w = 1mm$, $depth = 1mm$): measurement with a SQUID at $d = 12.8mm$ above the plate

Figure 9 represents the case of a defect free rectangular aluminum plate. The vertical component B_z of the resulting induction field, measured between A and A' is given at a vertical distance d of $1mm$ (measured with the Hall probe), and at a vertical distance d of $12.8mm$ (measured with the SQUID). Note that the difference between theoretical values and experimental values is so small that they cannot be distinguished.

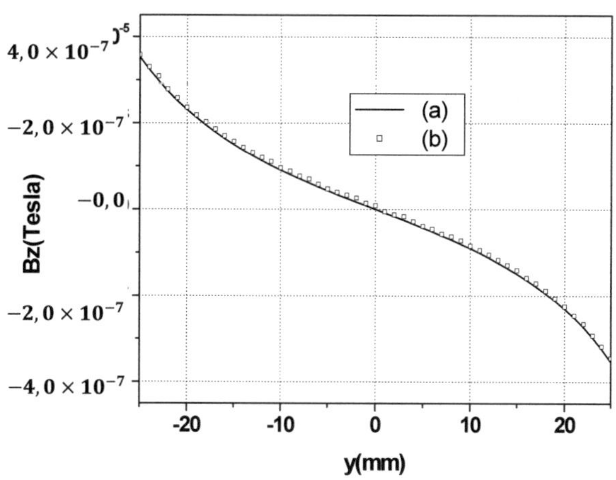

(A)

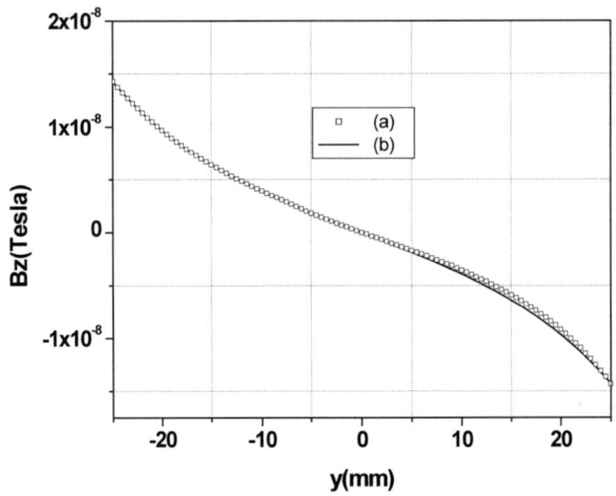

(B)

Fig. 9. Computed (a) and measured (b) values of B_z between A and A'in the case of a defect free rectangular aluminium plate: (A) measurement with a Hall probe at $d = 1mm$ above the plate, (B) measurement with a SQUID at $d = 12.8mm$ above the plate

6. Case of a crack with zero width showing metallic dislocations

We endeavored to repeat the same kind of experiment as above, but with a slot of zero width ($b_2 = w = 0mm, depth = c$) (figure 10).

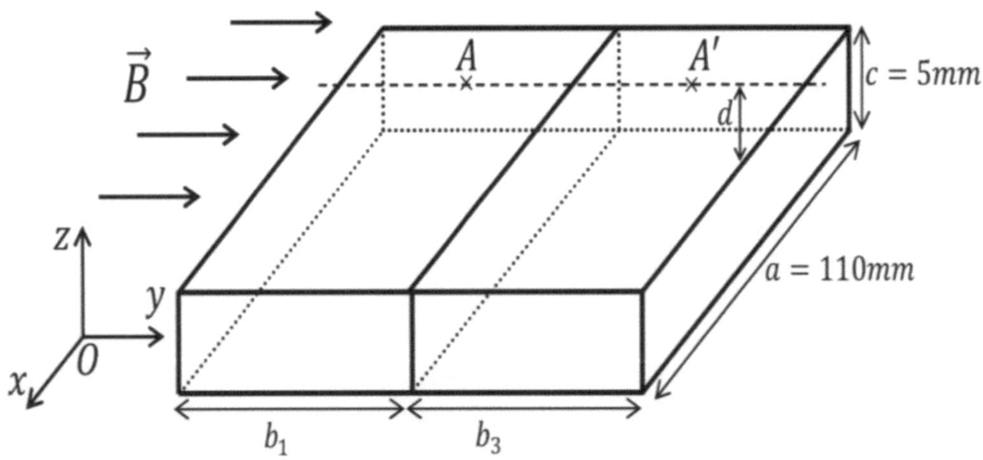

Fig. 10. Rectangular aluminium plate with a slot of zero width under consideration

We expected to find, along the line AA', a variation of the vertical component of the induced field depicted by curve (a) of figure 11, because all current lines are parallel to the Ox direction. In other words, we did not expect to detect the cut. We were extremely surprised to find the variation depicted by curve (b) of the same figure. In fact, we had considered that the two halves of the plate were equivalent to a homogeneous plate. What we neglected was that, when you cut a metallic plate, the cutting tool may damage the metallic microstructure of the plate, at least over small depth. Therefore, the intuitive representation of the situation is shown in figure 12. In this figure, the aluminum parts are themselves divided into two parts: one where the resistivity is normal ($\rho = 5.82 \times 10^{-8}\Omega.m$), and one where the resistivity is higher: $\rho^{'} > \rho$.

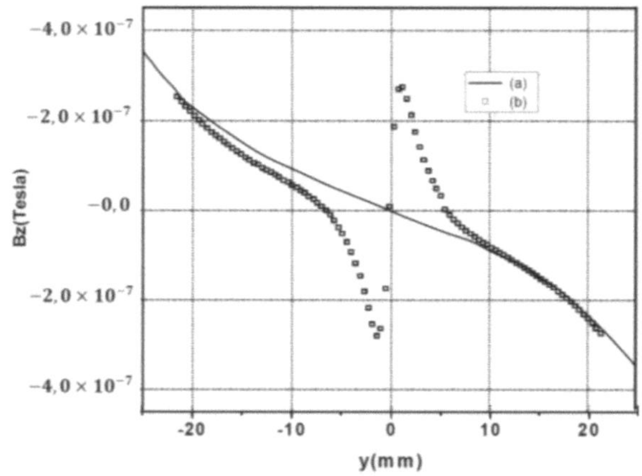

Fig. 11. Measured values of B_z, with a Hall probe at $d = 1mm$ above the plate, between A and $A^{'}$: (a) in the case of a defect free rectangular aluminium plate, (b) in the case of a rectangular plate with a slot of zero width

In figure 12, the thickness of the higher resistivity zone is considered $\frac{w^{'}}{2}$ at the edge of each half plate, so the total damaged edge area width is w'.

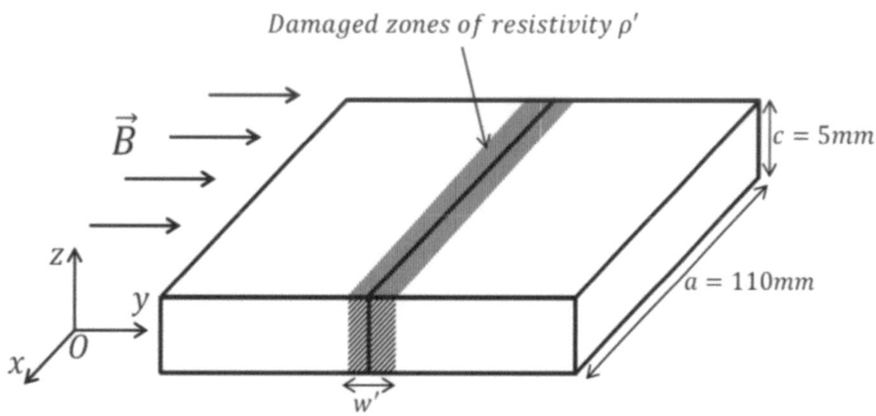

Fig. 12. Representation of the two half plates laid near each other, and of two assumed fatigued zones of resistivity ρ'

Therefore, the next problem is to determine the values of $w^{'}$ and $\rho^{'}$ which account conveniently for the experimental results. This is a simple classical optimization problem. In our case, we pick at random one arbitrary value of $w^{'}$ and $\rho^{'}$. This allows to evaluate the z component of the induced field. It is then necessary to define a global error $\epsilon(w^{'},\rho^{'})$ between the observed field, and the field corresponding to $w^{'}$ and $\rho^{'}$. Then we modify either

w' or ρ' to decrease ϵ and repeat this process until ϵ is as small as possible. This is developed in [Poloujadoff et al., 1994] and in the reference [Bettaieb et al., 2010].

This led us to determine a best value of ρ' $\rho = 5.82 \times 10^{-8}\Omega.m = 1.5 \times \rho)$ and of w' $(w' = 50\mu m)$ which provided a theoretical curve of B_z shown in figure 13 (a).

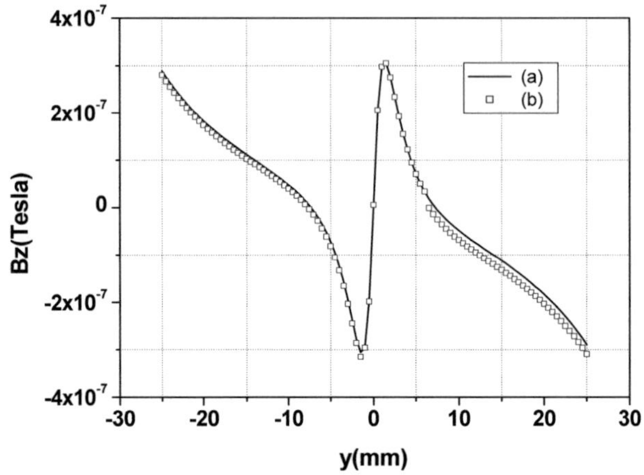

Fig. 13. (a) Theoretical curve with optimized values of $w' = 50\mu m$ and $\rho' = 1.5 \times \rho$, (b) experimental curve established with a Hall sensor at d=1mm above the aluminium plate in the case of a zero width crack

Since the publication of this reference, Dr. Denis GRATIAS suggested a further verification of this approach. This consisted in annealing the two half plates, then reputing the measurements. The annealing cycle is shown in figure 14.

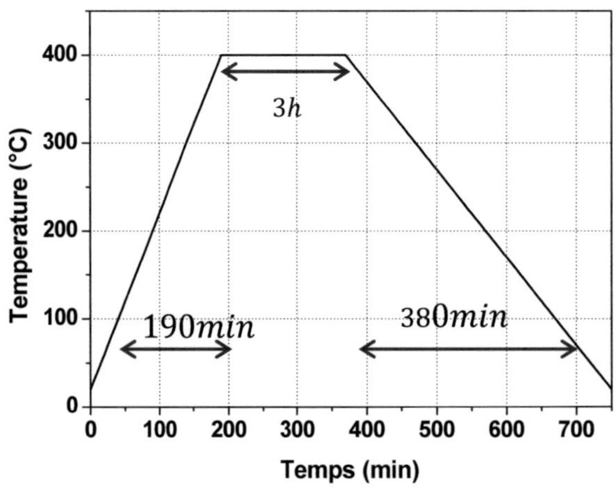

Fig. 14. Annealing cycle of the two half plates after cutting a plate

Then, we place again the two half plates in the same uniform magnetic excitation field. The curve of the vertical component of the induction field still shows a variation of w' and ρ', but much smaller than previously; this show *that annealing has been very effective, but has not been long enough* (figure 15).

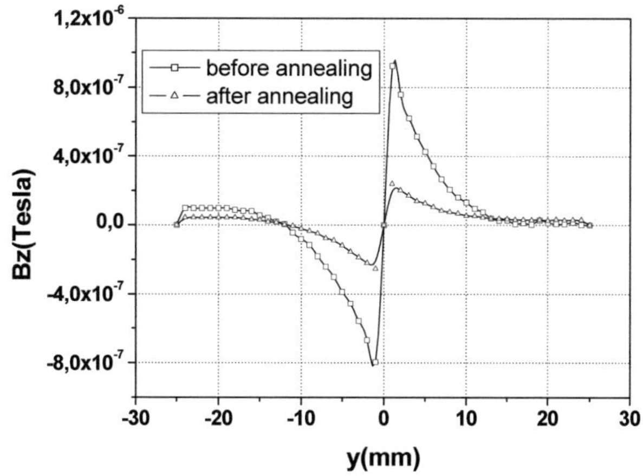

Fig. 15. Experimental curve established with a Hall sensor at d=1mm above the aluminium plate in the case of a zero width crack before and after annealing

7. Metallic dislocations created by a shock

The above study of dislocations can be carried on by creating small damaged zones by a direct or indirect shock with a hammer (figure 16). We have already reported a preliminary study of this phenomenon [Bettaieb et al., 2010], and we report some more recent progress below.

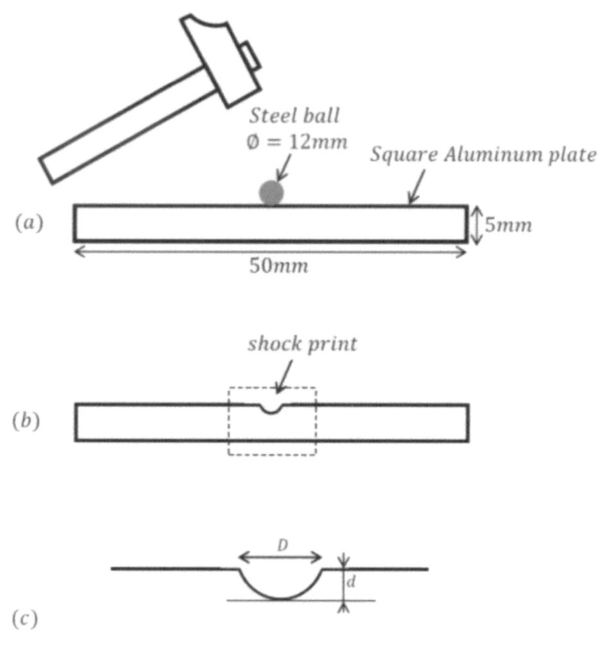

Fig. 16. (a) Creation of an impact zone by stroking a steel ball, (b) impact upon the square aluminium plate, (c) zoom of the shock print zone with modelling dimensions (D, d)

In our latest experiments, we stroke a steel ball in the middle of a square aluminum plate with a hammer. The dimensions of the damaged zones depend naturally on the radius of the steel ball and on the violence of the hammer shocks. We have considered only three cases; the depth d and the diameter D of some shock prints are given in the table I.

	Dimeter D	depth d
Impact 1	4mm	0.11mm
Impact 2	3mm	0.09mm
Impact 3	2mm	0.07mm

Table 1. Dimensions of the damaged zones

Consider the first impact, produced near the middle of the square plate. The Aluminum plate is laid upon the stage which has been already described in the figure 6. The excitation field being again at $180Hz$ with an amplitude of $0.15 \times 10^{-4}T$ if we use a SQUID and $2.45 \times 10^{-4}T$ if we use a Hall probe sensor. The stage is moved as described in §4, the variation of the vertical induction field component, measured by the Hall probe is shown in figure 17 (A). This proves that the impact may be easily detected, since the perturbation signal varies between $\pm 10^{-5}T$. It is also possible to scan a square domain $(-10mm \leq x, y \leq +10mm)$ around the center of the impact and to represent it as in figure 17 (B).

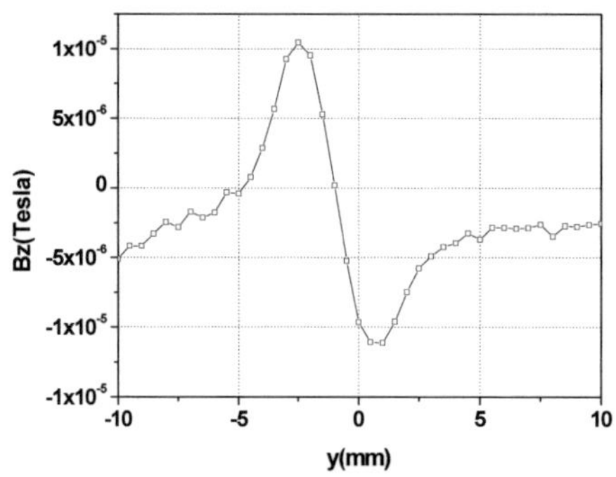

(A)

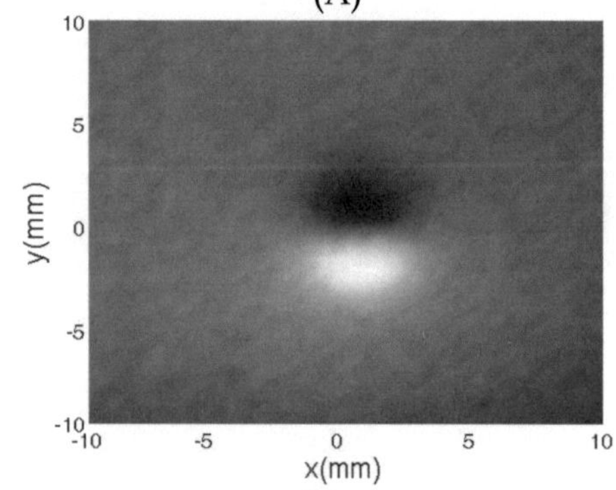

(B)

Fig. 17. (A) Measured values of B_z along the Oy axis for $x = 0mm$, (B) matrix representation of the values of B_z across a square domain $(-10mm \leq x, y \leq +10mm)$. Measurement with a Hall probe at d=1mm above the square aluminium plate ($D = 4mm$, $d = 0.11mm$)

Considering the second impact, we found similar results, but the signal varies within a smaller interval ($\pm 0.6 \times 10^{-5}T$) (see figures 18 (A) and 18 (B)). As for the third impact, it is not really detectable.

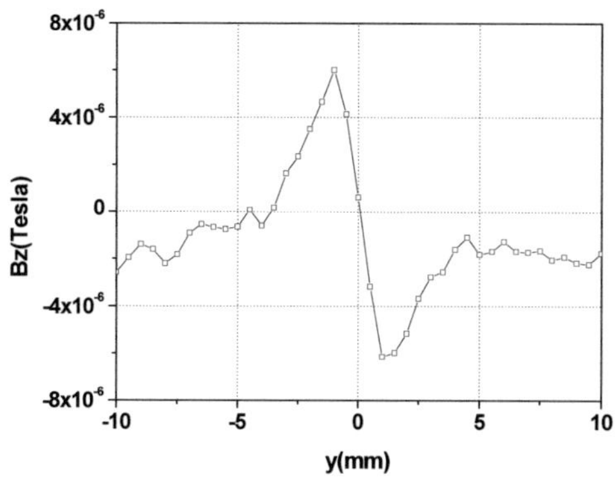

(A)

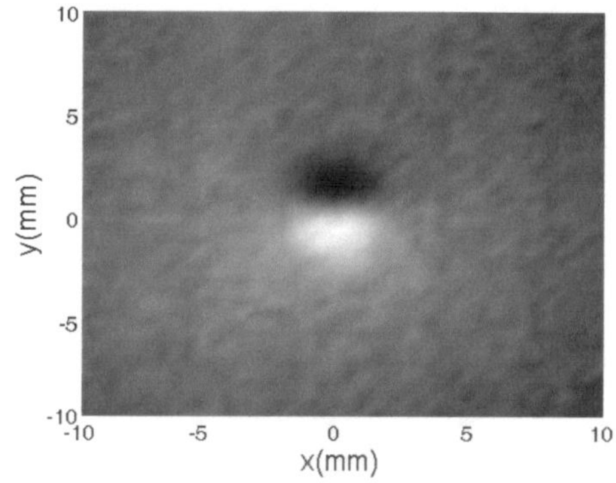

(B)

Fig. 18. (A) Measured values of B_z along the Oy axis for $x = 0mm$, (B) matrix representation of the values of B_z across a square domain ($-10mm \leq x, y \leq +10mm$). Measurement with a Hall probe at d=1mm above the square aluminium plate ($D = 3mm$, $d = 0.09mm$)

How may we interpret these results?

It is extremely important to consider that these results may be assigned to two different causes: the change of the shape of the aluminum square plate and/or a modification of the metallic microstructure inducing a change of the electric resistivity. To clarify this question, we have assumed that the effects of these two causes can be separated. Unfortunately, a rigorous mathematical analysis of such a structure would involve a cylindrical geometry and a uniform field, resulting in a very complicated 3D analysis. For this reason, we made the following approximations which happen to give a sufficient feeling of what happens.

This means that we first replaced the portion of a sphere, in figure 16 (c) by an empty parallelepiped with a square basis of side $w = D$ and height $h = d$ chosen to have the same volume than the portion of the half sphere (figure 19 (A)). This model accounts for the observed results (figure 19 (B)).

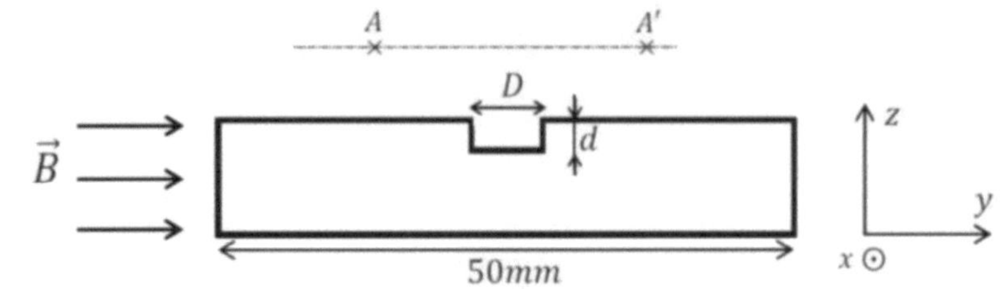

(A)

Fig. 19.A. Square aluminium plate with an empty space of same volume that the one in figure 16 (c)

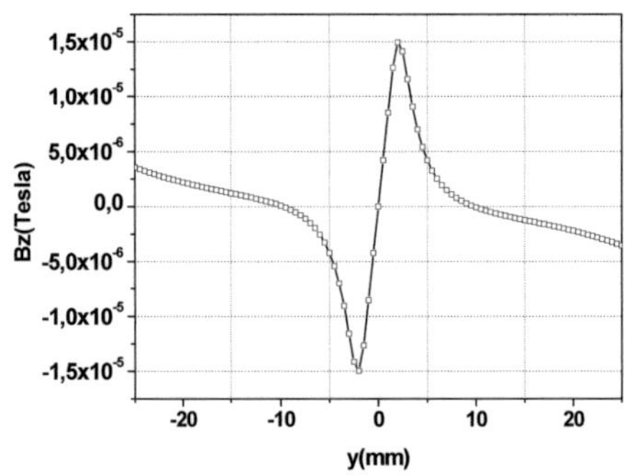

(B)

Fig. 19.B. Theoretical values of B_z corresponding to the figure 19 (A)

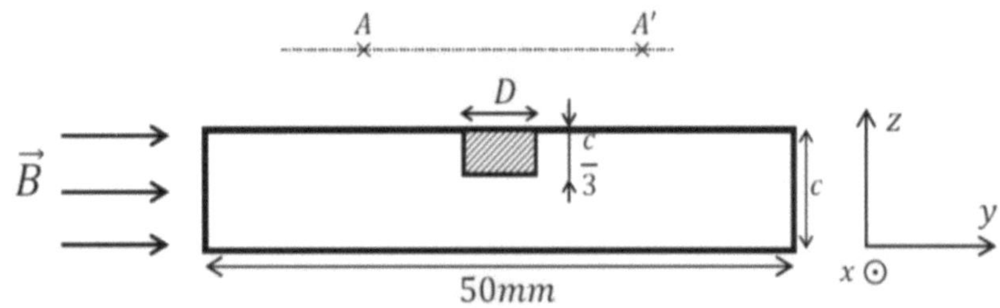

(A)

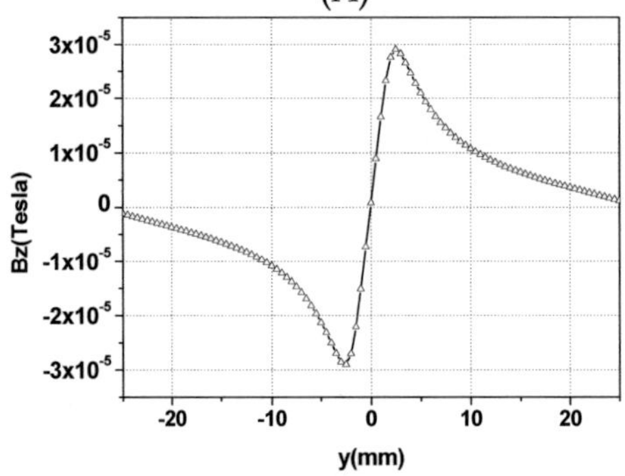

(B)

Fig. 20. (A) Square aluminium plate with an inner portion of resistivity $\rho' > \rho$, (B) theoretical values of B_z corresponding to the figure 20 (A).

In a second step, we assumed that the deformation of the plate was negligible, but the metallic microstructure has been changed over a depth as large as 1mm. We have considered that in this region the resistivity has been changed by 10% (figure 20 (A)). This model also accounts for the experimental results (figure 20 (B)).

Therefore, this experiment and the corresponding models show that the effect of the shock is partially explained by a deformation of the square aluminum plate and partially by dislocations created in the metal over same depth. We have proved that the effects of the dislocations are much more important [Bettaieb, 2009].

8. Metallic dislocations created by flexions

8.1 Experimental study

Consider a long aluminum strip (150mm) shown in figure 21 which has been bent by hand, so as to create a fatigue zone in the middle. In the already quoted previous paper [Bettaieb et al., 2010], we had shown that the fatigue of the metallic bent part could be detected. To this end, we had placed the strip in a uniform alternating horizontal field $\vec{B}$, and we have scanned its upper surface with a Hall probe. However, this first publication was limited to the feasibility of the detection.

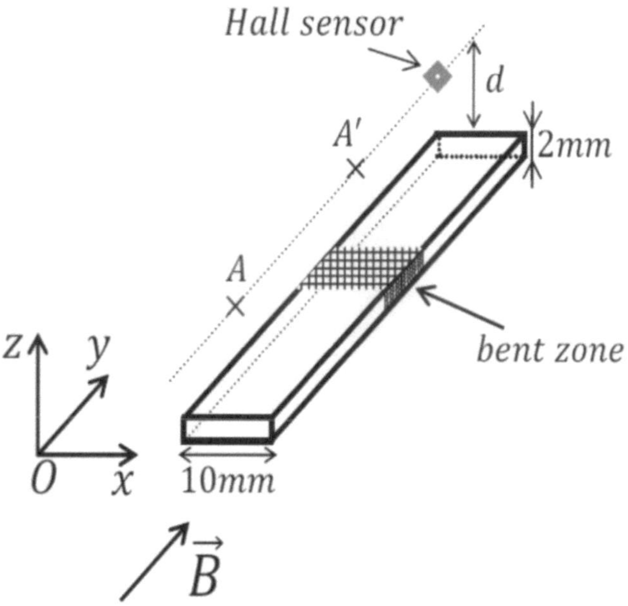

Fig. 21. Fatigued strip, the length of the damaged (bent) zone is approximately equal to 3mm

Since that time, we have greatly improved our experimental process, and have developed an inverse problem method for a better mathematical analysis of the results.

Our first effort has been to flatten the long strip carefully before measurement, as shown by figures 22 (figure 22 (A) is a side photo, and figure 22 (B) an upside down view of the strip during flattening). As a result, the signal obtained along the sensor track is symmetrical about its center (figure 23 (A)).

(A)

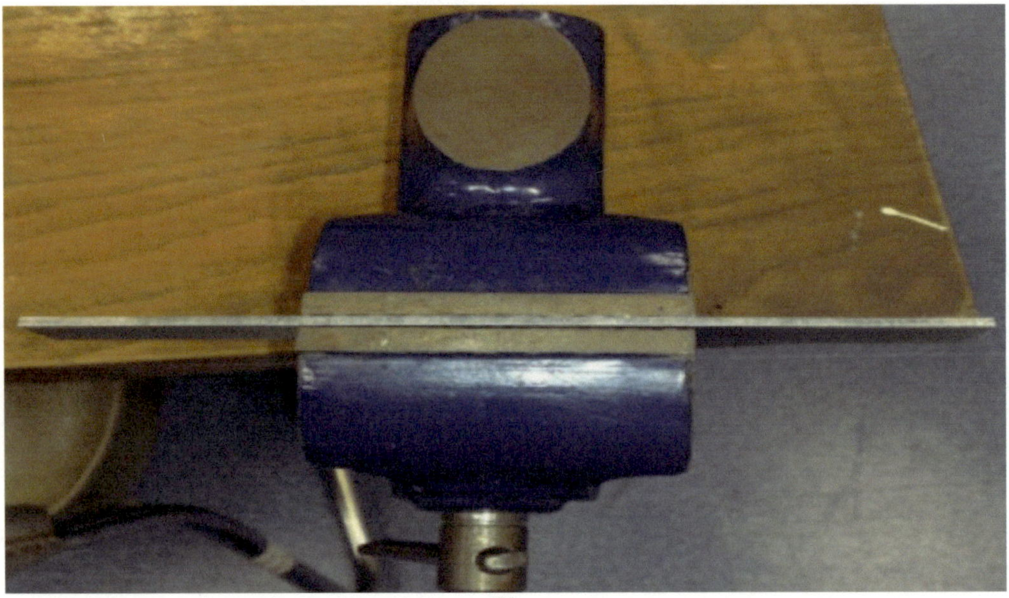

(B)

Fig. 22. Bent long strip being flattened in a shaper to eliminate the geometrical deformation before eddy current measurement (A: side view; B: upper view)

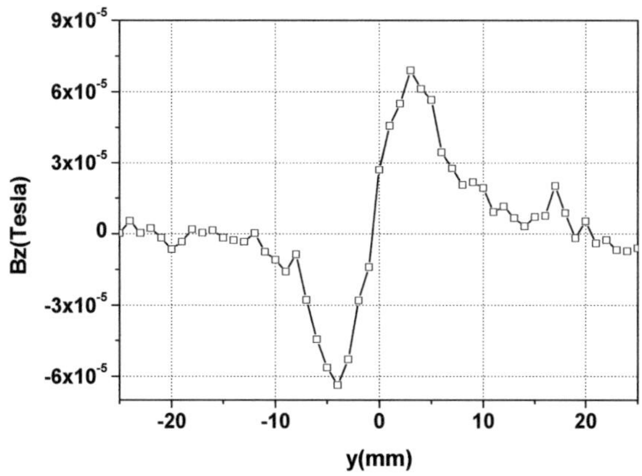

(A)

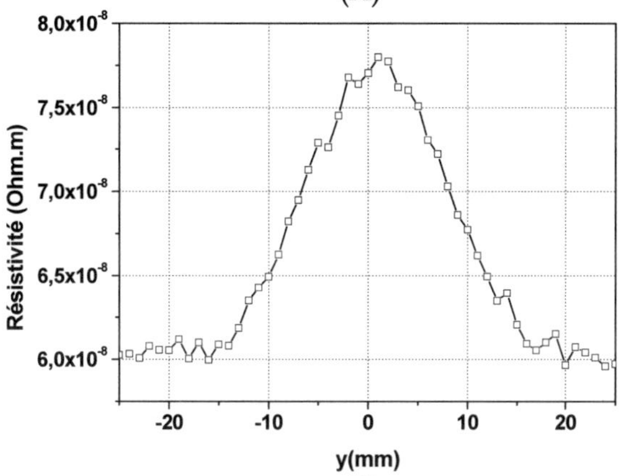

(B)

Fig. 23. (A) Values of B_z induced by the eddy currents within a flattened fatigued long strip (measurement with a Hall sensor at $d = 1mm$ above the strip), (B) Corresponding values of the estimated local resistivity along the damaged zone using the inverse solution

8.2 Theoretical evaluation of resistivity along the strip

The principle of the inverse problem is well known. The relationship between the induction field and the induced currents in the long strip is linear, the coefficients being the conductivity. Simultaneously, the relationship between the strip induced current and the outside induced induction field is also linear. Therefore, if the exciting induction field and the induced induction field are known, the resistivity can be determined [Bettaieb et al., 2010]. It is what has been done to estimate the resistivity (figure 23 (B)) along the strip. This latter result clearly shows the effect of the fatigue where the strip has been bent.

9. Conclusion

In this very original chapter, we have shown that a methodology generally used to detect cracks can be also used to measure the effects of a mechanical fatigue created by cutting tools, impact shocks or flexions. The natural continuation should be to study quantitatively most of the microstructure properties which influence the resistivity of the alloy. These include irradiations defects and phase transitions. We would certainly welcome collaboration offers from colleagues.

10. Acknowledgment

We would like to thank warmly Dr. Denis GRATIAS from ONERA/CNRS for several useful advices and discussions notably related to the nature of dislocations. He also suggested the check of our ideas by the annealing procedure.

11. References

Bettaieb, L. (2009). Contributions to Non Destructive Evaluation of amagnetic metlas using eddy currents. *Phd Thesis at University of Pierre and Marie Curie (UPMC)*, in French, (September 2009).

Bettaieb, L.; Kokabi, H.; Poloujadoff, M.; Sentz, A. & Krause, H. J. (2008). Non Destructive Testing (NDT) With High Tc RF SQUID. *Journal of Physics: Conference Serie 97*, (March 2008), 6pp.

Bettaieb, L.; Kokabi, H.; Poloujadoff, M.; Sentz, A.; Krause, H. J. & Coillot, C. (2010). Comparison of the use of SQUID an Hall effect sensors in NDE. *Materials Evaluation*, Vol.68, No.5, (2010), pp. 535-54.

Bettaieb, L.; Kokabi, H.; Poloujadoff, M.; Sentz, A. & Tcharkhtchi, C. (2010). Fatigue and/or crack detection in NDE. *Nondestructive Testing and Evaluation*, Vol.25, No.1, (March 2010), pp. 13-24.

Bettaieb, L.; Kokabi, H.; Poloujadoff, M.; Sentz, A. & Krause, H. J. (2009). Analysis of some Nondestructive Evaluation Experiments Using Eddy Currents. *Research in Nondestructive Evaluation*, Vol.20, No.3, (July 2009), pp. 159-177.

Poloujadoff, M.; Christaki, E. & Bergman, C. (1994). Univariant search: An opportunity to identify and solve conflict problems in optimization. *IEEE trans. Energy Conver.*, EC(9), (December 1994), pp. 659-664.

7

Magnetic Adaptive Testing

Ivan Tomáš[1] and Gábor Vértesy[2]
[1]*Institute of Physics, Praha,*
[2]*Research Institute for Technical Physics and Materials Science, Budapest*
[1]*Czech Republic*
[2]*Hungary*

1. Introduction

Tests of impending material degradation of engineering systems due to their industrial service make an indispensable part of any modern technological processes. *Destructive tests* are extremely important. Their considerable advantage consists in their *straightforwardness*. In great majority of cases they *directly test the very property*, which is at stake. For instance, the limiting endurable stress before the material yields is directly measured by a mechanical loading test, or the number of bending cycles before the system breaks is directly counted to find the fatigue limits, and so on. Besides, the destructive inspections are regularly used in situations of any materialized failure, when knowledge of the *final condition* of the material at the moment of the breakdown helps to avoid any next malfunction of the same or similar systems under equivalent circumstances. However, the destructive tests cannot be used on the finished or half-finished parts in process of their industrial production, simply because those parts - after having been destructively tested - cannot be used for their primary purpose any more. The only possible application of destructive tests in industry is to examine destructively every n^{th} produced piece only, which is not sufficient for reliability of goods presently required.

Nondestructive tests, do not suffer by those problems. Causing no harm, they can be used at each produced object, and they can be periodically applied even on systems in service. Judging by the vivid interest in the presently observed improvement of traditional nondestructive tests and in development of some recently discovered ones, the nondestructive evaluation of material objects attracts currently attention perhaps even more than the destructive assessment does. Evaluation of nondestructive tests keeps the user informed about the *actual condition* of the system, and should ensure avoidance of any failure *before* it ever happens. There are numerous methods of nondestructive tests, based on optical, acoustic, electrical, magnetic and other properties of the materials, which can be *correlated* with the watched quality of the whole system. The necessity of the *unambiguous correlation* between the nondestructively measured physical property and the guarded property of the system at stake is an *unreservedly required* claim. It calls for the nondestructive tests to be examined and re-examined to their *one hundred percent reliability* before they can be really applied in crucial cases. Multi-parametric output of a nondestructive testing method is therefore an extremely valuable and welcomed property, which enables to check and to cross-check validity of the necessary correlation and ensures reliability of single measurement with respect to the guarded physical quality of the tested system. The presently described method of Magnetic Adaptive Testing (MAT) is just one of the few *multi-parametric* nondestructive tests available.

Many industrial systems have construction parts, which are fabricated from the most common ferromagnetic construction materials, such as steel or cast iron. Nondestructive tests of their structural-mechanical quality can be very suitably based on magnetic measurements, because processes of magnetization of ferromagnetic materials provide detailed reflection of the material microstructure and its degradation.

The physical processes, which take place in the course of magnetization of ferromagnetic materials by an applied magnetic field, are well known. They are classified as reversible (mostly continuous changes of direction of the magnetization vector within magnetic domains) and irreversible (mostly discontinuous changes of volumes of magnetic domains, caused by jumps of domain walls from a position to a next one). These processes, the latter ones in particular, are closely correlated with structure of the sample, with its uniformity, texture, material defects, internal and external stresses and even with the sample shape. This can be disturbing, if we want to measure magnetic properties of the pure material. However, if we are interested just in the material structure, this correlation serves as a starting point of all magnetic techniques used for magnetic nondestructive tests of any ferromagnetic material. For detailed survey of such magnetic methods see e.g. papers (Jiles, 1988, 2001, Blitz, 1991, Devine, 1992) and many references to actual measurements quoted there.

Magnetic hysteresis methods belong to the most successful indirect ways of structural investigation of ferromagnetic construction materials. They are mostly based on detection of the material structural variations via variation of the traditional parameters of the *major*, saturation hysteresis loop, such as coercive field, H_C, remanent magnetic induction, B_R, maximum permeability, μ_{MAX}, and a few others. At first, these magnetic parameters are experimentally correlated with independently measured real structural / mechanical characteristics of the samples, and then from measurement of the former ones the latter can be determined. The point is that the magnetic parameters are measured *non-destructively and with less difficulty* than the real structural / mechanical characteristics, which in most cases can be learned destructively only. The few traditionally employed magnetic parameters are actually special points or slopes on the magnetic major hysteresis loop. These traditional parameters are very well suited for characterization of *magnetic properties* of ferromagnetic materials, but they were *never optimized* for magnetic reflection of modified *structural properties* of the measured samples. Besides, they are by no means the only available magnetic indicators of various nonmagnetic modifications of ferromagnetic materials. Correlation between alternative magnetic parameters and concrete structural changes of the studied material, can possibly be even *better adapted* to each specific task.

Structural aspects of ferromagnetic samples cover material non-uniformities such as local mechanical stresses, clusters of dislocations, grains, cavities, inclusions and many others. Their presence, distribution and magnitude control details of the magnetization processes, and there is no reason to expect that each type of structural defect affects all regions of each magnetization process with the same intensity. Measurement of a suitably chosen magnetic variable, e.g. magnetic induction, B, or differential permeability, μ, of differently degraded samples of the same material, while changing the applied field in a systematic way, confirms the fact. Analysis of a large *family of minor hysteresis loops* is a good example of such a systematic investigation of magnetization processes. Besides, as known from the Preisach model of hysteresis (Bertotti, 1998), such volume of data is able to make a complete magnetic picture of the sample.

A novel method, Magnetic Adaptive Testing (MAT), for magnetic nondestructive inspection of ferromagnetic construction materials (Tomáš, 2004) is introduced in this chapter. The method is based on systematic measurement of minor magnetic hysteresis loops. The essential difference between material testing by the traditional hysteresis- and by the MAT-approach is shown in Fig. 1 schematically. The left hand part of the figure represents the traditional measurement of the *single major* (saturation) hysteresis loop. The major loop is measured for each of the degraded samples and the material degradation can be described through variation of values of any *of the few major loop parameters*, e.g. H_C, B_R, ... as functions of an independent degradation variable, ε, (e.g. as functions of mechanical tension). The right hand part of the figure depicts schematically volume of the measured data for MAT. The large family of minor hysteresis loops is measured for each of the degraded samples and the material degradation can be then described through variation of values *of any of the point (and/or slope) on any of the minor loops*, i.e. $B(F_i, A_j)$ (and/or $\mu(F_i, A_j)$), as functions of an independent degradation variable, ε, (e.g. as functions of mechanical tension).

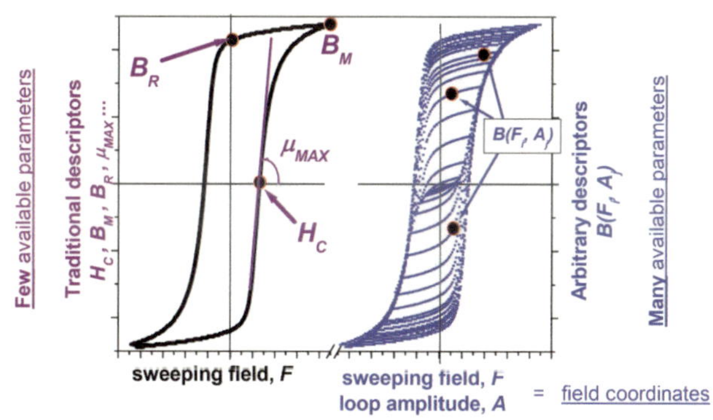

Fig. 1. Schematic comparison of the traditional magnetic hysteresis testing (left) and Magnetic Adaptive Testing (right). The traditional testing uses only a few parameters of the major loop for description of the material. Magnetic Adaptive Testing has the choice to pick up the best from many available parameters indexed by the field coordinates.

The method of Magnetic Adaptive Testing utilizes systematic measurement of large families of minor hysteresis loops, from minimum amplitudes up to possibly the maximum (major) ones on degraded ferromagnetic samples/objects. From the large volume of the recorded data, those are applied for *evaluation* of the degradation, which reflect the material degradation in the most sensitive or otherwise the most convenient way. Such - *best adapted for the investigated case* - data are used as the MAT-parameter(s) and its / their dependence on an independent variable accompanying the inspected degradation is referred to as the

MAT degradation function(s). The following text is dedicated to description of basic principles of MAT, to specific advantages and disadvantages of the method and to a number of cases when MAT was successfully applied.

2. Description of the method

The basic features and processes of use of the method of Magnetic Adaptive Testing are probably most instructive and easy to understand, if they are demonstrated on a typical concrete example of MAT application. Therefore, throughout Section 2, properties are described of samples manufactured from low carbon steel, which was degraded (plastically deformed) by previous application of 7 different magnitudes of mechanical tension. The seven steel samples are ring-shaped with the magnetizing and the pick-up coils wound on each of them. All the figures presented in Section 2 refer to this example.

The magnetic induction method appears to be the easiest way of the systematic measurement for MAT. A Permeameter described in Fig. 2 makes a practical example of the possible experimental arrangement. The driving coil wound on the magnetically closed sample gets a triangular waveform current with stepwise increasing amplitudes and with a fixed slope magnitude in all the triangles (see Fig. 2, right). This produces a triangular variation of the magnetizing field with time, t, and a voltage signal, U, is induced in the pick-up coil for each kth sample:

$$U(\mathrm{d}F/\mathrm{d}t,\ F,\ A_j,\ \varepsilon_k) = K^*\partial B(\mathrm{d}F/\mathrm{d}t,\ F,\ A_j,\ \varepsilon_k)/\partial t = K^*\mu(\mathrm{d}F/\mathrm{d}t,\ F,\ A_j,\ \varepsilon_k)^*\ \mathrm{d}F/\mathrm{d}t, \quad (1)$$

where K is a constant determined by geometry of the sample and by the experimental arrangement. As long as $F=F(t)$ sweeps linearly with time - i.e. $|\mathrm{d}F/\mathrm{d}t|$ is (the same) constant for measurement at each of the samples, Eq.(1) states, that the measured signal is simply proportional to the differential permeability, μ, of the measured magnetic circuit, as it varies with the applied field, F, within each minor loop amplitude, A_j, for each kth measured sample. If correct results without influence of any previous remanence should be obtained, it is evident that each sample has to be thoroughly demagnetized before it is measured.

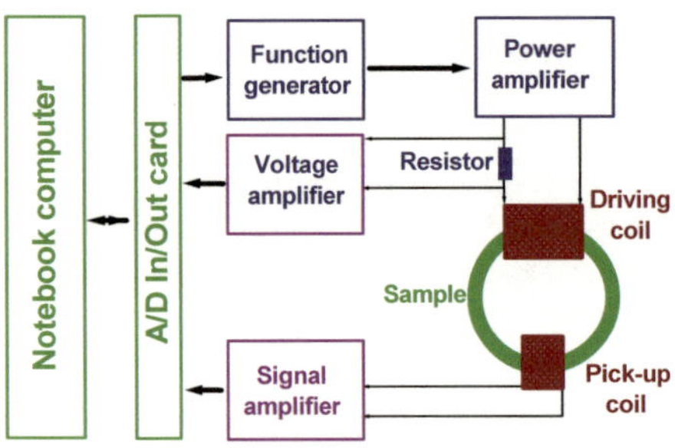

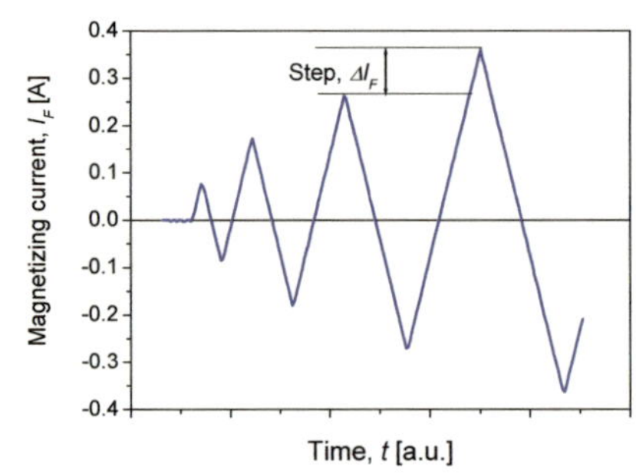

Fig. 2. Left: Block-scheme of the Permeameter. The driving coil produces triangular variations of the applied magnetic field. The signal coil picks-up the induced voltage proportional to differential permeability of the sample. This scheme shows arrangement when a magnetically closed sample is measured and both the coils are wound directly on the sample. Right: Triangular variation of the magnetizing current with time.

The Permeameter works under control of a notebook PC, which sends the steering information to the function generator and collects the measured data. An input/output data acquisition card accomplishes the measurement. The computer registers actually two data files for each measured family of the minor μ-shaped loops. The first one contains detailed information about all the pre-selected parameters of the demagnetization and of the measurement. The other file holds the course of the voltage signal, U, induced in the pick-up coil as a function of time, t, and of the magnetizing current, I_F, and/or field, F. A typical example of one family of the μ-shaped loops (the reference, unstrained sample with $\varepsilon_0=0\%$) is shown in Fig. 3a, and Fig. 3b presents the seven families of all the seven ring samples of the steel degraded by the previously applied mechanical tension to the strain values ε_k = 0%, 1.7%, 3.5%, 5.8%, 7.8%, 9.8%, 17.9%, respectively. Evidently it is a lot of data and our task is to compare them and to find the most suitable ones for description of the investigated material degradation.

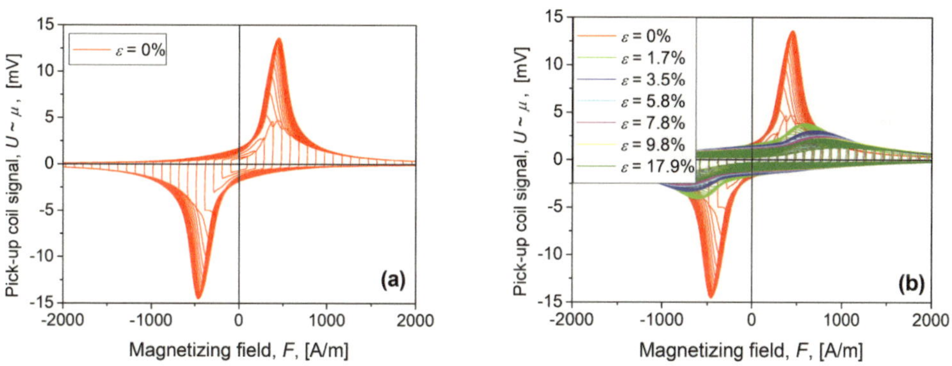

Fig. 3. Examples of families of the μ-shaped loops vs. magnetizing field, F. (a) Measured on the unstrained - $\varepsilon_0=0\%$ - low carbon steel sample. The positive and negative parts of the signal correspond to the increasing and decreasing parts of the triangular waveform of the current (field), respectively. (b) Measured on all the seven ring samples, manufactured from low carbon steel, which was degraded by mechanical tension down to strain values ε_k = 0%, 1.7%, 3.5%, 5.8%, 7.8%, 9.8%, 17.9%. The curves are plotted in dependence on the magnetizing field, F, within the limits ± 2000 A/m.

Instead of keeping the signal and the magnetizing field in shapes of continuous time-dependent functions, it is practical to interpolate the family of data for each ε_k-sample into a discrete square (i, j)-matrix, $U(F_i, A_j, \varepsilon_k)$, with a suitably chosen step, $\Delta A = \Delta F$. (Because dF/dt is a constant, identical for all measurements within one experiment, it is not necessary to write it explicitly as a variable of U.) MAT is a relative method (practically all the nondestructive methods are relative), and the most suitable information about degradation of the investigated material can be contained in variation of *any* element, of such matrices as a function of ε, relative with respect to the corresponding element of the reference matrix, $U(F_i,A_j,\varepsilon_0)$. Therefore all $U(F_i, A_j, \varepsilon_k)$ elements will be divided by the corresponding elements $U(F_i,A_j,\varepsilon_0)$ of the reference sample matrix and *normalized* elements of matrices of relative differential permeability $\mu(F_i,A_j,\varepsilon_k) = U(F_i, A_j, \varepsilon_k)/U(F_i, A_j, \varepsilon_0)$, and their proper sequences

$$\mu(F_i, A_j, \varepsilon) = U(F_i, A_j, \varepsilon)/U(F_i, A_j, \varepsilon_0) \quad (2)$$

as *normalized μ-degradation functions* of the inspected material will be obtained. Throughout the text of this chapter, all the degradation functions are considered as *normalized* with respect to the corresponding values of the reference sample.

Beside the μ-matrices and the corresponding *μ-degradation functions*, also matrices of the integrated, $B=\int \mu dF$, or the differentiated, $\mu'_F=d\mu/dF$, and/or $\mu'_A=d\mu/dA$ values can be computed and used for definition of the *B-degradation functions*, $B(F_i,A_j,\varepsilon)$, of the *μ'_F-degradation functions*, $\mu'_F(F_i,A_j,\varepsilon)$, and/or of the *$\mu'_A$-degradation functions*, $\mu'_A(F_i,A_j,\varepsilon)$. In fact the *B*-, μ'_F-, and μ'_A-degradation functions do not contain more or other information than the μ-degradation functions, but their use is sometimes more helpful and they are occasionally able to show certain material features with higher sensitivity or in different relations. This will be shown later in the application examples. (Note: As the μ'_A–degradation functions are actually not used in any of the examples here, μ'_F is simply denoted as μ' thoughout this chapter.) In some cases it turns out, that degradation functions of *reciprocal* values, such as 1/μ-degradation functions and the similar others, are more convenient than the direct ones. Application of the reciprocal degradation functions proves effective especially in situations when - with the increasing parameter ε - the direct degradation functions approach kind of a "saturation".

An example of the μ-degradation functions and *1/μ*-degradation functions of the low carbon steel degraded by the mechanical tension is shown in Fig. 4.

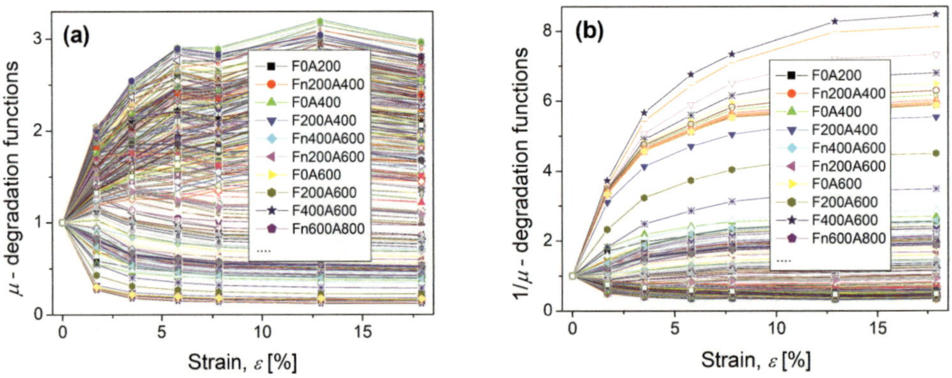

Fig. 4. The μ-degradation functions (a) and *1/μ*-degradation functions (b) of the low carbon steel degraded by the mechanical tension. The functions were calculated for the step $\Delta A = \Delta F$ = 200A/m and they are plotted up to the maximum amplitude A_j = 3600 A/m.

Number of the degradation functions obtained from the MAT measurement depends on magnitude of the maximum minor loop amplitude, A_j, up to which the measurement is done, and on choice of the step value $\Delta A = \Delta F$ which is used for computation of the interpolated data matrices. The maximum amplitude A_j = 3600 A/m and the step $\Delta A = \Delta F$ = 200 A/m were applied in the presented example of investigation of the degraded steel. Under these conditions there are altogether 289 exploitable μ-degradation functions (all the unreliable "boundary" ($F_i=\pm A_j$) degradation functions were excluded from the total number of $_{n=0}\Sigma^{n=17}(2n+1)=324$) and the same number of $1/\mu$-degradation functions, as they are plotted in Fig. 4.

Once the degradation functions are computed, the next task is to find the *optimum* degradation function(s) for the most sensitive and enough robust description of the investigated material degradation. A 3D-plot of sensitivity of the degradation functions can substantially help to choose the optimum one(s). The μ- and $1/\mu$-sensitivity maps corresponding to the degradation functions of Fig. 4 are plotted in Fig. 5. Slope of the linear regression of each degradation function is here defined as the function sensitivity. Thus the *sensitivity map* is a 3D-graph of these slope values plotted against the degradation functions field coordinates (F_i, A_j). As it follows from the presented sensitivity maps, the most sensitive μ-degradation functions are those with field coordinates around (F_i=-1400 A/m, A_j=3600 A/m), and/or (F_i=2200 A/m, A_j=3600 A/m), whereas the most sensitive $1/\mu$-degradation functions evidently have the F-coordinate equal to F_i = 400 A/m and the optimum amplitudes start from A_j = 600 A/m (the most sensitive $1/\mu$-degradation function) and the sensitivity decreases (but keeps rather high anyway) for larger minor loop amplitudes. The most sensitive μ- and $1/\mu$-degradation functions are plotted in Fig. 6.

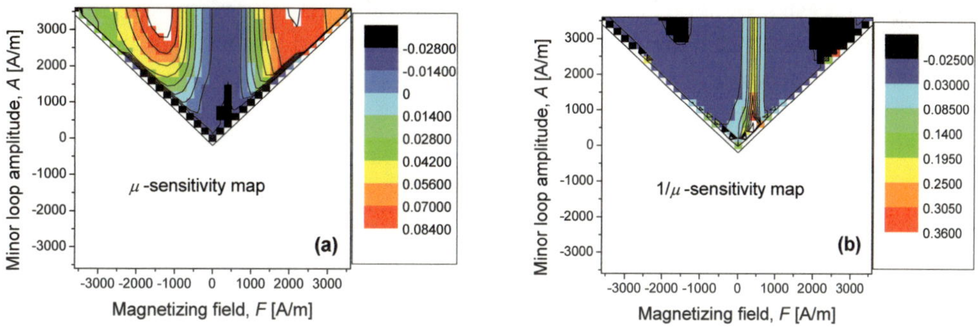

Fig. 5. The μ-sensitivity map (a) and $1/\mu$-sensitivity map (b) of the low carbon steel degraded by the mechanical tension. The maps were calculated for the step $\Delta A = \Delta F$ = 200 A/m and they are plotted up to the maximum amplitude A_j = 3600 A/m.

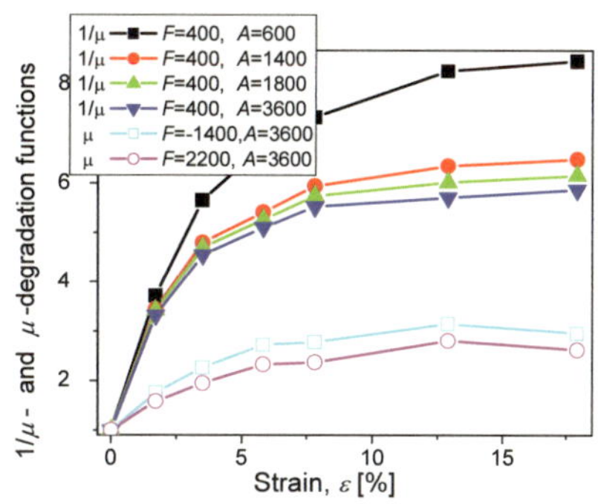

Fig. 6. The most sensitive degradation functions for plastic deformation of a steel for the step $\Delta A = \Delta F$ = 200 A/m. The top sensitive is the $1/\mu$-degradation function with field parameters F_i = 400 A/m, A_j = 600 A/m. The most sensitive μ-degradation functions are more flat.

3. Methodological hints

During measurement of the induced signal in the example presented in previous Section, samples shaped like thin rings were used, and the magnetizing and the pick-up coils were wound directly on them. Relatively *small field* was able to magnetize the material substantially, as there was *no demagnetization* effect, and *homogeneous magnetization* of the sample material was rightfully assumed. Signal measured on such *magnetically closed* samples is proportional to differential permeability of the samples *material*, and magnetizing field, *F*, *inside* the samples can be easily calculated as

$$F = N\, I_F/L \qquad [\mathrm{A/m, 1, A, m}], \tag{3}$$

where N is number of turns of the magnetizing coil, I_F is the magnetizing current and L is circumferential length of the magnetic circuit (e.g. of the sample ring).

If the measured samples are *magnetically open*, the situation is more complicated. "Long and narrow" specimens can be also successfully magnetized by coils placed *around* their body, as the demagnetization of such shapes is still acceptable for the MAT measurement. However, demagnetization field comes here into picture and the total inside field cannot be calculated as easily as suggested by Eq.(3). And what to do with short stout or large flat shapes? The magnetic circuit of such samples can be artificially closed, and the samples can be magnetized and measured with the aid of a magnetically soft yoke. The yoke can be either *passive* with the coils wound around body of the sample, or the yoke can be *active* (we speak then about an active *inspection head*) with the magnetizing coil wound around the bow of the yoke and the pick-up coil around one or both legs of the yoke.

Problems may appear in cases of rough or uneven surfaces of samples, where quality of magnetic contact between the sample and the yoke fluctuates from sample to sample. This introduces fluctuation of quality of the magnetic circuit and therefore also of the pick-up signal, which can be mistaken for variation of the samples material. The most rigorous solution of such situation is then simultaneous measurement of the tangential field on the sample surface (Stupakov, 2006, Stupakov et al., 2006, Perevertov, 2009), which substantially complicates the experiment, however. Another approach, which is able to solve the problem satisfactorily in majority of situations is to insert a non-magnetic spacer between the sample / yoke touching surfaces. The spacer makes the fluctuation relatively smaller and allows successful MAT measurement even on rough surfaces of magnetically open samples, see (Tomáš et al., 2012) and Section 3.5. However, fortunately, in great majority of investigated samples the different pieces of sample series have more-or-less similar quality of surface, which ensures reliable and repeatable measurements, even if the surface is rough, considering that MAT is a relative measurement, and degradation functions are normalized with the corresponding value of the reference sample.

In MAT measurements with open samples and attached yokes we deal with non-uniform magnetization. As a consequence, the magnetizing field inside the sample cannot be easily calculated like (3), we cannot use the magnetizing *field* coordinates (F_i, A_j), but we use the magnetizing *current* coordinates (I_{Fi}, I_{Aj}) instead. Besides, with the non-uniform magnetic circuits we cannot speak about the signal U to be proportional to the differential permeability of the *material*, but evidently we deal with an *effective* differential permeability of the existing *circuit* instead. As MAT is a *relative* method, the current coordinates are also well applicable for identification of the mutually corresponding magnetic states of the samples to be related and compared.

Measurements on closed samples and on open ones, either in a solenoid or helped by attached yokes, evidently yield quantitatively different data. If measured on identical materials, are they comparable? Do they show similar trends? Next two Sections of this part give answers to such questions and describe MAT measurements performed on equivalent series of closed and open samples (Section 3.1) and on the same series measured in a solenoid and with the aid of yokes (Section 3.2). The remaining Sections of this part present several other points of the methodological discussion, namely the potential of multi-parametric character of MAT (Section 3.3), the influence on MAT sensitivity of speed of magnetization (Section 3.4) and of nonmagnetic spacers (Section 3.5), and eventually the influence of sample material temperature on the MAT results (Section 3.6). The last Section of this part, Section 3.7, refers to several traditional hysteresis and Barkhausen measurements of degraded samples and compares their results with those obtained by MAT.

3.1 Magnetically closed and open samples

Structural degradation of samples manufactured from compressed steel are here studied by MAT and their optimal $1/\mu$-degradation functions are compared with their rolling reduction, and with their independently measured Vickers hardness (*HV*). The material is low-carbon steel with 0.16 wt.% of C, 0.20 wt.% of Si, 0.44 wt.% of Mn, with the rest of Fe, and the samples were produced from steel plates cold rolled down to 0, 5, 10, 20 and 40% reduction. Three sample geometries are used, namely picture-frames, rectangular plates and rectangular bars. Dimensions and shapes of the samples are shown in Fig. 7.

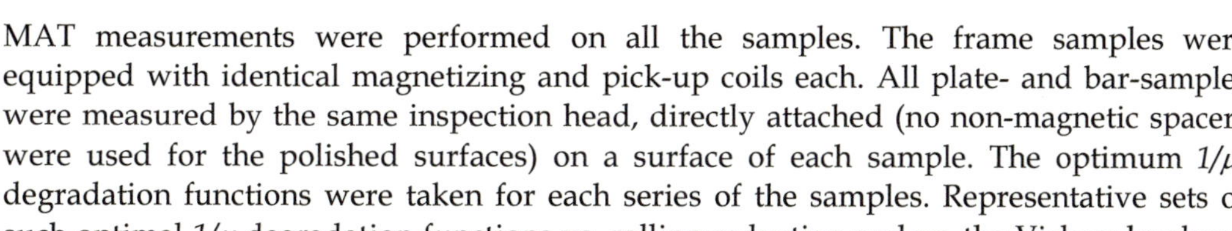

Fig. 7. Dimensions and shapes of the frames, of the plates and of the bar-samples.

MAT measurements were performed on all the samples. The frame samples were equipped with identical magnetizing and pick-up coils each. All plate- and bar-samples were measured by the same inspection head, directly attached (no non-magnetic spacers were used for the polished surfaces) on a surface of each sample. The optimum $1/\mu$-degradation functions were taken for each series of the samples. Representative sets of such optimal $1/\mu$-degradation functions vs. rolling reduction and vs. the Vickers hardness are plotted in Fig. 8 for the frame-, for the plate- and for the bar-shaped samples. In each measurement series, all the degradation functions are normalized by data of the virgin (not rolled) sample of the relevant shape. As seen from Fig. 8, consecutive series of the optimal MAT degradation functions well reflect the material modifications, regardless of the actual sample shapes. The results on series of differently shaped samples agree with each other *qualitatively*, and the frame-shaped magnetically closed samples show the highest sensitivity, as expected. For making the measurements also *quantitatively* comparable and repeatable, it is necessary to perform the tests always under the same experimental conditions (speed of magnetization, steps of increasing current, etc.), to use the same measuring head (and/or the same coils), and evidently also the same shape of samples (see also Vértesy et al., 2008a).

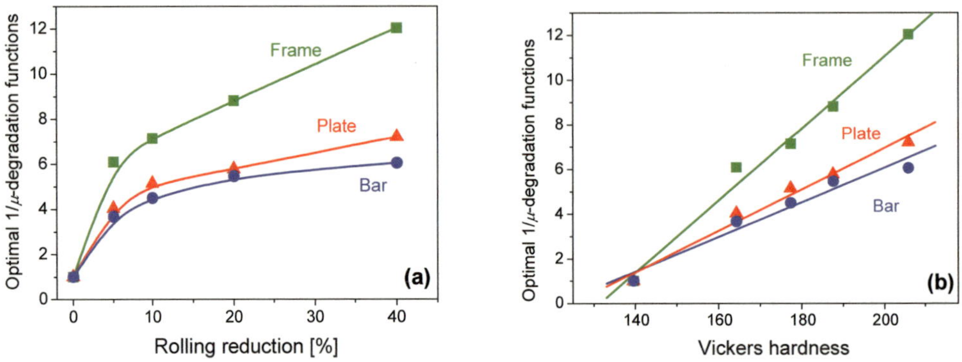

Fig. 8. Optimal *1/μ*-degradation functions for the three series of differently shaped samples. (a) Dependence on the rolling reduction. (b) Dependence on Vickers hardness.

3.2 Open samples between yokes and in a long solenoid

Low carbon commercial steel CSN 12050 was plastically elongated in an Instron loading machine up to final strain values 0, 0.1, 0.2, 0.9, 1.5, 2.3, 3.1, 4.0, 7.0 and 10.0%. A single series of identically shaped long flat samples (115x10x3 mm^3) was manufactured from the degraded steel pieces and it is here studied by MAT in two different ways, both applicable for these magnetically open samples. In one measurement series the samples were magnetized by a short coil, placed around middle of each sample and the magnetic circuit was artificially closed from both sides of the sample by two symmetrically positioned passive yokes (no non-magnetic spacers). The arrangement is shown in Fig. 9a, while Fig. 10a shows pick-up coil signals of the investigated samples. In the other measurement series the same samples were magnetized as open, in a long solenoid. This solenoid, with one of the samples inside, is shown in Fig. 9b. Fig. 10b demonstrates the pick-up coil signals originating from this experimental arrangement.

Dramatic quantitative differences can be observed between signals of Figs. 10a and b, even though the double-peak character of the signal is qualitatively preserved in both. And, in spite of difference between the directly measured signals, character of the MAT degradation functions vs. plastic strain is again qualitatively the same. It is shown in Fig. 11, where the optimal 1/μ-degradation functions of the discussed two cases are compared. Evidently, sensitivity of measurements is significantly higher if the samples are at least artificially closed by the yokes, but even in the worse case of the open long thin samples in a solenoid the material degradation is well measurable.

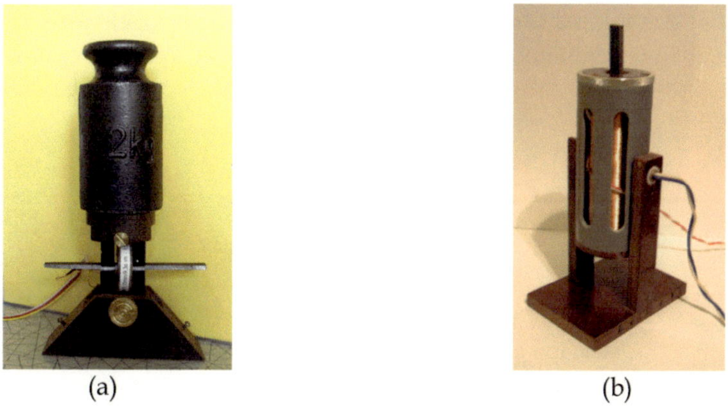

Fig. 9. The arrangements, used for the measurements. (a): The samples are magnetized by the short coil placed around middle of each sample and a pair of soft passive yokes closes the magnetic circuit from bottom and top. (The 2 kg weight is used only for pressing the yokes better on the sample surfaces.) (b): The samples are magnetized inside the long solenoid.

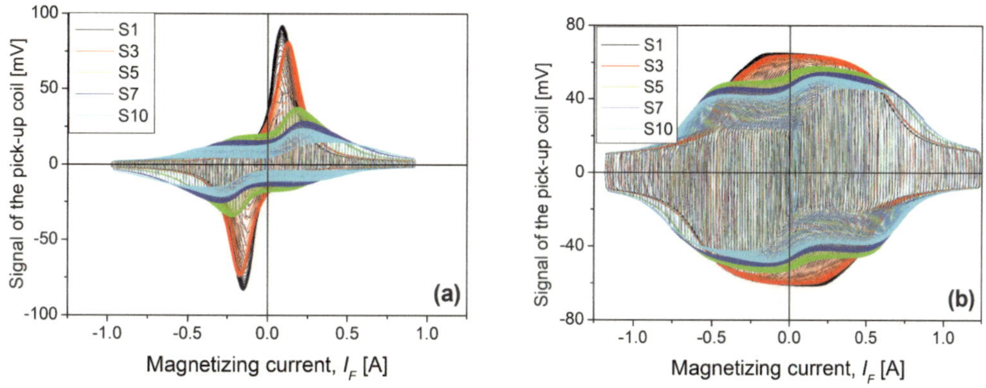

Fig. 10. Several pick-up coil signals of the investigated samples. (S1 is the reference, not deformed sample, S10 is the sample with 10 % strain). (a): Samples with artificially closed magnetic circuit by a pair of passive yokes. (b): Open samples in the solenoid. Remark: Scales in the (a) and (b) figures are not mutually comparable as neither the magnetizing solenoids nor the pick-up coils are the same in the two experimental arrangements.

3.3 Full use of the multi-parametrical character of MAT

Magnetic Adaptive Testing offers a great number of degradation functions. Usually *those* degradation functions are used as calibration curves for assessment of unknown samples, which are *monotonous and most sensitive* to the investigated material degradation. However, in some cases, magnetic reflection of the degraded material yields *only non-monotonous* degradation functions, see e.g. study of low cycle fatigue in (Devine et al., 1992, Tomáš et al., 2010). A single non-monotonous degradation function does not allow to decide whether a measured unknown sample belongs to the ascending or to the descending part of it. It will be shown in this Section, that the multi-parametric behavior of MAT can solve this question and that a suitable combination of two or more non-monotonous degradation functions obtained from a *single* MAT measurement makes such a decision possible, see also (Vértesy & Tomáš, 2012a).

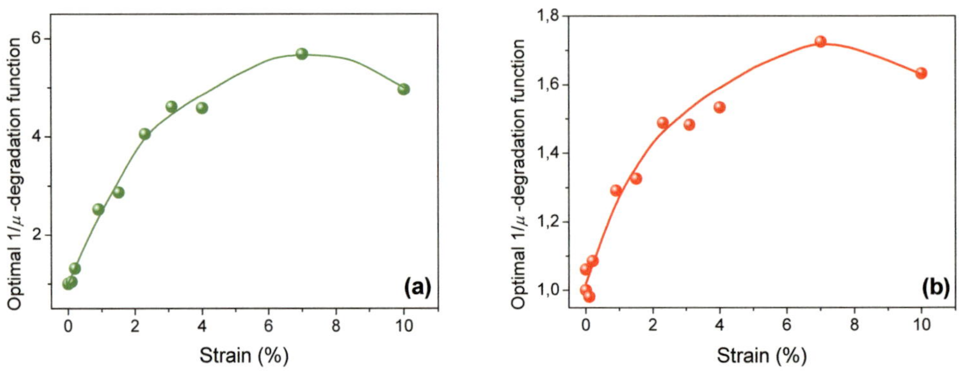

Fig. 11. MAT degradation functions vs. the plastic strain obtained on the tensile stressed steel samples. (a): Samples were artificially closed by a double yoke. (b): Open samples in the solenoid.

As an illustration of such a solution again the series of long flat samples (115x10x3 mm^3) is presented, which were manufactured from the commercial steel CSN 12050, plastically deformed by tensile stress up to the ten strain values 0, 0.1, 0.2, 0.9, 1.5, 2.3, 3.1, 4.0, 7.0 and 10.0%. The same samples were used in Section 3.2 already for demonstration of equivalency between results of two different ways of measurement. Data from the arrangement with two passive yokes (see Fig. 9a) are described here.

The non-monotonous character of the *1/μ*-degradation functions is shown in Fig. 11, and the same is true about the *1/μ'*-degradation functions, which are presented in Fig. 12. The most sensitive *1/μ'*-degradation functions come from the area of coordinates around I_F=40 mA, I_A=150 mA values (see the red area in the *1/μ'*-sensitivity map plotted in Fig. 13), but as shown in Fig. 12a, they are non-monotonous at the range of large strains. If *1/μ'*-degradation functions from other areas of the sensitivity map are plotted, all of them show the non-monotonous shape as well. However, not all of them have the peaks in the same range of strains as the most sensitive ones do. An example of such a *1/μ'*-degradation function is plotted in Fig. 12b. The function in Fig. 12b comes from the area around I_F =400 mA, I_A =600 mA (see the yellow area at the right-hand-side in the *1/μ'*-sensitivity map in Fig. 13.), it is substantially less sensitive than the optimal *1/μ'*-degradation function of Fig. 12a, but it is nicely monotonous at large strains (and non-monotonous at low strains, where the sensitive function does not have any problem). Evidently, the sensitive *1/μ'*-degradation function of Fig. 12a can be used as the *main* calibration function, but if an unknown sample falls into its non-monotonous part, the auxiliary *1/μ'* -degradation function of Fig. 12b answers the question about whether the unknown sample belongs to the ascending or to the descending part of the Fig. 12a-curve.

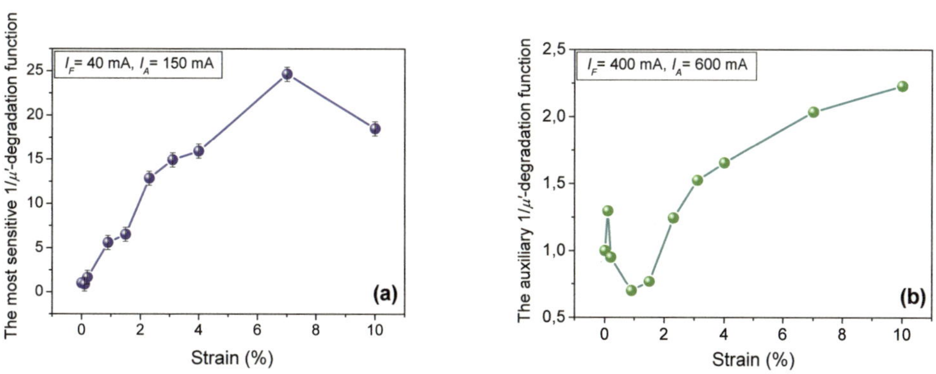

Fig. 12. (a): The optimum 1/μ'-degradation function from the (I_F=40 mA, I_A=150 mA) area. (b): The auxiliary 1/μ'-degradation function from the (I_F =400 mA, I_A =600 mA) area.

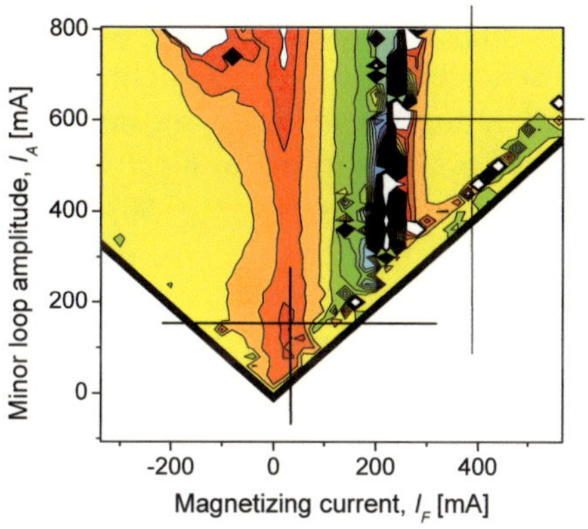

Fig. 13. The 1/μ'-sensitivity map of the steel degraded by tensile stress. Crossing lines indicate the areas from where degradation functions of Fig. 12 were taken.

3.4 Speed of magnetization

The most frequent way of MAT measurement proceeds by the electromagnetic induction method in a Permeameter similar to the scheme in Fig. 2 under the condition $|dF/dt|$=const. This condition guarantees that the magnetization processes in the samples continue always at the constant speed, which is desirable. As according to Eq.(1) the measured signal is directly proportional to dF/dt, this field-slope is set *high enough* with the intention of getting a sufficient signal-to-noise ratio. At the same time, however, it is advisable to set it *low enough* in order to minimize eddy currents and other dynamic effects, which can adversely influence shape of the signal and of the degradation functions via their implicit dependence on the value of dF/dt. In this Section influence of the rate of change of the magnetization processes on sensitivity of degradation functions in Magnetic Adaptive Testing will be investigated, see also (Tomáš et al., 2009).

The samples employed for this investigation were the same series of circular rings of low carbon steel plastically deformed by uniaxial tension, which was introduced as an illustrative example in Section 2. Arrangement of the experiment coincided with Fig. 2, so that the varying magnetization of each sample was achieved by application of a time-dependent magnetic field, $F(t)$, due to triangular waveform current, $I_F(t)$, in the magnetizing coil, with step-wise increasing field-amplitudes, A_j, corresponding to the step-wise increasing current-amplitudes, I_{Aj}. The rate of change in all the triangles was constant (but for its sign), namely dI_F/dt=const. for the current-slope and/or dF/dt=const. for the magnetizing field-slope, in each measured family of the minor loops and in any measurements mutually compared. Three measurements of the same series of samples were carried out. The applied current-slopes and the corresponding field-slopes in the three measurements were dI_F/dt = 0.5 A/s, 4 A/s, 32 A/s, and dF/dt = 0.8 kA/m/s, 6.4 kA/m/s, 51.2 kA/m/s, respectively.

The sensitivity maps of the μ- and of the $1/\mu$-degradation functions computed for the measurement with the lowest field-slope dF/dt = 0.8 kA/m/s were plotted in Fig. 5 for the field-coordinates $-A_j \le F_i \le +A_j$, $0 < A_j \le 3.6$ kA/m, with the step $\Delta A = \Delta F = 0.2$ kA/m. The μ- and of the $1/\mu$-sensitivity maps for dF/dt = 6.4 kA/m/s, and 51.2 kA/m/s were qualitatively similar to those in Fig. 5. Three interesting regions of the field-coordinates (F_i, A_j) can be seen in each of the sensitivity maps, with slight shifts of the extreme regions due to different field-slopes. The white regions in Fig. 5a indicate the most sensitive *increasing* μ-degradation functions, the black areas indicate the most sensitive *decreasing* μ-degradation functions. For the $1/\mu$-degradation functions (see Fig. 5b) it naturally is just reversed. The most sensitive degradation functions for the field-slope dF/dt = 0.8 kA/m/s are plotted in Section 2. in Fig. 6. The same degradation functions, together with their counterparts for higher values of the field slope, namely for dF/dt = 6.4 kA/m/s, and 51.2 kA/m/s are drawn in Fig. 14 in the present Section.

As the curves in Fig. 14 show, there is very little influence of the varied magnetizing field-slope from 0.8 till 51.2 kA/m/s (i.e. 64 times larger speed of magnetization) on sensitivity of the best μ-degradation functions within their regions of monotonous *increase* (i.e. in the two "white" areas of the sensitivity map of Fig. 5a). However, the best degradation functions from the "white" region of Fig. 5b, (i.e. those with a monotonous *increase* of the $1/\mu$-degradation functions) are influenced substantially as it is seen on Fig. 14c. As the degradation functions are created by the signals ratio shown in Eq. (2), explanation of this behavior can be found through a closer look at the recorded induced voltage signals both for the degraded samples, ε_k, and for the normalizing reference sample, ε_0, as was discussed in details in (Tomáš et al., 2009).

There is no universal advice as for what magnetizing field-slope to chose and what region of degradation functions to use for the most successful MAT nondestructive tests in individual cases. As the name of the method suggests, it is recommended to *adapt* optimally the choice both of the degradation functions field-coordinates region and the magnetizing field-slope as to obtain the satisfactory signal-to-noise ratio and the best degradation function sensitivity at the same time. The concrete adaptation is completely governed by properties of the investigated material degradation and by the level of noise and the available rate of change of the magnetizing field of the used measuring technique.

However, generally speaking it can be stated, that as long as the MAT degradation functions are picked-up from localities of the field coordinates where the local differential permeability is *high* (frequently close to the maximum permeability at the used minor loop), then the (usually "reciprocal") degradation functions possess *very high sensitivity*, but level of their sensitivity is *strongly field-slope-dependent*. On the other hand, if the MAT degradation functions are chosen from localities of the field coordinates where the material is closer to saturation and the local differential permeability is *low*, sensitivity of such degradation functions is *not extremely high*, but it is *little dependent on the applied magnetizing field-slope*.

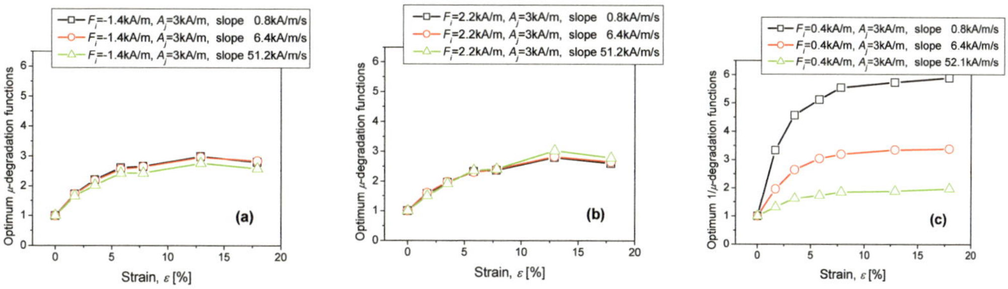

Fig. 14. The steepest degradation functions measured at the three magnetizing field-slopes dF/dt = 0.8 kA/m/s (□), 6.4 kA/m/s (O), and 51.2 kA/m/s (Δ), at the three important coordinate regions of the sensitivity maps. For easy comparison all the degradation functions are determined at the minor loop with amplitude A_j = 3kA/m: $\mu(F_i,A_j,\varepsilon)$ at F_i = -1.4 kA/m (a), $\mu(F_i,A_j,\varepsilon)$ at F_i = +2.2 kA/m (b), and $1/\mu(F_i,A_j,\varepsilon)$ at F_i = +0.4 kA/m (c).

3.5 Non-magnetic spacers

Magnetic measurement of flat samples, which cannot be magnetized (and/or closed magnetically) in any better way than by an attached magnetizing/sensing soft yoke, set into direct contact with the sample surface, suffers often by fluctuation of quality of the magnetic sample/yoke contact. This is a well-known problem, with unpolished surfaces in particular, which can be improved by using as large yoke as possible and/or by application of a spacer between the yoke and the sample. However, size of the sample frequently does not allow a very large yoke to be used and even a thin nonmagnetic spacer decreases and distorts the measured signal substantially, so that measurement of basic *magnetic parameters* of the sample material in this way is very difficult, if not impossible.

However, spacers are quite applicable for magnetic "structuroscopy", i.e. for magnetic measurement of relative structural changes of ferromagnetic construction materials, especially if the measurement is carried out by a method analyzing the measured signal like it is done e.g. in Magnetic Adaptive Testing. The following measurement explains and illustrates use of spacers, see also (Tomáš et al., 2012).

Samples of gradually increasing brittleness were prepared from ferromagnetic steel 15Ch2MFA in the shape of rectangular prisms 10 x 10 x 30 mm^3. Material for the samples was embrittled in the same way as described in Section 4.3.2. Three samples of each grade of brittleness (of the same tempering temperature T_T of their preparation, see Section 4.3.2) were used for the reported measurement. Quality of surfaces of the samples corresponded to their ordinary machining (milling) and grooves from the milling or even scratches were visible on some of them. No polishing of the surfaces was implemented, some surfaces were evidently worse than others.

The measurement was done inductively on the magnetically open samples using a pair of passive yokes. Magnetization of the samples was carried out by a short driving coil, which was each of the samples before the measurement inserted into. The induced voltage signal was recorded from a pick-up coil, placed co-axially with the driving one. In order to be able to magnetize the short thick samples by the small driving coil appreciably, two soft yokes of laminated Si-Fe sheets (cross sections of the legs 10x20 mm^2, maximum height in the bow 30 mm) were used to short-cut the magnetic flux during magnetization of the samples. The yokes were attached to surfaces of the samples either directly, or over thin spacers (paper stickers) glued to each contact face of the two yokes. Thickness of the applicable spacers was from zero up to 0.9 mm for the starting examination, spacers with thickness 0.08 mm and 0.23 mm were chosen as optimal for the reported measurements. Arrangement of the sample, yokes and the driving and pick-up coils are shown in Fig. 15 (left) and photo of the sample holder can be seen in Fig. 15 (right), showing a sample, the driving coil (the pick-up coil is covered by the driving one), the yokes, and a system of levers insuring a reproducible pressure of the yokes on each sample.

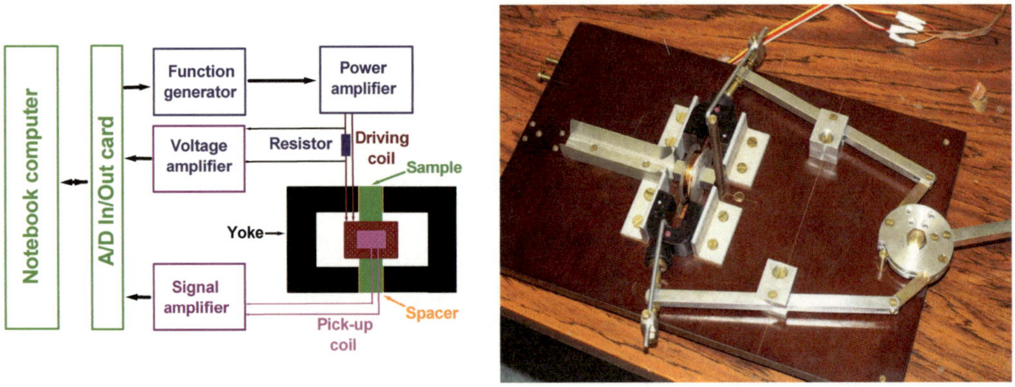

Fig. 15. Left: Sketch of the Permeameter with a magnetically open sample, coils, yokes and spacers. Right: The sample holder with the coils and the attachable yokes. The prismatic sample is placed into the coil system (by shifting it in the metal tray rightwards down to the brass stop) and the black yokes are pressed against the sample by the system of levers with the handle on the right in the picture.

The speed of change of the magnetizing current in its linear region was 1 A/s for the measurements without spacers and with spacers 0.08 mm thick, and it was 3 A/s for the measurement with the spacers 0.23 mm thick. (Increase of the speed of change of current in the latter case - as compared with the former ones - was used in order to make the signal conveniently larger.) Also the maximum amplitudes and the constant steps were different for different measurements, their applied values can be approximately read from Figs. 16a,b,c. Their choice was rather arbitrary, guided by the intention to get the sample appreciably magnetized and to fill up the range between the minimum and maximum amplitudes of the magnetizing current with a representative number of minor loops. The voltage signals measured in the pick-up coil were proportional to the effective differential permeability of the used magnetic circuits and their characteristic shapes can be seen in Fig. 16.

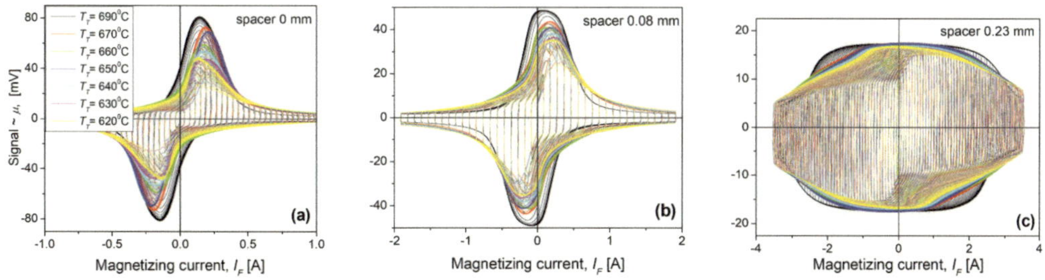

Fig. 16. (a): Measured signals from the most ductile (T_T = 690⁰C) till the most brittle (T_T = 620⁰C) samples, with the yokes without any spacer.
(b): Signals of the same samples measured with non-magnetic spacers 0.08 mm thick, glued on the yokes. Notice the range of the vertical axis is twice smaller and of the horizontal axis is twice larger than in figure (a).
(c): Signals of the same samples measured with non-magnetic spacers 0.23 mm thick, glued on the yokes. Notice the range of the vertical axis is four times smaller and of the horizontal axis four times larger than in figure (a).

Decision about optimum thickness of spacers can be found empirically, following measurement of peak values of signals of "good" and "bad" samples at a convenient (the same for all spacers) speed of change of the magnetizing field, using a series of available spacers. Fig. 17 shows result of such a measurement on two most ductile (T_T = 690⁰C) samples, one with the best surfaces and the other with the worst ones. As it can be seen in Fig. 17, magnitudes of the peak values of the two samples approach each other with increasing spacer thickness. Spacers with thickness 0.08 mm and 0.23 mm were chosen as optimum ones for the illustrative measurement.

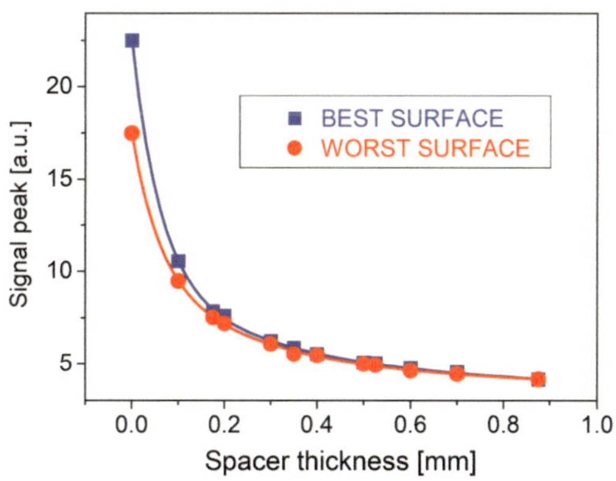

Fig. 17. Peak values of signals of two most ductile (T_T = 690⁰C) samples, one with the best surfaces and the other with the worst ones. The two curves approach each other with increasing spacer thickness.

The optimum μ-degradation functions and the μ'_F-degradation functions for the measurement without spacer and with spacers 0.08 mm and 0.23 mm thick are plotted in Fig. 18.

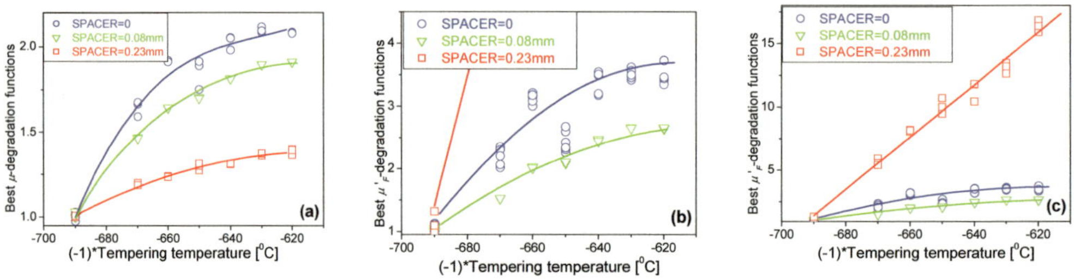

Fig. 18. (a): The best μ-degradation functions measured without any spacer (O) on the yoke faces, with spacers 0.08 mm (∇) and with spacers 0.23 mm (□).
(b): The best μ'_F-degradation functions measured without any spacer (O) on the yoke faces, with spacers 0.08 mm (∇) and with spacers 0.23 mm (□). Notice the vertical scale is doubled in comparison with figure (a).
(c): The same curves as in figure (b), just the vertical scale is four times larger than in (b).

Influence of the non-magnetic spacers inserted between the sample and the magnetically soft yokes is clearly seen in Figs. 16a,b,c: The increased distance between surface of the sample and faces of the yokes increases demagnetization within the magnetic circuits. This means a decrease of the magnetizing *field* in the sample with spacers at the use of the same magnetizing *current* as for the circuit without the spacers. However, the decreased magnetizing field fluctuates less. Beside the control of fluctuation of the magnetizing field, presence of well-defined spacers (their thickness should be found empirically by trial-and-error approach) modifies shapes of the signals measured on the series of degraded samples. Generally, with application of a spacer, substantially decreased scatter of the experimental points accompanied by a small decrease of sensitivity of the degradation functions can be expected. Such a regular behavior is presented in Figs. 18, for the μ- and the μ'_F-degradation functions measured with thin spacers (0 and 0.08 mm) in particular. A substantially thicker spacer (in our case 0.23 mm), however, is able even to modify shape of the signals considerably (see Figs. 16c), which can result in a qualitatively new shape of the degradation functions with quantitatively increased sensitivity (see Fig. 18c).

Use of the spacers for testing of material structural degradation can be recommended as a simpler alternative to more sophisticated and more demanding measurement with a direct "on-line" recording of the magnetic field at the sample surface. Especially, when it is measured by a series of field sensors and the resulting field value is obtained by extrapolation of their readings down to the sample surface, see e.g. (Stupakov, 2006, Perevertov, 2009), the outcome can be very reliable and allows compensation of fluctuating quality of the sample-yoke magnetic contact by determination of the immediate value of the magnetizing field inside the samples. Such measurements are even able of a reasonable determination of absolute magnetic parameters of the samples, which evidently cannot be done with the spacers only.

Successful use of the non-magnetic spacers can be applied for relative indication of the samples' micro-structural conditions. The spacers modify shapes of the measured signals, they substantially reduce scatter of experimental points accompanied by a slight decrease of the degradation functions sensitivity. Sometimes, thick spacers in particular, are able to modify shapes of the measured signals qualitatively and to bring about considerable increase of sensitivity, especially in the degradation functions computed from the signal derivatives. However, spacers hardly can be employed (without a simultaneous measurement of the sample surface field) for determination of purely magnetic parameters of the samples.

3.6 Temperature dependence

An important feature of any NDT technique is the permissible range of temperatures within which the method can be successfully applied. For *magnetic* methods, the tested objects are mostly ferromagnetic iron-based construction materials, namely various kinds of steel or cast iron. These materials are electrical conductors, their electric conductivity is substantially reduced not far above room temperature already, and this is the reason, why e.g. the widely successful eddy current inspections loose above room temperature a great portion of their efficiency. Not so the magnetic methods. Magnetic properties of iron-based materials do not change around the room temperature significantly yet, as they are still pretty far from their Curie point. Therefore, magnetic methods and relative magnetic NDT methods in particular, give results rather independent of temperature, if the tests are carried out within the working temperatures of most industrial objects, let us say up to about 200^{0}C (see Vértesy et al. 2010a).

Such behavior of MAT is demonstrated on a series of 5 ductile cast iron specimens, solidified with different cooling rates, which in its turn lead to five different microstructures with different values of Brinell hardness (see Vértesy et al. 2010b for details). MAT measurements were performed on each specimen at two different temperatures (20^{0}C and 180^{0}C) by application of an inspection head. Different rates of change ("slopes") of the magnetizing current were used. The specimens were heated by a hot plate and temperature of surface of the specimens was measured by a thermocouple. The obtained optimally chosen μ-degradation functions are shown in Fig. 19. As it can be seen there, linear

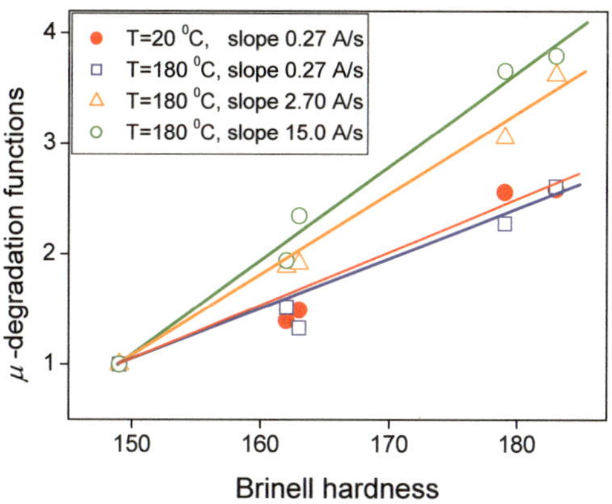

Fig. 19. The μ-degradation functions vs. the Brinell hardness measured at 20^{0}C and at 180^{0}C temperatures. Different rates of change of magnetizing current (slopes) were used in the latter case.

correlation was found between the nondestructively determined optimal μ-degradation functions and the Brinell hardness of the investigated material, and this correlation does not depend on the temperature as long as slope of the magnetizing current is kept low enough (see the red and blue lines in Fig. 19). If higher slopes are used, the influence on the μ-degradation functions is *negligible at room temperature*: the μ-degradation functions at 20^{0}C at rates 2.7 and 15.0 A/s coincide with the red line (not shown in Fig. 19, but kindly compare with Figs. 14a,b). At temperature 180^{0}C another behavior is seen: the μ-degradation functions at slopes 2.7 and 15.0 A/s increased their sensitivity with increase of the slope (see the orange and green lines).

3.7 Comparison of MAT descriptors with traditional magnetic parameters

As mentioned in Section 2, Magnetic Adaptive Testing is a magnetic hysteresis method, which - in contrast to the traditional hysteresis ones - makes use of data taken not only from the saturation-to-saturation *major* hysteresis loops of a series of degraded ferromagnetic materials, but collects information from systematically measured *families of minor* hysteresis loops. Such a large system of data is supposed to be able to completely characterize (see e.g. Preisach, 1935, Mayergoyz, 1991, Bertotti, 1998) each of the measured material, and therefore it is natural to expect it to contain also those special magnetic data, which vary with degradation of the investigated material with top sensitivity. The top-sensitive data *can* be those, utilized by the traditional hysteresis approach, but experience shows that the top sensitive data are usually different from the traditional ones.

For making a comparison between results of MAT and of a number of other widely used nondestructive magnetic methods, measurements are considered on two series of differently shaped samples of different materials plastically deformed by two different ways. The first was a series of TRIP steel magnetically open flat long bar samples (the same series as in Section 4.2.2) plastically elongated by uniaxial tension. The results of this first series are plotted in Fig. 20a. The other series were low carbon steel magnetically closed frame shaped samples (the same series as in Section 3.1) plastically compressed by cold rolling. The results of this second series are plotted in Fig. 20b. Each of the series was measured by several magnetic methods. For making the comparison possible, the results of each measurement were normalized by the corresponding value of the reference (not strained) sample. Fig. 20 shows data of the two applied ways of plastic deformation as they are reflected by the traditional major loop parameters (H_C and, $1/\mu_{MAX}$ and hysteresis losses, W), by Barkhausen noise measurement (*1/RMS*), and by the top sensitive MAT degradation functions ($1/\mu$ and $1/\mu'$). It shows that in both these illustrative cases the traditional hysteresis parameters, and also the Barkhausen data result in *qualitatively the same correlations* between the investigated magnetic characteristics and the material degradation. It is the first important conclusion of these comparative measurements, which shows in general the usefulness and equivalency of magnetic measurements.

On the other hand, this comparison clearly demonstrates that properly chosen MAT degradation functions - the $1/\mu$ '-degradation functions in particular - reflect the material degradation (at least that due to plastic deformation, as shown in our example) with substantially higher sensitivity. *The MAT-degradation functions are typically much more sensitive than the traditional parameters.* The evident reason for this statement is the fact, that the MAT parameters are optimized just for the investigated material and for the studied way of degradation. It is worth of mentioning that similarly good results were achieved on other materials and other ways of degradation, as well.

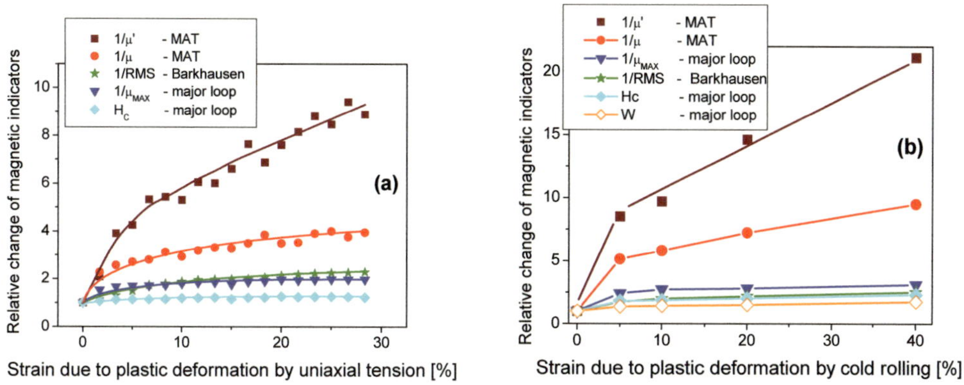

Fig. 20. Comparison of the available MAT most sensitive degradation functions picked up from the families of minor loops, with other available magnetic parameters. (a): TRIP steel *open* samples *elongated* by uniaxial tension. (b): Low carbon *closed* samples *compressed* by cold rolling.

Higher sensitivity, however, is not the single advantage of the MAT approach as compared with the traditional hysteresis and other magnetic ways of NDT. Another good point for MAT is its *multi-parametric output*, which allows not only re-check and cross-check the measured results, but in cases of non-monotonous dependencies it is often possible to use one set of data for clarification of another one, as it is shown e.g. in Section 3.3 or in (Vértesy & Tomáš, 2012). Another, experimentally very friendly feature of MAT is the fact, that there is *no need of saturation* of the measured samples. This is a very important advantage especially with magnetically open samples, where to reach saturation is usually an extremely difficult task, and the traditional methods are often forced to use minor instead of major loops anyway and just to assume the differences from the real major loops are not substantial (which is often justified, however).

Speaking about the magnetically open samples, one more interesting property must be mentioned, namely the unrestricted application of non-magnetic ***spacers*** for MAT measurement with magnetizing/sensing inspection heads or with magnetic-flux-short-cutting yokes. The point is, that for purposes and needs of MAT actually ***any magnetic value***, which can be parametrized by its field- and/or current-coordinates and which varies in a definite way with the investigated degradation of the material, is acceptable and applicable for indication of the studied degradation. Use of non-magnetic spacers brings substantial simplification of attitude and easiness of evaluation of testing, especially on samples with not ideal surfaces. See Section 3.5 and (Tomáš et al., 2012).

Last but not least, systematic measurement of the family of minor loops, with the constant rate of change of the sample magnetization in particular, makes it possible to use special ways of extrapolation for rather precise *determination of quasi-static magnitudes of magnetic parameters* of the material even on magnetically open samples. This simple idea is thoroughly described by Ušák in his recent paper (Ušák, 2010).

4. Examples of application

4.1 Elastic deformation

Application of MAT on elastic deformation of steel samples was investigated only to a limited extent, so far. There is still plenty of space for getting experience in this field. An unpublished example is shown of a preliminary investigation of the μ-degradation functions reflecting uniaxial tension applied on a sample of Mn-Si steel (CSN 13240: Fe, 0.33-0.41%C, 1.1-1.4%Si, 1.1-1.4%Mn). Fig. 21 shows sketch of a magnetically *closed* sample shape, which made it possible to clamp the sample by its top and bottom pads into jaws of a loading machine, and to apply equal tensional stress to the two legs of the sample window. Stress-strain loading curve of the sample is shown in Fig. 22a. The magnetizing and the pick-up coils were wound on the legs and magnetization of the material was thus looped in the closed path around the sample window.

Fig. 21. Magnetically closed steel sample for testing influence of the actively applied tensional stress on the shape of the MAT degradation functions. The red rectangles show positions of the magnetizing and pick-up coils on legs of the sample window.

The result of the MAT measurement and evaluation is given in Fig. 22b. The complicated shape of the μ-degradation function reflects presence of at least three effects influencing the magnetic results: appearance of magnetic anisotropy with the pressure-induced *easy axis along* direction of the positive pressure (pulling), appearance of magnetic anisotropy with the pressure-induced reorientation of the grains in the polycrystalline material (probably reorientation of the material crystalline grains so that they get more and more oriented by their (111) axis along the direction of the pull; this should very probably induce an anisotropy with a *hard axis along* direction of the pull, and appearance of dislocations and variation of the dislocation distribution.

4.2 Plastic deformation

4.2.1 Plastic deformation of austenitic (paramagnetic) steel

4.2.1.1 Cold rolled austenitic steel

The aim of the example presented in this Section is to show application of MAT for characterization of cold rolled austenitic stainless steel specimens (Vértesy et al., 2007). Titanium stabilized austenitic stainless steel, 18/8 type, was studied. The specimens were annealed at 1100°C for 1 hour. Then they were quenched in water in order to prevent any carbide precipitation, and to achieve homogeneous austenitic structure as the starting material. The as-prepared stainless steel specimens were cold-rolled at room temperature down to about 60% strain. The compressive plastic deformation of the material increased its hardness and at the same time the originally paramagnetic austenite specimens became more and more ferromagnetic, due to appearance of bcc α′-martensite. Vickers hardness, *HV*, was measured destructively and nondestructive MAT measurements were performed on each specimen by attaching a magnetizing/sensing inspection head on surface of the stripe-shaped samples.

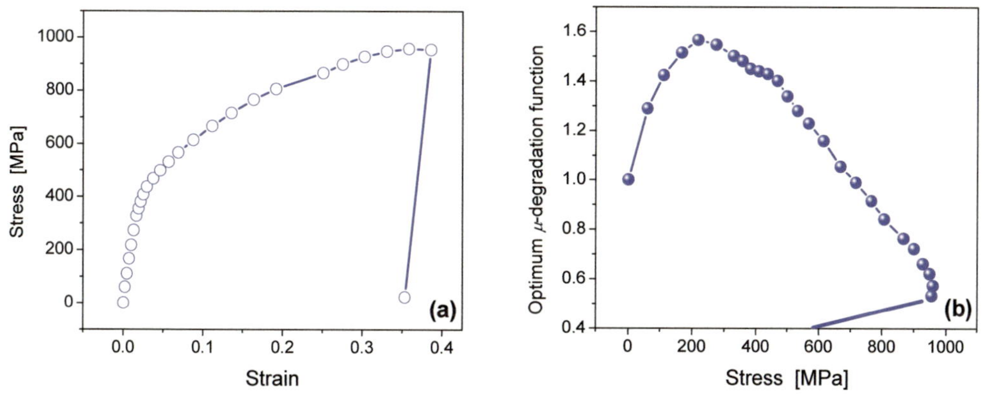

Fig. 22. (a): Stress-strain loading curve of the sample of the CSN 13240 steel, (b) The most sensitive μ-degradation function (F_i=280, A_j=560 A/m) as it depends on the applied stress. Experimental points on the two curves show stress values at which the sample was magnetized and measured.

Values of the measured Vickers hardness, *HV*, together with the most sensitive μ ′-degradation function are shown in Fig. 23 in dependence on the plastic deformation strain. Evidently the MAT results offer substantially higher sensitivity and reliability of indication of the state of the stainless steel after the deformation than the traditionally used destructive hardness measurements.

4.2.1.2 Austenitic steel strained by tensile stress

Austenitic stainless steel SUS316L investigated in the present example was also plastically deformed as that in the previous Section 4.2.1.1, however this time not by compression, but by a tensile stress instead. Flat samples with original dimensions 100x70x3mm^3 were loaded up to the 50% strain. In contrast to the compressed samples, the tensile deformation did not introduce such a large percentage of the ferromagnetic phase into the deformed samples and neither the signals measured by an attached inspection head nor the μ-degradation functions gave much hope for sensitive enough magnetic indication of the strain values. However, the *B*-degradation functions (and also the μ ′-degradation functions - not shown here) gave very acceptable results. The magnetic indication of the strain was substantially more efficient (up to about 30%) if the samples were magnetized along the direction of the material prolongation than normal to it (up to a few % only), as it can be seen in Fig. 24. However, it is evident, that MAT was applicable even in such an unpromising case and that it was also able to reflect anisotropy induced into the material by the stress. For more details see (Vértesy et al., 2011).

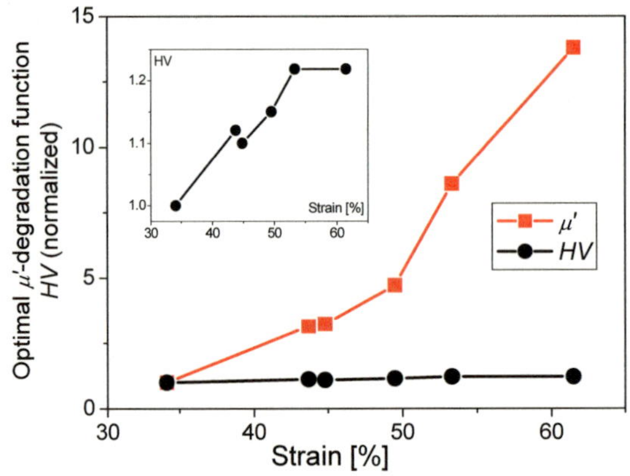

Fig. 23. The optimally chosen μ'-degradation function and the Vickers hardness, *HV*, vs. plastic strain of the samples. The inset shows the same values of *HV* but with more detailed vertical scale than that of the large plot.

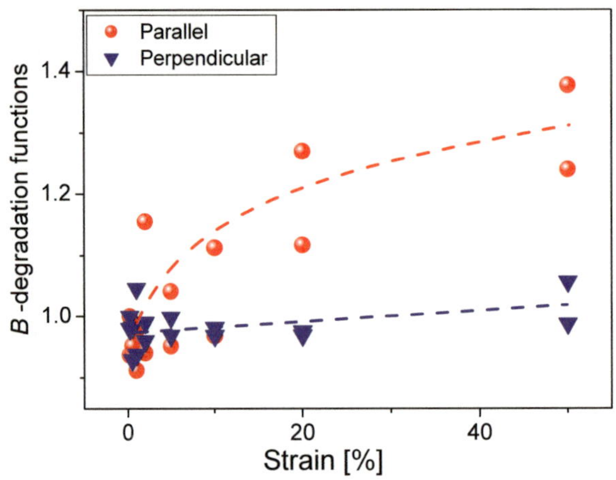

Fig. 24. The optimal *B*-degradation functions vs. residual strain of stainless steel for parallel and perpendicular magnetization with respect to the elongation.

4.2.2 Plastic deformation of ferromagnetic steel

Plastic deformation of ferromagnetic steel was investigated by MAT many times, always with excellent results. Papers (Stupakov & Tomáš, 2006, Tomáš et al., 2006, Tomáš et al., 2009a, 2009b, Vértesy et al., 2008c) and others can be quoted here, and results of a typical measurement on low carbon steel plastically deformed by uniaxial tension were used for the description of the MAT method in Section 2 and for investigation of the influence of the rate of change of the magnetizing field on sensitivity of the MAT degradation functions in Section 3.4. The results described in those two Sections are connected with the same material and similar experiments. MAT description of plastic deformation was also studied by measurements performed on magnetically open and closed samples in Section 3.1. At this place also a concise information about MAT study of a series of Transformation Induced Plasticity (TRIP) steel samples as of a specific material is added.

Transformation Induced Plasticity steels are modern materials for car industry with an excellent combination of ductility and high tensile strength. Typical mechanical parameters of TRIP materials are the upper yield strength about 540 MPa, the fracture strength about 750 MPa, and the fracture strain up to 30%. Our samples of TRIP 700 type steel with the original dimensions 150x20x1.18 mm^3 were loaded by tensile stress along their length (the production rolling direction of the material) and they were magnetized also along this direction by an attached inspection head in the MAT tests.

Eighteen samples were deformed to their relaxed strain values from 0% up to 28.33%. Vickers hardness (*HV*) measurements showed a significant increase of hardness due to the samples elongation. MAT experiments revealed a sensitive linear correspondence between the optimal μ- and μ'-degradation functions and HV (see Fig. 25) (and/or the strain - not presented here), and also a typical drop of the peak signal value (similar to that in Fig. 3b) between the reference (not deformed) sample and the very first deformed (with strain 1.7% only) sample. The scatter of points observed in Fig. 25 is evidently due to the untreated industrial surface of the commercial material together with application of the attached inspection head for the MAT measurement. No non-magnetic spacers (by which the scatter of points would certainly be reduced) were used in those experiments yet.

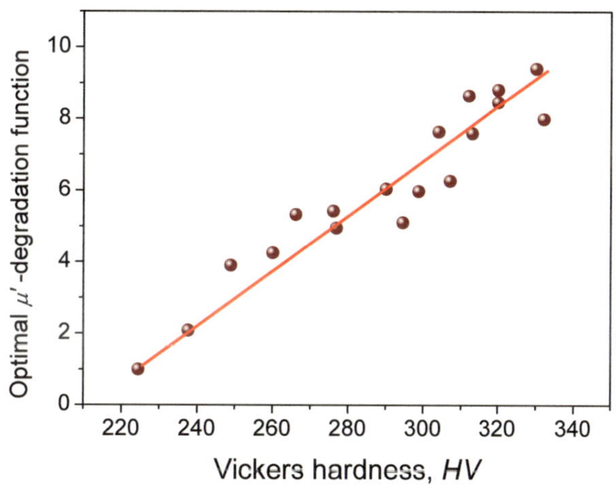

Fig. 25. The linear dependence of the optimal μ'-degradation function and Vickers hardness of the TRIP-700 steel plastically deformed by tensile loading.

4.3 Steel embrittlement

Decrease of ductility accompanied by increase of brittleness makes steel construction elements less reliable and more fault-prone. Number of processes exists, which can induce such a (usually adverse) change. Among examples of so far tested targets of MAT it is considered expedient to mention especially measurements of embrittled steel aiming at possible application of the Magnetic Adaptive Testing as a nondestructive method for checking the state of ductility/brittleness of materials used as steel coats of pressure vessels in nuclear reactors. The operational conditions of nuclear pressure vessel steel, that lead to such quality degradation include long-term bombardment by neutrons combined with intense material ageing in the elevated working temperature. As experimenting with irradiated material requires protected environment of "hot" laboratories, first investigation of applicability of potential methods is usually performed on steel samples, which underwent *simulated not radioactive treatment* by artificial thermal ageing and artificial mechanical or thermal embrittlement.

Abilities of MAT to reflect structural modifications of thermally aged steel (Vértesy et al., 2008b), of steel embrittled by cold rolling (Vértesy et al., 2008a, Tomáš et al., 2009a, 2010) and also of steel embrittled by special thermal processing (Tomáš et al., 2011) were investigated. Only after these experiments with the not radioactive simulated materials, the measurements proceeded to irradiated pressure vessel steel in a hot laboratory. Concise information about these experiments is presented in the following text in Section 4.3.1 (embrittlement by thermal ageing), Section 4.3.2 (embrittlement by thermal processing) and Section 4.3.3 (embrittlement by neutron irradiation). Successful MAT tests of several kinds of steel embrittled by cold rolling was described in Sections 3.1, 3.7 and 4.2.1.1.

4.3.1 Steel embrittlement by thermal ageing

Cu precipitates are known to be formed in the Fe metal matrix by thermal ageing of Fe-1 wt.% Cu samples at 500^{0}C and these precipitates have equally large impact on mechanical properties of this material as they have in the real pressure vessel steel. Two series of this Fe-Cu alloy were annealed at 850^{0}C for 5 hours, followed by water quenching. Then one part of the samples was cold-rolled down to 10% deformation and isothermally aged at 500^{0}C for 0, 50, 500 and 5000 min. The other part of samples was aged similarly, but without the previous cold rolling (0% deformation). Vickers hardness, *HV*, of the deformed material was measured with the standard Vickers indentation technique. Then 2 mm thick ring-shaped samples were prepared having 18 mm outer and 12 mm inner diameter, respectively. The samples were equipped with a magnetizing coil and a pick-up coil each. MAT measurements were performed on those rings.

The most sensitive μ-degradation functions of the two sample series (with 10% and with 0% deformation) were determined and plotted together with the normalized Vickers hardness values in Fig. 26. The definite variation of mechanical hardness, *HV*, of the aged material testifies about the structural modifications, leading to observable change of the ductile/brittle properties. However, as it is shown in Fig. 26, MAT reflects those changes with substantially higher sensitivity than *HV* and in contrast to *HV*, MAT is monotonous. Details about the experiment and detailed discussion on the material changes can be found in (Vértesy et al., 2008b).

4.3.2 Steel embrittlement by thermal processing

One of available technologies simulating radiation embrittlement of steel is a special thermal processing described in (PHARE, 1995). At this place (see (Tomáš et al., 2011) or details) investigation by MAT is shortly presented of a large series of samples of the 15H2MFA reactor steel (widely used in nuclear plants pressure vessels VVER 440), which were artificially embrittled in this way. The magnetic measurement was accompanied by destructive Charpy impact tests done on the same samples, so that the resulting MAT degradation functions could be plotted in direct dependence on the critical ductile-brittle-transition-temperature (*DBTT*) of the material. The samples were prepared (30 pieces of each embrittled grade) in the shape of rectangular prisms 10 x 10 x 55 mm³, with a V-notch in the middle, applicable for the mechanical Charpy tests. However, before they were sacrificed for the mechanical destruction, they were nondestructively tested magnetically by MAT with the use of soft yokes short-cutting the magnetic flux and with nonmagnetic spacers (see Section 3.5 and Fig. 15) suppressing scatter of experimental results. The most sensitive degradation functions proved to be the $1/\mu$ '- and μ '-degradation functions (as usual) and the optimal ones are plotted in Fig. 27 as functions of the *DBTT*.

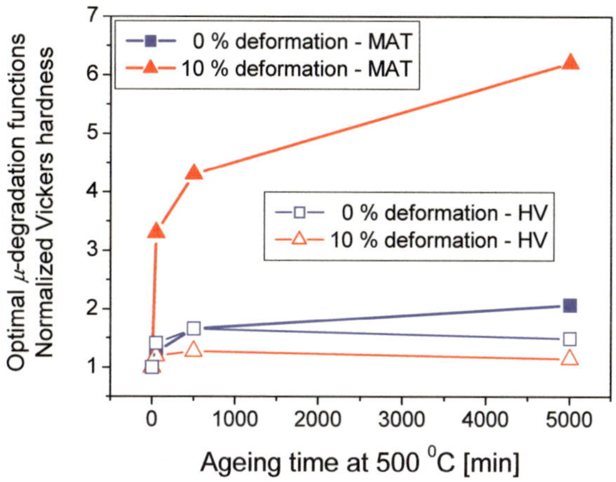

Fig. 26. The most sensitive μ-degradation functions and the averaged *normalized* Vickers hardness values of the thermally aged Fe-Cu alloy in dependence on the ageing time. The blue curves (■, □) describe the samples without previous deformation, the red ones (▲, Δ) belong to the samples compressed by 10% before start of the ageing.

MAT is shown in this Section to be sensitive enough to reflect variation of brittleness of the nuclear pressure vessel steel. It is essential, however, to keep in mind that this test of MAT on the steel, which was *embrittled by thermal processing*, reports about ability of MAT to reflect presence and density and size of *defects, which were introduced into the material **just and only** by the applied thermal processing*. The thermal processing shifted *DBTT* and Vickers hardness in this steel, but microstructure of this processed steel was certainly *different* from that of the same kind with *DBTT* shifted about the same (i.e. of the same mechanical brittleness), but embrittled by combination of the real fluence of neutrons and thermal ageing in a nuclear reactor. *The material defects after different treatments are substantially different even though the Charpy impact tests may show identical brittleness.* Magnetic Adaptive Testing is called "adaptive" just for the reason that it is able to react selectively and differently to different types of defects; it can be optimally *adapted* to them. This is considered to be one of the advantages of the method, and it was many times confirmed. Therefore, the present investigation proved MAT to "see" defects causing modification of brittleness by the thermal processing, similarly as it "saw" defects modifying brittleness of low carbon steel by cold rolling as reported in (Tomáš et al., 2009a) or in (Takahashi et al., 2006), and here in Sections 3.1 and 3.7. We can *anticipate* - by analogy - MAT to be able to see also defects caused by combination of neutron irradiation and thermal ageing, but this is all. In order to *prove* it, real experiments on irradiated/aged samples will have to be carried out. However, magnetic "structuroscopy" and its various methods were already successfully tested at least on some irradiated samples, see e.g. (Takahashi et al., 2006, Dobman et al., 2008, Vandenbossche, 2009, Gillemot & Pirfo Barroso, 2010) and the next Section shows, that also measurements by MAT on irradiated samples were equally successful.

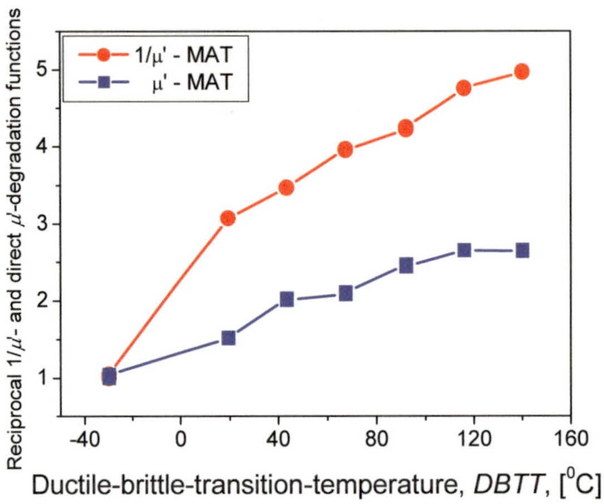

Fig. 27. Optimal reciprocal 1/μ '-degradation function and direct μ '-degradation function of the 15H2MFA nuclear pressure vessel steel embrittled by the special thermal treatment. Values of the DBTT were taken from destructive Charpy impact tests carried out on the same samples.

4.3.3 Steel embrittlement by neutron irradiation

For studying influence of neutron irradiation, block samples with dimensions 10x10x30 mm^3 were prepared from three types of nuclear reactor pressure vessel steel: JRQ is the "western", 15H2MFA is the "eastern" type reactor steel, 10ChMFT is the welding metal for steel 15H2MFA.

The samples were irradiated by neutron fluence in the 1.58×10^{19} - 1.19×10^{20} n/cm^2 range by E>1 MeV energy neutrons. (The value 1.19×10^{20} n/cm^2 corresponds to the neutron fluence, which a pressure vessel coat of a VVER 440 reactor obtains in approximately 20 years.) Always two samples of each material were irradiated with the same fluence, two not-irradiated ones of each material were available as reference samples. For investigation of irradiated samples by MAT technique a specially designed sample holder was built, very similar to that in Fig. 15. The samples were magnetized by a short solenoid, placed around the sample and the magnetic circuit was closed *from one side* of the sample by a passive soft magnetic yoke. A plastic dummy yoke was used from the other side, just to keep the sample fixed in the holder. Application of a *single* yoke made it possible to measure each sample independently four times from all four sides of the block. Thin non-magnetic spacers (0.08 mm) were placed between the sample and the yoke, in order to reduce fluctuation of quality of magnetic contact between the sample surface and the yoke. Measurements were performed in a protected "hot" laboratory, the samples were never touched by naked hand.

The resulting optimal MAT degradation functions are plotted in Fig. 28a for each of the three different materials. The $1/\mu$ '-degradation functions characterized the material modification the best, and very satisfactory correlation was found between the degradation functions and the neutron fluence. The type of correlation is similar for all the investigated materials: the $1/\mu$ '-degradation functions monotonously increase with increasing amount of neutron irradiation. Fig. 28b presents how *DBTT* of the 15H2MFA steel can be determined from the $1/\mu$ '-degradation function of Fig. 28a if a power law formula between the neutron fluence and *DBTT* (corresponding to typical conditions of a VVER 440 reactor) is applied for recalculation. $DBTT_0$ is the ductile-brittle transition temperature of the 15H2MFA steel before irradiation.

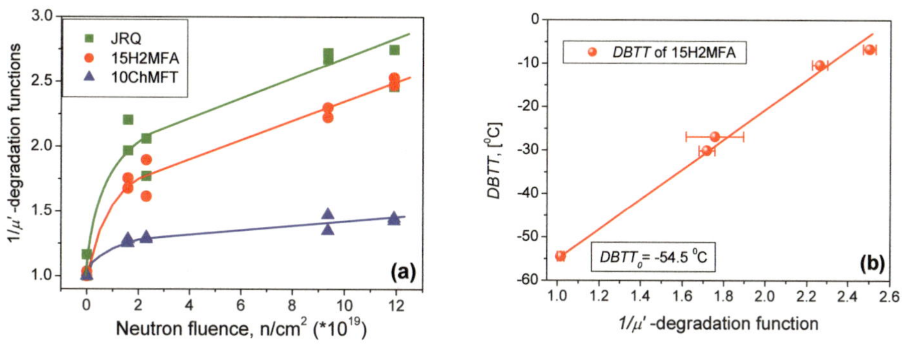

Fig. 28. (a): The optimally chosen $1/\mu$ '-degradation functions for the three investigated steels. (b): *DBTT* of the 15H2MFA steel in dependence on the $1/\mu$ '-degradation function of Fig. 28a.

In conclusion, the applicability of MAT was also demonstrated for inspection of neutron irradiation embrittlement of nuclear reactor pressure vessel steels. It should be noted at the same time, that geometry of the investigated samples was rather unfavorable for any magnetic measurement. For next measurements, and for industrial application in particular, another geometry of the samples is recommended. For instance long thin rods (e.g. 30-50 mm length and 2-3 mm diameter) could be used as surveillance samples for magnetic nondestructive tests. Such samples should be measured *contactless* in a solenoid, and much better results, not complicated by the severe problems with bad surfaces of samples having spent long periods inside a nuclear reactor, can be expected in such case.

4.4 Low cycle fatigue

Here MAT response is described to varied structures of a series of commercial steel CSN 12021 (used for high pressure pipelines), which were modified by application of low cycle fatigue. Twelve large flat steel samples were fatigued by ± cycles with the constant strain value ±0.25% and the constant strain rate of change 0.3%/s. The numbers of kilocycles (kc) applied to different samples were 0, 2, ..., 20, 22 kc. The samples would be destroyed near to 24 kc. After application of the relevant numbers of cycles, magnetically closed samples (rectangles 38x20x5 mm^3 with a hole ∅10 mm in the middle) were cut-out from the fatigued middle parts of the plates. The magnetizing and pick-up coils were wound directly on the rectangular samples through the middle hole. An optimum $1/\mu$-degradation function is presented in Fig. 29.

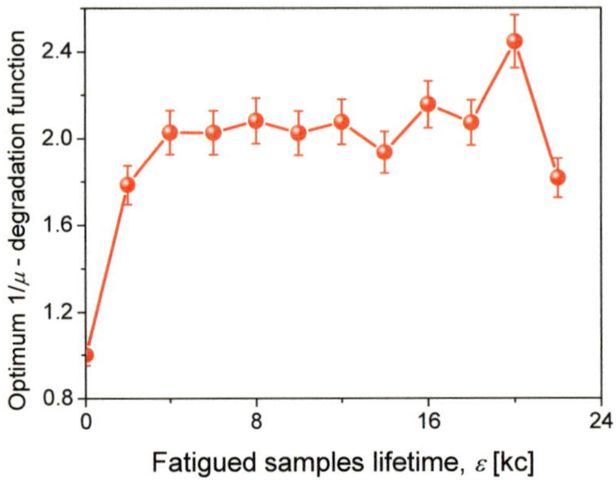

Fig. 29. The most sensitive *1/μ*-degradation function of the low cycle fatigued degraded steel. The error bars show the ±5% range, which corresponds to the scatter of magnetic properties measured on nominally identical samples of the used commercial steel.

Shape of the degradation functions (both of the *1/μ*-ones plotted here and of the *μ*-ones in (Tomáš et al., 2010) confirm the expected profile published earlier in (Lo et al., 2000), but suggesting perhaps more details and certainly more sensitivity. Magnetic reflection of the microstructure reveals radical change in the first 15% of the samples lifetime, corresponding probably to redistribution of dislocations into a volume configuration of areas with higher and lower dislocation densities, which create kind of a "cellular" pattern. In the second phase, up to about 70% of the lifetime, the situation is stabilized, which is manifested by a constant level of the degradation functions: the cellular structure of dislocation bands persists in a dynamic equilibrium with nucleation and annihilation and with some periodical shifting. During the last part of the samples lifetime, microscopical displacements of the material lead to appearance of micro-cracks, mostly at the samples surface. High local stresses at ends of the micro-cracks cause significant changes of the magnetic degradation functions again.

These last part variations of the degradation functions are the most interesting/important from the practical point of view. If indicated in time, they could serve as a warning of the approaching end of lifetime of the sample, or better of an industrial object. Shape of the degradation function in Fig. 29, suggests possible existence of such a warning. The limited number of the individual degraded samples did not allow investigation of this area in as much details as necessary within the described experiment. Behavior of the degradation functions suggests possibly even more substantial variations than in (Lo et al., 2000). The problem certainly deserves more attention and more experiments in future, with focus on the last part of the samples lifetime. Beneficial assistance of nonmagnetic spacers together with use of inspection heads on magnetically open fatigued samples should yield satisfactory results.

4.5 Decarburization of steel surface

If a steel object is exposed to normal air atmosphere at elevated temperature for an appreciable period of time, unwanted softening of the steel surface by decarburization can appear. The surface is modified into a layer of practically pure iron - the ferrite - and fatigue lifetime of such an object is dramatically shortened. Existence and also thickness of the magnetically soft surface decarburized layer can be detected by the classical MAT approach, of course, as it can be found in (Skrbek et al., 2011). It will be shown in this Section, however, that - as suggested by (Perevertov et al., 2011) - in the case of decarburized surfaces a *modified MAT line of attack* is possible with advantage. It is based on the idea that the decarburized surface ferrite layer gets magnetically *saturated* in substantially lower applied fields than rest of the sample, and that value of the saturated magnetic flux of each ferrite layer should be proportional to its thickness. Principles and straightforward results of this MAT attitude will be shortly described here, whereas its details can be found in (Tomáš et al., 2011).

The modified MAT tests were carried out by magnetization of ring samples of carbon spring steel 54SiCr6 ($D/d/w$ = 60/54/3 mm, D - outer diameter, d - inner diameter, w - axial width). All the samples were annealed at 800^{0}C in air atmosphere for T = 0, 1, 4, 8, and 20 hours, respectively. A magnetizing and a pick-up coil were wound on each of the samples. The measured voltage signals of each sample were numerically integrated, and the integral at a minor loop amplitude A_j, measured on a sample annealed in air for the time-period, T, was denoted as the corresponding maximum magnetic flux $\Phi_T(A_j)$. Indicating $\Phi_0(A_j)$ such maximum magnetic flux of the reference sample, differences $(\Phi_T - \Phi_0)(A_j)$ were plotted as functions of the loop amplitudes, A_j, in Fig. 30a.

These reduced magnetic fluxes, $(\Phi_T - \Phi_0)$, of the decarburized surfaces were plotted for each loop with amplitude A_j in dependence on values of the optically measured thickness, t, of the surface ferrite (t was measured independently destructively on auxiliary samples sections). It turned out, that the plots $(\Phi_T - \Phi_0)(t)$ were very linear within surprisingly broad range of A_j. Fig. 30b gives an example of such a linear relation for the minor loop amplitude A_j = 1 kA/m, and similar linear fits were obtained in all the investigated range of the minor loop amplitudes. Fig. 30c shows the correlation coefficient, R, of such linear fits.

In conclusion it can be stated that modified MAT measurement of suitable minor hysteresis loops of ferromagnetic steel objects can serve as a fast, convenient and nondestructive method for determination of the surface decarburization level. Provided, of course, that a reference sample and at least one object of the same material, with an independently measured appreciable thickness of the decarburized layer is available so that the necessary calibration *line* can be prepared. Range of the appropriate minor loop magnetizing amplitudes was shown to be broad, the measurement simple and not requiring any sophisticated equipment.

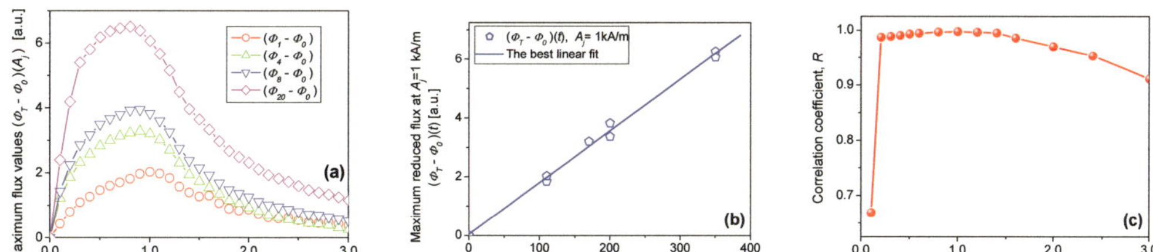

Fig. 30. (a): The maximum magnetic flux values of the samples reduced by the maximum flux value of the reference sample, $(\Phi_T - \Phi_0)(A_j)$, at minor loops with amplitudes A_j. (b): The reduced maximum magnetic flux values, $(\Phi_T - \Phi_0)$, of the air-annealed samples, at the minor loop with amplitude A_j=1 kA/m, as a function of thickness, t, of the surface decarburized ferrite. (c): Correlation coefficient, R, of the linear regression fit with $(\Phi_T - \Phi_0)(t)$, for the used range of the minor loop amplitudes, A_j.

4.6 Welding quality tests

Point welding is a frequently used technology in industry, and checking quality of welding is very important. Welding process produces serious phase transformation in the welded area of the material, which is accompanied by modification of magnetic characteristics. It is shown in this Section how MAT can be applied for inspection of point welding of car-body sheets.

Thickness of the point-welded sheets was 1.2 mm, the material was cold rolled low carbon steel. Two series of welded points were produced by 2.4 bar pressure welding and the welding current was modified from 15.6 up to 21.8 kA. Diameter of the welded points was around 7 mm. A MAT-checked row of welded points is shown at the bottom of Fig. 31a, a destructively tested control series of welded points can be seen at the top of Fig. 31a. Result of the destructive tests confirmed, that welded points made by the welding current between 18.5 and 20 kA were proper. The points with lower current were not welded strongly enough, the points above 20 kA were over-welded (over-welding damaged the material).

MAT measurements were performed on each welded point with a magnetizing yoke. The result of the measurement is shown in Fig. 31b, where the optimally chosen $1/\mu$-degradation function vs. the welding current can be seen. It is worth to mention, that even a small change of the welding current results in a measurable modification of the MAT degradation functions at the beginning of the curve, and this becomes more pronounced in the critical region from 18.5 to 20 kA. All the MAT degradation functions were normalized by the corresponding value of a reference spot (where no welding was taken). More than 100% difference was found between the MAT degradation functions corresponding to the lowest and to the highest energy of welding, which makes possible to characterize quality of the welding quite easily.

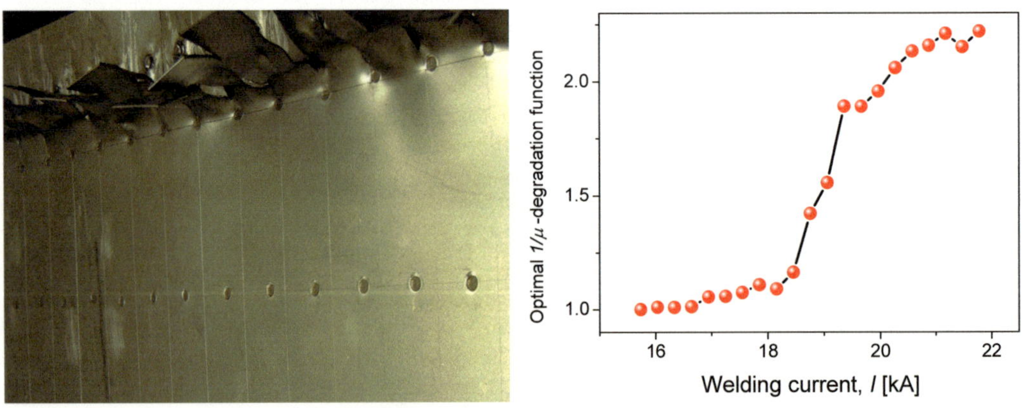

Fig. 31. (a): Two rows of welded points, each produced by a different welding current. The upper row shows results of the destructive test. (b): The optimally chosen $1/\mu$-degradation function vs. the welding current, measured on the bottom row of welded points shown in figure (a).

Choice of the optimal degradation functions is not critical, the top sensitivity area of $1/\mu$-degradation functions in the sensitivity map is very large. For the next possible industrial application also tolerance towards of the results with respect to exact positioning of the magnetizing yoke was investigated. The same welded spot was measured several times after each other, the position of the yoke with respect to the welded spot was modified, and MAT evaluation was performed. The results are seen in Fig. 32, where three characteristic cases are shown. In the left case the welded spot was in the central area of the yoke, in the middle case only half area of the welded spot was under the yoke, and in the right case the welded spot was completely outside. The figure shows how the *1/μ*-degradation functions values were changing compared with the central position of the yoke. Even the most unfavorable case caused 3% difference in the MAT degradation function value only.

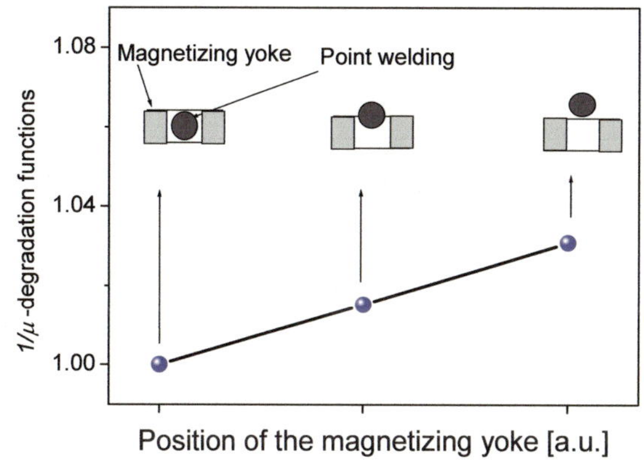

Fig. 32. Variation of the MAT *1/μ*-degradation functions, if the measurement was performed on the same welded spot at three characteristic positions of the yoke with respect to the welded spot, as indicated in the figure.

4.7 Search for flaws

Detection and evaluation of thickness reduction of pipes are very important issues for prediction of lifetime of pipelines in order to avoid severe accidents (Ju, 2007). Recently many nondestructive testing techniques have been used for the measurement of pipe wall thinning, but none of them gives satisfactory result in the case of ferromagnetic carbon steels, especially if the steel is thick or if it is *layered*. In this Section it is shown how Magnetic Adaptive Testing can be applied for inspection of wall thinning in layered ferromagnetic materials. Experiments were performed on a system of ferromagnetic carbon steel plates, one of them containing an artificial slot (see also (Vértesy et al., 2012b)).

Preliminary experiments verified, that samples with different total thickness give different signals from the MAT inspection head. Evidently similar modifications of the signal can be expected if thickness of the sample or of one of the layered samples is modified only locally. It was simulated by three plates (180x170x3 mm^3) made of a carbon steel, with one of them containing a 9x2 mm^2 slot in the middle. Fig. 33 shows the case if all the three sheets are put together. The samples were measured in such a way, that they were magnetized by an inspection head from the top side, while the slot was placed at the bottom. The measurements were performed on both "double" and "triple" configurations (Fig. 34), and the head was attached in both parallel and perpendicular orientations with respect to the axis of the slot. The head was moved on the top sample surface along a direction perpendicular to the axis of the slot. Distance, x, of the center of the head from the middle of the slot was the independent parameter. In this case cross-section of the magnetic yoke of the inspection head was 10.5x16 mm^2, the total outside length 31 mm, and the total outside height of the bow 31 mm. The distance between the legs was 10 mm.

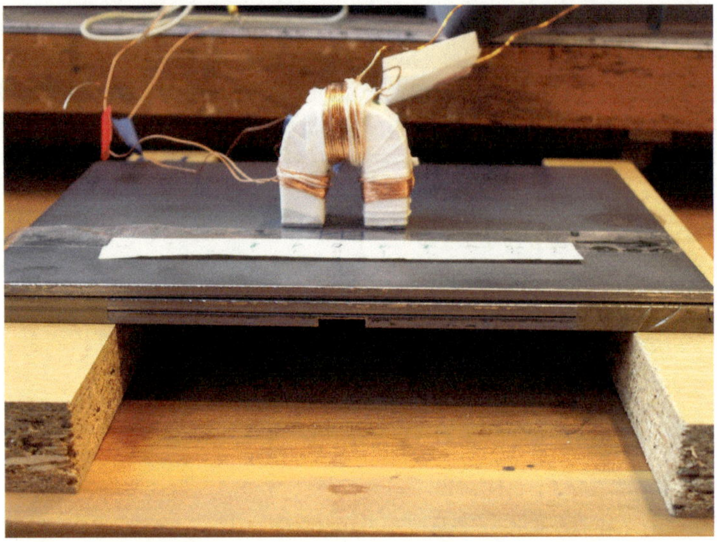

Fig. 33. Measurement of the triple sheet with perpendicular orientation of the yoke.

Results of the measurements, performed on the double sheet system and sketch of the experimental arrangement are shown in Fig. 34a. The yoke was oriented perpendicular to the axis of the slot in this case. The optimally chosen μ-degradation functions were normalized by the value, measured at the largest distance from the center of the slot (x = - 60 mm). A very similar result was obtained if the yoke was oriented parallel with the slot axis and moved similarly as in the former case. The measurement was performed in a triple sheet configuration, too, with the results shown in Fig. 34b. The experiments revealed that Magnetic Adaptive Testing is a promising tool for nondestructive detection of thinning of steel slabs from the other side of the specimen, and that it is able to give good results even for a layered ferromagnetic material.

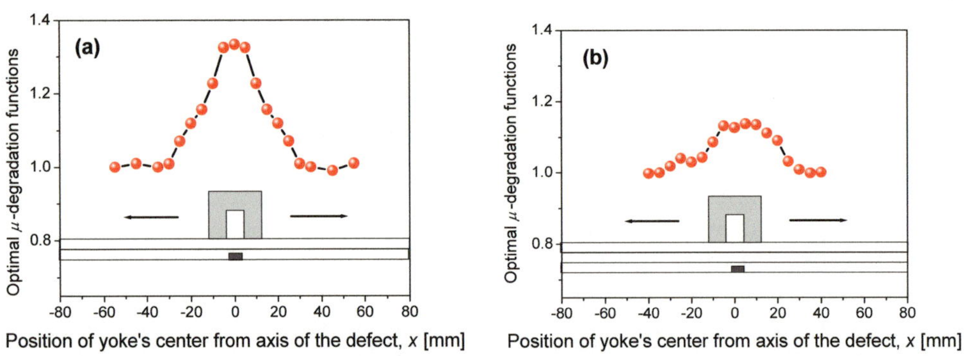

Fig. 34. Optimal MAT degradation functions in dependence on the inspection head position for perpendicular configuration in a (a) double sheet system, (b) triple sheet system.

4.8 Cast iron

4.8.1 Ductile cast iron

Special series of ductile cast iron samples was investigated (see also (Vértesy et al., 2010c)) Staircase-shaped samples were prepared, where different cooling rates resulted in different structures. Correlations were studied of the *nondestructively* measured MAT degradation functions with the *destructively* determined Brinell hardness and also with the electrical conductivity of the samples. Four staircase-shaped block specimens with 5 steps each were prepared. (One of these samples was used here earlier for temperature dependence investigations, see Section 3.6.) Chemical composition of the cast iron samples is shown in Table 1. The amount of ferrosilicon and of commercial pure iron in the melt was controlled, in order to get different carbon equivalents. The targeted carbon equivalent of sample 1 was set to a low value, in order to have cementite in the matrix. The targeted chemical compositions of samples 2, 3 and 4 were the same, but they were given different heat treatments, in order to have different matrices.

Sample	C	Si	Mn	P	S	Mg
1	3.05	2.041	0.166	0.049	0.016	0.024
2	3.58	2.592	0.127	0.065	0.016	0.025
3	3.46	2.575	0.128	0.066	0.015	0.025
4	3.43	2.523	0.130	0.064	0.016	0.026

Table 1. Chemical composition of the ductile cast iron samples (values in wt%).

These staircase-shaped block specimens were used for magnetic measurement: each step was measured. The samples were magnetized by a magnetizing inspection head, attached on the surface. The optimally chosen *B*-degradation functions of the samples are shown in Fig. 35 in dependence on the Brinell hardness and on the electric conductivity. All the steps of all the measured (four) samples are taken into account within this figure. (Fig. 19 in Section 3.6 shows the dependence of a μ-degradation function on the hardness for sample 4.) The degradation function values of different samples were normalized by the sample with the lowest hardness.

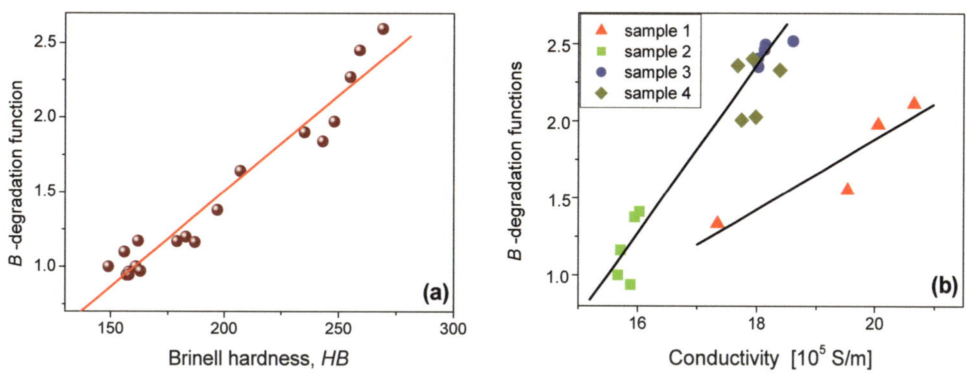

Fig. 35. The most sensitive *B*-degradation functions in dependence on Brinell hardness (a) and on conductivity (b).

The Brinell hardness and the MAT degradation functions correspond to each other very well, linear correlation was found. It should be emphasized that the presented relationship was obtained when *all* the samples (with different chemical compositions and preparation conditions) were considered for the evaluation. (Considering the steps within each single sample, the scatter of points is significantly smaller, as shown in Fig. 19). In any case the reliability and reproducibility of the MAT degradation functions are good, as determined from the sensitivity maps. A good correlation was also found between the MAT degradation functions and conductivity, but in this case the samples with different chemical composition should be handled separately, as can be seen in Fig. 35b. The optimally chosen MAT degradation functions of samples 2, 3 and 4, which have rather similar compositions, lie more-or-less on a straight line as a function of conductivity. This is also valid for sample 1, but with a different slope.

4.8.2 Flake graphite cast iron

Flake graphite cast iron materials with various graphite structures and matrices were prepared and their magnetic properties were measured and evaluated systematically. Correlation of the *nondestructively* measured MAT parameters has been studied with the *destructively* determined Brinell hardness, the pearlite/ferrite ratio and the graphite morphology. Three flake graphite cast iron materials with different chemical compositions listed in Table 2 were prepared. Disks, prepared of these materials were subjected to two kinds of heat treatment: annealing (AN) to obtain a *ferrite-based* matrix and normalization (NR) to obtain a *pearlite-based* matrix. To treat these annealed and normalized disks, they were heated at 1123 K in a furnace for one hour and then either cooled in the furnace or in air for AN and NR, respectively. Altogether 9 flake graphite cast iron materials with various matrices and graphite shapes were thus produced. After grinding the specimen surfaces, their Brinell hardness, *HB,* was measured confirming that the normalizing and annealing treatments were successful in producing the fully ferritic and fully pearlitic matrices, respectively. Metallographic examination also revealed, that the as-cast samples (samples without the heat treatments) had differently mixed pearlite/ferrite matrices.

Material	Chemical composition (mass%)						
	C	Si	Mn	P	S	Cr	Ti
CE4.7	3.77	2.78	0.78	0.025	0.015	0.029	0.015
CE4.1	3.36	2.15	0.69	0.018	0.010	0.014	0.011
CE3.7	3.13	1.66	0.72	0.017	0.020	0.038	0.010

Table 2. Chemical composition of flake graphite cast iron samples (values in wt%).

The samples were magnetized by an attached inspection head for the MAT investigation and degradation functions of all the investigated samples were evaluated. The optimal $1/\mu$-degradation functions are plotted against the Brinell hardness in Fig. 36.

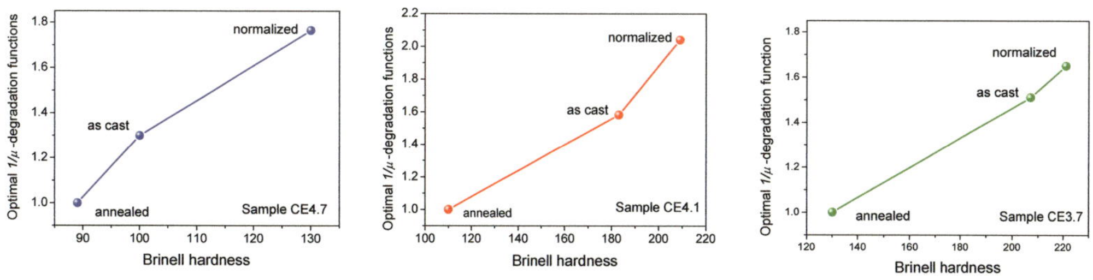

Fig. 36. Optimally chosen *1/μ* -degradation functions of the sample series CE4.7, CE4.1 and CE3.7 vs. Brinell hardness.

The applied annealing resulted in completely ferrite-based matrices (the 0% pearlite/ferrite ratio) and the normalization resulted in completely pearlite-based ones (the 0% pearlite/ferrite ratio). The pearlite/ferrite ratio of the as-cast samples was different for different sample series, and it was determined by metallographic examination. The pearlite/ferrite ratio and the average length of graphite flakes were evaluated using an image processing software. Correlation between the optimal *1/μ*-degradation function and the pearlite/ferrite ratios is given in Fig. 37.

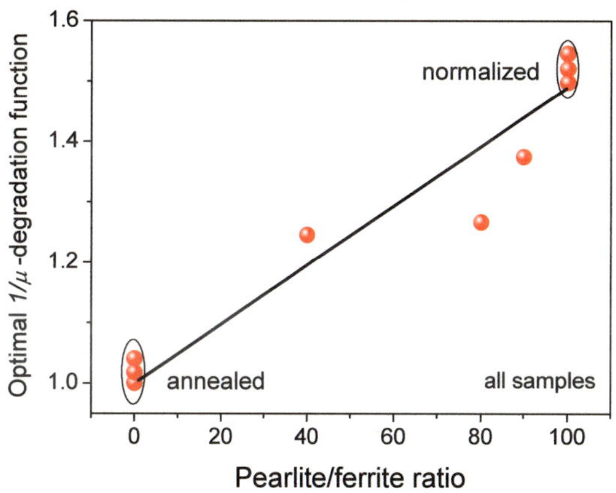

Fig. 37. Optimally chosen *1/μ*-degradation function of all the samples vs. perlite/ferrite ratio. The annealed and the normalized samples are indicated.

Graphite area and graphite length characterize the graphite structure of the investigated samples. MAT degradation functions were evaluated as functions of these quantities for the three as-cast samples, and the corresponding values were compared with each other. Correlation between the magnetic parameters and the graphite area can be seen in Fig. 38a, while correlation between the magnetic parameters and the graphite length is plotted in Fig. 38b.

Concluding the observations, closely linear correlation was found between the MAT degradation functions and the Brinell hardness. It is valid if influence of the heat treatments was investigated within the series of samples with the same chemical composition (as shown in Fig. 36), but also if influence of chemical composition was studied for different as-cast samples, and also if even all the samples were considered within the same graph (these last two cases are not shown here). This confirms the expectation that magnetic hardening correlates with the mechanical hardening of cast iron very well. This is why MAT is also an excellent tool for determining the pearlite/ferrite ratio. Degradation functions of all 0% pearlite/ferrite ratio samples are very close to each other, the difference is not larger than the error of measurement, and the same holds for the normalized samples (100% pearlite/ferrite ratio case). The difference of about 50% in magnitude of the shown $1/\mu$-degradation function is more than enough for determination of variations between the pearlite/ferrite ratios from 0% to 100% (see Fig. 37). The larger scatter in Fig. 37 of the degradation functions evaluated on the as-cast samples can be probably attributed to inaccurate determination of the pearlite/ferrite ratio, which was rather estimated from microphotographs of their structure. MAT was also found to be a useful tool for measurement of parameters of graphite morphology. Very satisfactory linear correlation between optimally chosen MAT degradation functions and both the graphite length and the graphite area was found on the as-cast samples in Fig.38. The correlation between MAT degradation functions and the graphite morphology is evident even if all the samples are considered, however in this case the scatter of points is larger.

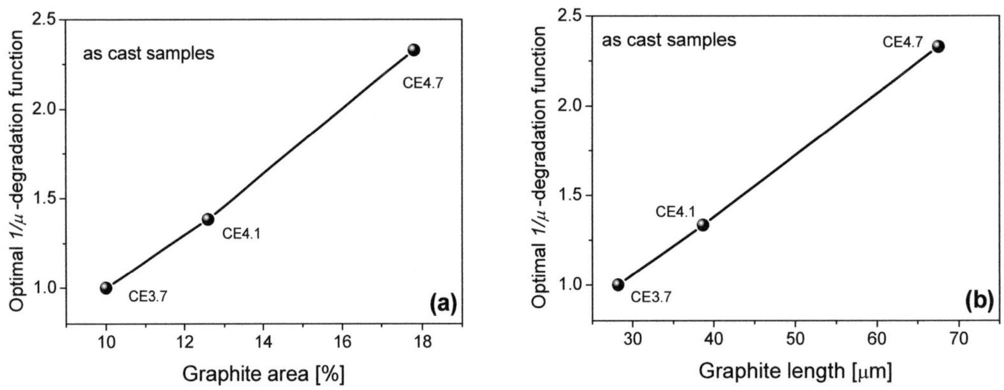

Fig. 38. Optimally chosen $1/\mu$ -degradation functions of the as-cast samples vs. the graphite area (a) and vs. the graphite length (b).

5. Conclusions

An original magnetic nondestructive method of Magnetic Adaptive Testing is presented in this chapter as a sensitive option to other ways of investigation of modification of properties of ferromagnetic construction materials. Physical basis, principles and ways of application of the method are explained in the first and second Sections, suggestions with respect to its efficient use are given in Section 3, and Section 4 is dedicated to several examples of measurements, where MAT was utilized, usually with very satisfactory results.

MAT belongs to the family of nondestructive *magnetic hysteresis* material testing methods, which generally investigate structural modifications of *bulk* of the tested objects, and it is shown to *surpass sensitivity* of the traditional hysteresis methods mainly due to its specific selection from the recorded large data-pool of that part of information, which is *optimal with respect to the investigated material and to the way of its degradation.*

The most advantageous features of Magnetic Adaptive Testing were demonstrated through many successful measurements on many different materials, which underwent different types of degradation (fatigue, annealing, plastic deformation, neutron irradiation, etc.). MAT generally offers significantly larger sensitivity - with respect to any degradation parameter - than any of other magnetic nondestructive and probably even of any existing destructive method. A very important, advantageous feature of the MAT technique is that it *does not require magnetic saturation* of investigated samples. The most appropriate MAT descriptors of each type of degradation are chosen from a big data-pool, and this multi-parametric choice frequently makes possible to characterize the material degradation in a more complex way, than other methods. It was also shown that the optimally chosen degradation functions are reliable and repeatable. Magnetic Adaptive Testing is a comparative measurement, it does not offer any absolute magnetic quantities. Inspection of any unknown sample is based on a previous "teaching" process.

The most serious drawback of this technique is that in great majority of practical applications the MAT inspection is required to operate on magnetically open industrial objects, where shape of the object (through the demagnetization factor) and condition of its surface can introduce uncertainty into the measurement. However, it was shown that qualitatively the same correlation exists between the MAT degradation functions and the degradation variables both for open and for closed magnetic circuits (Section 3.1, and/or 3.2), and also that the surface problem can be treated rather successfully (Section 3.5).

Main directions promising to contribute to further development of the MAT application are probably a next improvement of MAT inspection heads for contact measurement on surfaces of industrially interesting objects, namely in finding a satisfactory elimination of fluctuation of quality of magnetic contact between the inspection head and surface of the object, which is frequently uneven, rough and deteriorated. An important step in this direction was achieved by application of non-magnetic spacers, which was described in Section 3.5 and successfully applied in several latest measurements. Perhaps a *simple* measurement of a parameter proportional to the real value of the magnetizing field inside the object magnetized from its surface could do the trick.

As for the most promising next industrial application of MAT, the nondestructive surveillance program for irradiated steel of nuclear pressure vessel reactor steel (Section 4.3.3) should be pointed out, and quantitative measurement of steel surface decarburization (Section 4.5). Probably also fatigue tests of periodically loaded objects could give very well applicable results, which, however, still require broad next experimental study, substantially more detailed than the first attempt mentioned in Section 4.4.

6. Acknowledgements

The authors if this chapter would like to express their gratitude to all the colleagues, who were co-authors of the quoted papers and also to those who took part at preparation of the presented experiments. Special thanks aim at the members of the Universal Network for Magnetic Nondestructive Evaluation, the annual workshops of which made a very fruitful and supporting platform for discussions and inspiration. The work was financially supported by the Czech Institute of Physics, v.v.i. of ASCR (project AVOZ 10100520) and by the Hungarian Research Institute for Technical Physics and Materials Science of HAS, by the

Hungarian-Czech and by the Hungarian-Japanese Bilateral Intergovernmental S&T Cooperation projects, by the project No.101/09/1323 of the Grant Agency of the Czech Republic, and by the Hungarian Scientific Research Fund (projects T-035264, K-62466 and A08 CK 80173).

7. References

Bertotti G. (1998) *Hysteresis in Magnetism*, Academic Press, San Diego

Blitz J. (1991) *Electrical and Magnetic Methods of Nondestructive Testing*, Adam Hilger IOP Publishing, Ltd, Bristol

Devine M.K. (1992) Magnetic detection of material properties (1992) *J. Min. Met. Mater. (JOM)*, pp. 24-30

Devine M.K., Kaminski D.A., Sipahi L.B. Jiles D.C. (1992) Detection of fatigue in structural steels by magnetic property measurements, Journal of Materials Engineering and Performance, Vol. 1, pp. 249-253

Dobmann G., Altpeter I., Kopp M., Rabung M., Hubschen G. (2008) ND-materials characterization of neutron induced embrittlement in German nuclear reactor pressure vessel material by micromagnetic NDT techniques, *Electromagnetic Nondestructive Evaluation (XI)*, p.54, IOS Press, ISBN 978-1-58603-896-0

Gillemot F., Pirfo Barroso S. (2010) Possibilities and difficulties of the NDE evaluation of irradiation degradation, *Proceedings of 8th International Conference on Barkhausen Noise and Micromagnetic Testing (ICBM8)* (ISBN 978-952-67247-2-0)

Jiles D.C. (1988) Review of magnetic methods for nondestructive evaluation, *NDT International* Vol. 21, p. 311

Jiles D.C. (2001) Magnetic methods in nondestructive testing, In: *Encyclopedia of Materials Science and Technology*, K.H.J. Buschow et al., Eds, p.6021, Elsevier Press, Oxford

Ju, Y. (2007) Remote measurement of pipe wall thinning by microwaves, *Advanced Nondestructive Evaluation II*, Volume 2 pp. 1128-1133

Lo C.C.H., Tang F., Biner S.B., Jiles D.C. (2000) Effects of fatigue-induced changes in microstructure and stress on domain structure and magnetic properties of Fe-C alloys, *J. Appl. Phys.*, Vol. 87, p. 6520

Mayergoyz I.D. (1991) *Mathematical Models of Hysteresis*, Springer-Verlag, Berlin PHARE PH2.01/95 Project "Reactor Pressure Vessel" 1995

Preisach F. (1935) Uber die magnetische Nachwirkung, *Zeit. fur Physik* Vol. 94 p. 277

Perevertov O. (2009) Increase of precision of surface magnetic field measurements by magnetic shielding, *Meas. Sci. Technol.*, Vol. 20, p. 055107

Perevertov O., Stupakov O., Tomáš I., Skrbek B. (2011) Detection of spring steel surface decarburatization by magnetic hysteresis measurements, *NDT&E International* Vol. 44 pp. 490–494

Skrbek B., Tomáš I., Kadlecová J., Ganev N. (2011) NDT Characterization of Decarburization of Steel after Long-Time Annealing, *Kovove Materialy – Metallic Materials*, Vol. 49, No.6, pp.401-409

Stupakov O. (2006) Investigation of applicability of extrapolation method for sample field determination in single-yoke measuring setup, *J. Magn. Magn. Mater.* Vol. 307 pp. 279-287

Stupakov O., Tomáš I., Kadlecová J. (2006) Optimization of single-yoke magnetic testing by surface fields measurement, *J.Phys. D* Vol. 39, pp. 248-254

Stupakov O., Tomáš I. (2006) Hysteresis minor loop analysis of plastically deformed low-carbon steel, *NDT&E Int.* Vol. 39 pp. 554-561

Takahashi S., Kobayashi S., Kikuchi H., Kamada Y. (2006) Relationship between mechanical and magnetic properties in cold rolled low carbon steel, *J. Appl. Phys.*, Vol. 100, p.113908

Takahashi S., Kikuchi H., Ara K., Ebine N., Kamada Y., Kobayashi S., Suzuki M. (2006) In situ magnetic measurements under neutron radiation in Fe metal and low carbon steel, *J. Appl. Phys.* Vol. 100, p.023902

Tomáš I. (2004) Non-Destructive Magnetic Adaptive Testing of Ferromagnetic Materials, *J. Magn .Magn. Mater.* Vol. 268/1-2, pp. 178-185

Tomáš I., Stupakov O., Kadlecová J., Perevertov O. (2006) Magnetic Adaptive Testing - low magnetization, high sensitivity assessment of material modifications, *J. Magn. Magn. Mater.* Vol. 304 pp.168-171

Tomáš I., Vértesy G., Kobayashi S., Kadlecová J., Stupakov O. (2009a) Low-carbon steel samples deformed by cold rolling - analysis by the magnetic adaptive testing, *J. Magn. Magn. Mater.*, Vol. 321 pp. 2670-2676

Tomáš I., Vértesy G., Kadlecová J. (2009b) Influence of rate of change of magnetization processes on sensitivity of Magnetic Adaptive Testing, *J. Magn. Magn. Mater.* Vol. 321 pp. 1019-1024

Tomáš I., Vértesy G., Kadlecová J. (2009c) Influence of rate of change of magnetization processes on sensitivity of Magnetic Adaptive Testing, *J.Magn. Magn. Mater.* Vol. 321 pp. 1019-1024

Tomáš I., Kadlecová J., Vértesy G., Skrbek B. (2010) Investigation of Structural Modifications in Ferromagnetic Materials by Magnetic Adaptive Testing, *Proceedings of 8th International Conference on Barkhausen Noise and Micromagnetic Testing (ICBM8)*, pp.153-162 (ISBN 978-952-67247-2-0)

Tomáš I., Kadlecová J., Konop R., Dvořáková M. (2011) Magnetic nondestructive indication of varied brittleness of 15Ch2MFA steel, *Proceedings of 9th International Conference on Barkhausen Noise and Micromagnetic Testing (ICBM9)* ISBN 978-952-67247-4-4 (paperback), ISBN 978-952-67247-5-1 (CD-ROM), pp. 55-63

Tomáš I., Kadlecová J., Vértesy G. (2012) Measurement of flat samples with rough surfaces by Magnetic Adaptive Testing, *IEEE Trans. Magn.*, in press

Ušák E. (2010) A new approach to the evaluation of magnetic parameters for non-destructive inspection of steel degradation, *Journal of Electrical Engineering*, Vol. 61. pp. 100-103

Vandenbossche L. (2009) Magnetic Hysteretic Characterization of Ferromagnetic Materials with Objectives towards Non-Destructive Evaluation of Material Degradation, Gent University, Gent, Belgium, PhD Thesis

Vértesy G., Tomáš I., Mészáros I. (2007) Nondestructive indication of plastic deformation of cold-rolled stainless steel by magnetic adaptive testing, *J. Magn. Magn. Mater.* Vol. 310 pp. 76-82

Vértesy G., Tomáš I., Takahashi S., Kobayashi S., Kamada Y., Kikuchi H. (2008a) Inspection of steel degradation by Magnetic Adaptive Testing, NDT & E INTERNATIONAL, Vol. 41. pp. 252-257

Vértesy G., Tomáš I., Kobayashi S., Kamada Y. (2008b) Investigation of thermally aged samples by Magnetic Adaptive Testing, *Journal of Electrical Engineering,* Vol. 59. pp.82-85

Vértesy G., Tomáš I., Takahashi S., Kobayashi S, Kamada Y, Kikuchi H., (2008c) Inspection of steel degradation by Magnetic Adaptive Testing, NDT & E INTERNATIONAL, Vol. 41. pp. 252-257

Vértesy G., Uchimoto T. Tomáš I., Takagi T. (2010a) Temperature dependence of magnetic descriptors of Magnetic Adaptive Testing, *IEEE Trans..Magn.,* Vol. 46. pp. 509-512

Vértesy G., Uchimoto T. Tomáš I., Takagi T. (2010b) Nondestructive characterization of ductile cast iron by Magnetic Adaptive Testing, *J. Magn. Magn. Mater.* Vol. 322 pp. 3117-3121

Vértesy G., Uchimoto T., Takagi T., Tomáš I. (2010c) Nondestructive inspection of ductile cast iron by measurement of minor magnetic hysteresis loops, *Materials Science Forum,* Vol. 659. pp. 355-360

Vértesy G., Ueda S., Uchimoto T., Takagi T., Tomáš I., Vértesy Z. (2011) Evaluation of Plastic Deformation in Steels by Magnetic Hysteresis Measurements, In: *Electromagnetic Nondestructive Evaluation (XIV),* T. Chady et. al., Eds., pp. 371-378, IOS Press, Amsterdam

Vértesy G., Tomáš I. (2012a) Complex characterization of degradation of ferromagnetic materials by Magnetic Adaptive Testing, *IEEE Trans. Magn.,* in press

Vértesy G., Tomáš I. Uchimoto T., Takagi T. (2012b) Nondestructive investigation of wall thinning in layered ferromagnetic material by magnetic adaptive testing, NDT & E INTERNATIONAL, Vol 47. pp. 51-55

Part 3

Concrete Nondestructive Testing Methods

8

Elastic Waves on Large Concrete Surfaces for Assessment of Deterioration and Repair Efficiency

D. G. Aggelis[1], H. K. Chai[2] and T. Shiotani[3]
[1]*Materials Science & Engineering Department, University of Ioannina,*
[2]*Department of Civil Engineering, Faculty of Engineering, University of Malaya*
[3]*Graduate School of Engineering, Kyoto University*
[1]*Greece*
[2]*Malaysia*
[3]*Japan*

1. Introduction

Most of society's infrastructure supporting certain sectors of human activity is based on cementitious materials. Bridges, highways, water intake facilities and other structures are made of concrete. These structures sustain external function loads, own weight, as well as deterioration by temperature cycles and attack of environmental agents during their useful life span. The number of civil infrastructures built more than 50 years ago may be estimated to several hundreds of thousands worldwide (Chai et al. 2010). The operational efficiency of these structures is of primary importance for economic reasons but mostly for human safety. Concrete structures have ceased to be considered maintenance-free. They should be inspected in regular intervals, their damage level should be evaluated and when necessary, repair action should be applied. Maintenance or repair projects should be based on prioritization as to the importance of the structure and its damage status. Therefore, economic, fast and reliable characterization schemes are highly demanded. In most of the currently available methods of assessment, elastic modulus of material and strength characteristics are the primary criteria of consideration. In order for the information to be obtained, it often becomes necessary that mechanical tests are conducted on samples acquired from the target structure, through extraction of cores or other similar exercises, which turn out to be inflicting further damage to the already-degraded target structure. In addition, considering the nature of these exercises, which are usually adopted at selected locations, the assessment results are local and often not representative for the overall structure. Recently, extensive research has been reported in elastic wave-related techniques resulting in reliable assessment of the structural condition of actual concrete materials and structures. The characterization of concrete structures by Non Destructive Testing (NDT) and specifically stress wave methods produces mainly qualitative but very important results concerning damage. The surface layer of concrete suffers the most of environmental degradation as well as maximum stresses particularly by flexural loading. It is reasonable therefore, that considerable deterioration initiates from the surface. The deterioration is often observed in the form of small-scale cracking or macroscopic surface breaking cracks which can propagate to the interior and accelerate the deterioration of the whole structure, especially if the embedded metal reinforcement is exposed to various environmental agents. According to the results, proper repair action can be taken and therefore, the functionality of a structure can be extended. This is extremely important bearing in mind that the cost of building new infrastructure is much higher compared to a focused repair project aiming at extending the service life of the structure for decades.

The most widely used feature in elastic wave NDT is pulse velocity, which is correlated with the damage degree. Despite its rough results, pulse velocity monitoring of large concrete structures is of great significance since it offers a general estimation of the condition and enables proper repair action. Empirical correlations between pulse velocity and damage or strength have been exploited for a long time (Jones 1953, Kaplan 1959, Anderson and Seals 1981, Popovics 2001). It is accepted that pulse velocity, even though not extremely sensitive to damage (reduced prior to final failure by a percentage of not more than 20%), is very indicative of the internal conditions of the material (Van Hauwaert et al. 1998, Shiotani and Aggelis 2009). It possesses essential characteristics making it the most widely accepted and used parameter in concrete NDT. One of these advantages is its direct relation to the elastic constants through well known elasticity relations (Sansalone & Streett 1997, Naik et al. 2004). This way, elastic waves measurements enable the estimation of the modulus of elasticity and consequently projection to the strength of the material through empirical relations (Kaplan 1960, Keating et al. 1989, Shah et al. 2000 , Monteiro et al. 1993, Philippidis & Aggelis 2003).

Another essential feature of pulse velocity is its independence to the propagation distance. Pulse velocity is firmly fixed to the elastic constants and nominally its value does not exhibit changes with distance like other parameters, e.g. amplitude and frequency which are continuously decaying due to attenuation (Landis and Shah 1995, Owino and Jacobs 1999, Jacobs & Owino 2000). In addition to the well known use and interpretation of pulse velocity, frequency and dispersion features have recently been studied with the aim of more accurate characterization concerning the damage content and characteristic size (Aggelis & Philippidis 2004, Philippidis & Aggelis 2005, Punurai et al 2006, Aggelis & Shiotani 2007a, In et al. 2009, Chaix et al. 2006). Dispersion originates from inhomogeneity and hence the velocity dependence on frequency should be much more evident in damaged than in healthy media. Concrete is inhomogeneous by nature due to porosity and aggregates and exhibits moderate dispersive trends (Philippidis & Aggelis 2005, Punurai et al. 2006). Nevertheless, cracks due to size and severe impedance mis-match with the matrix material, are stronger scatterers of elastic waves and influence wave propagation more effectively than the inherent inhomogeneity of the material.

As aforementioned, concrete structures suffer deterioration in the form of distributed micro-cracking as well as macroscopic surface breaking cracks. Cracking allows water and other chemical agents to penetrate into the material, further deteriorating the structure, oxidizing the reinforcement and finally compromising its load bearing capacity (Ohtsu & Tomoda 2008). It is desirable to mitigate these deteriorating actions or even strengthen the structure by repair. Repair can take place in the form of cement injection either targeted to seal specific macroscopic cracks (Aggelis & Shiotani 2007b, Thanoon et al. 2005, Issa & Debs 2007, Yokota & Takeuchi 2004), as well as a measure to reinforce the whole surface applying injection in a pattern of boreholes on the surface of the structure (Shiotani et al. 2009, Shiotani & Aggelis 2007). Concerning specific macro-cracks, their depth can be evaluated by ultrasound either by the travel time of the longitudinal wave refracted from the tip of the crack, or the amplitude of the surface Rayleigh wave that survives beneath the crack (Doyle & Scala 1978, Hevin et al. 1998, Liu et al. 2001, Pecorari 2001, Song et al. 2003, Zerwer et al. 2005, Aggelis et al. 2009). The same measurements can be repeated after application of grouting agent in order to estimate the efficiency of repair. However, although this can be done for specific surface breaking cracks it would be not be practical for a large concrete surface that normally has numerous cracks. Therefore, although specific cracks may be indicatively targeted for detailed measurements, it is also desirable to establish a way to evaluate the general conditions of a large area of the structure both before and after repair. The relative change of wave parameter values before and after maintenance actions offers a deterministic measure of repair efficiency (Kase & Ross 2003, Aggelis & Shiotani 2007b, Shiotani et al. 2009, Aggelis et al. 2009) which otherwise should be based solely on empirical criteria. This way the effectiveness of the repair can be quantified by the change of wave features, as elastic waves propagate in long distances through the material and gather information from different parts. In the present chapter, distributed micro-cracking on large concrete surfaces is examined by elastic waves before and after repair. Results are presented from an actual old and deteriorated structure, as well as a large concrete block with simulated defects inside. It is exhibited that complementary use of Rayleigh and longitudinal waves is appropriate for the evaluation of initial condition and repair efficiency, while combined use of wave velocity and major frequency, as well as their dependence (dispersion) enhances the evaluation of repair. This kind of dispersion effects are attributed to inhomogeneity (Tsinopoulos et al. 2000) and are diminished after repair. Additionally, development in computer and electronic engineering enables tomography reconstruction of the interior characterized by the value of the propagation velocity of either longitudinal or Rayleigh waves, which can be related directly to the mechanical properties of the material (Sassa 1988, Kepler et al. 2000, Kobayashi et al. 2007, Shiotani et al. 2009, Aggelis et al. 2011). For the full field measurement of a large concrete surface, the newly developed Rayleigh wave tomography methodology is discussed showing the potential to successfully map the subsurface defects, exploiting the connection between penetration depth and applied wavelength (Chai et al. 2010). The chapter comprehensively reviews the current practice in structural health monitoring of real structures using elastic waves, in addition to discussing new trends and features for more detailed assessment of the damage condition.

2. Longitudinal and Rayleigh waves

Wave propagation is strongly affected by the elastic properties and density of constituent materials. Longitudinal wave velocity is given in Eq. (1):

$$C_P = \sqrt{\frac{E}{\rho}\frac{(1-v)}{(1+v)(1-2v)}} \tag{1}$$

where E is the elasticity modulus, ρ is density, and v is the Poisson's ratio.

It has been correlated with strength (Kaplan 1959, Kaplan 1960, Kheder 1999, Qasrawi 2000, Naik et al. 2004) and damage (Van Hauwaert et al. 1998, Ono 1998, Mikulic et al. 1999, Aggelis & Shiotani 2008a, Shiotani & Aggelis 2009) of concrete materials, offering rough but valuable estimations because the damage condition influences the mechanical properties and, hence, the wave speed. Employing a number of sensors, the velocity structure of the material can be constructed and the internal condition can be visualized, highlighting the existence of voids or cracks (Sassa 1988, Kepler et al. 2000, Kobayashi et 2007, Aggelis & Shiotani 2007b, Shiotani et al. 2009). Furthermore, the use of Rayleigh (or surface) waves seems quite suitable for surface opening cracks investigation, since they propagate along the surface of the structure. Additionally, they occupy higher percentage of energy than the other types of waves. For example, as mentioned in (Graff 1975) a point source in a homogeneous half space radiates 67% of its energy in the form of Rayleigh waves, while only 7% in compressional ones. Moreover, since they are essentially two-dimensional, their energy does not disperse as rapidly as the energy associated with three-dimensional dilatational and shear waves. Specifically, their amplitude is inversely proportional to the square root of propagation distance while for a longitudinal wave the amplitude is inversely proportional to the distance (Owino & Jacobs 1999). This makes them more easily detectable than other kinds of waves, as will also be discussed in the present text. The distinct difference from longitudinal waves concerns the particle motion which is elliptical with the vertical component greater than the horizontal, as opposed to the unidirectional oscillation of the particles in longitudinal waves (Graff 1975). The Rayleigh motion decreases exponentially in amplitude away from the surface (Jian et al. 2006). Practically, the penetration depth of these waves is considered to be similar to their wavelength (Sansalone and Streett 1997, Jian et al. 2006). The velocity of surface Rayleigh waves, C_R, which will be also discussed in this chapter, is given by (2):

$$C_R = \frac{(0.87 + 1.12v)}{1+v}\sqrt{\frac{E}{2\rho(1+v)}} \tag{2}$$

From equations (1) and (2) it is derived that by measuring both longitudinal and Rayleigh velocities, apart from the Young's modulus, the Poisson's ratio can be calculated as well.

3. Experimental part

3.1 Repair

On the surface of the old concrete structure (dam) numerous cracks were observed, as is typical due to long exposure to freezing-thawing cycles, and water attack. Repair was applied by cement injection in three different ways. First, cement injection was conducted from the opening of the thicker cracks using syringes, see Fig. 1. As to thin cracks, they were repaired by surface application of cement. Finally, cement was injected in a pattern of boreholes on the surface using a constant pressure, as described in (Shiotani & Aggelis 2007). The actual result is that empty pockets in the structure created by cracks or extensive porosity were filled with cementitious material. This material is initially liquid in order to penetrate to the thin crack openings, but due its cementitious nature, hydration reaction transforms it into a stiff inclusion, sealing the crack sides. It results in considerable decrease of permeability which is crucial for water intake facilities and restores some of the load bearing capacity since voids are replaced by a stiff material reinforcing the structures cross section. Concerning initial crack depth evaluation, some of the cracks can be interrogated by longitudinal or Rayleigh waves (Doyle et al. 1978, Liu et al. 2001, Tsutsumi et al. 2005, Aggelis et al. 2009). However, given the large number of cracks this is not practical. A measure of the quality of the entire surface must be obtained in a time-effective manner.

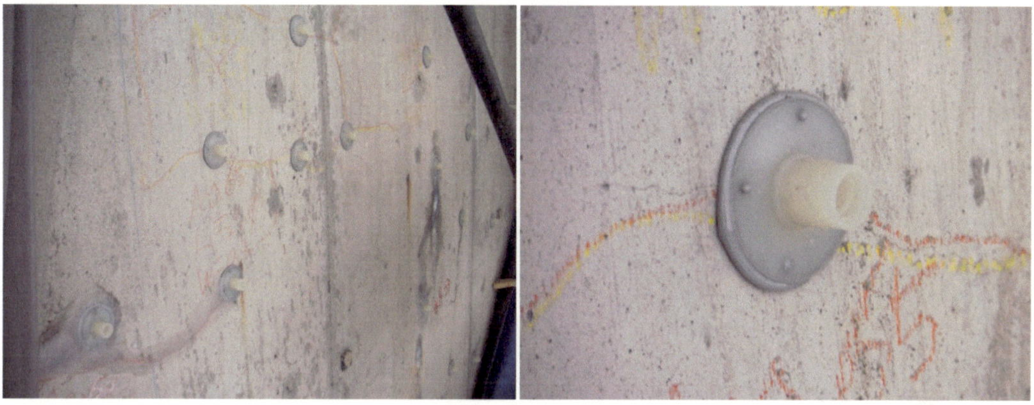

Fig. 1. (a) Multiple locations of injection on the surface, (b) detail of the injection point.

3.2 Wave measurements

In order to estimate the condition of the material elastic wave measurements were applied before and after repair. In the specific case, three vertical arrays of four sensors were attached using a rectangular pattern to record the surface response, as seen in Fig. 2. All three dimensions of the concrete block (dam pier) were much larger than the monitored area and the wavelength, and therefore, no influence was expected to either longitudinal or Rayleigh waves. Excitation was conducted by a steel ball (35 mm in diameter) resulting in a frequency peak of approximately 10 kHz (see Fig. 3(a)) and a longitudinal wavelength of approximately 400 mm while the Rayleigh wavelength is of the order of 200 mm. The excitation was consecutively conducted near each sensor (which acted as a trigger) and the rest acted as receivers. Therefore, several paths (all possible combinations between two individual sensors) with different lengths were examined. The horizontal spacing was 1.2 m, the vertical 1.5 m, while the longest paths corresponding to the largest diagonals of the whole monitored area were 5.1 m long. The sensors were acoustic emission transducers, namely Physical Acoustics, PAC, R6 sensitive at frequencies below 100 kHz. The acquisition system was a 16 channel PAC, Mistras operated on a sampling frequency of 1 MHz. The sampling time of 1 μs, resulted in an error lower than 0.3%, since even for the shortest distances of 1.2 m, the transit time was approximately 300 μs. The sensors were attached on the surface using electron wax.

Pulse velocity was measured by the time delay of the first detectable disturbance of the waveform. As to the Rayleigh velocity, the reference point used for the measurement was the first peak of the Rayleigh burst that stands much higher than the initial, weaker longitudinal arrivals (Sansalone & Streett 1997, Qixian & Bungey 1996, Aggelis & Shiotani 2007a, 2008b) (see Fig. 3(b)).

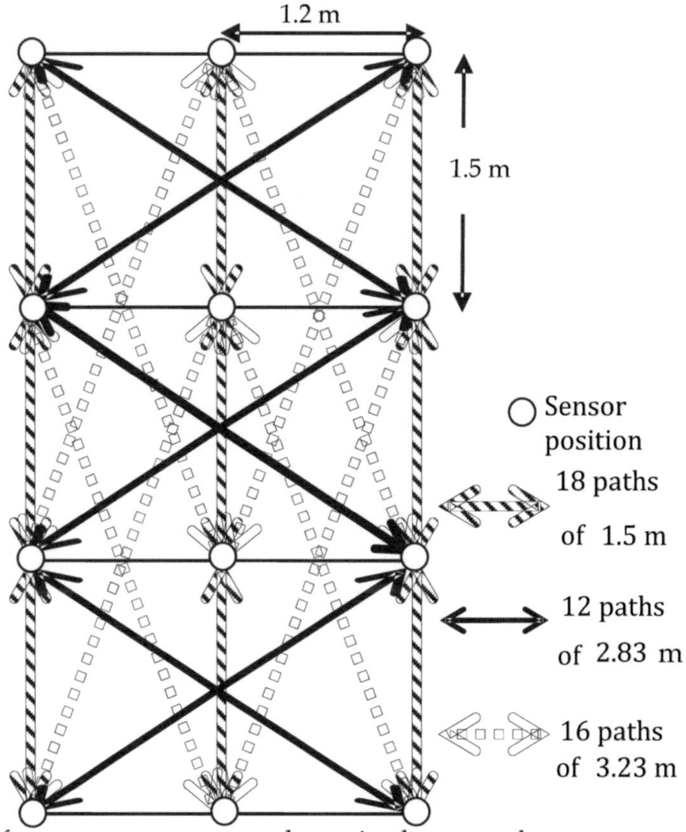

Fig. 2. Pattern of wave measurements and examined wave paths on concrete surface.

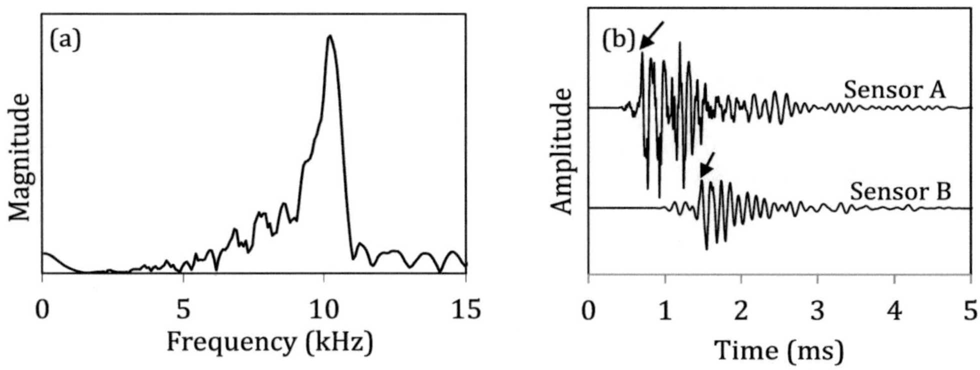

Fig. 3. (a) Frequency spectrum of the excitation, (b) Typical waveforms for sensors at different distances from the excitation.

4. Results

4.1 Wave velocity

Figure 4 depicts the longitudinal velocity as a function of the propagation distance. It is clear that a strong dependence on the distance exists. As already mentioned, nominally wave velocity depends on the elastic constant and not on the travel path. However, there are several reasons that impose this phenomenon. One is the typical crack density compared to the length of the wave path. The possibilities that a short path is crack-free, are higher than a longer one. Therefore it is reasonable that the highest velocities are exhibited at short quarters. Additionally, attenuation effects are accumulated reducing the signal amplitude for longer propagation. This hinders the correct identification of the waveform's leading edge for signals collected at large distances and thus the velocity is underestimated. This can be considered as an indirect effect of damage on the velocity, through the difficulty on interpretation that the attenuation imposes.

The average velocity measured for the short paths of 1.2 m before repair averages at 4500 m/s, while for the longest paths of 5.1 m it is below 4000 m/s. After repair and allowing for a period of two weeks for the injection cement to hydrate properly, the curve is elevated by about 250 m/s, as seen in Fig. 4. This corresponds to an increase of more than 5%, from 4228 m/s to 4455 m/s.

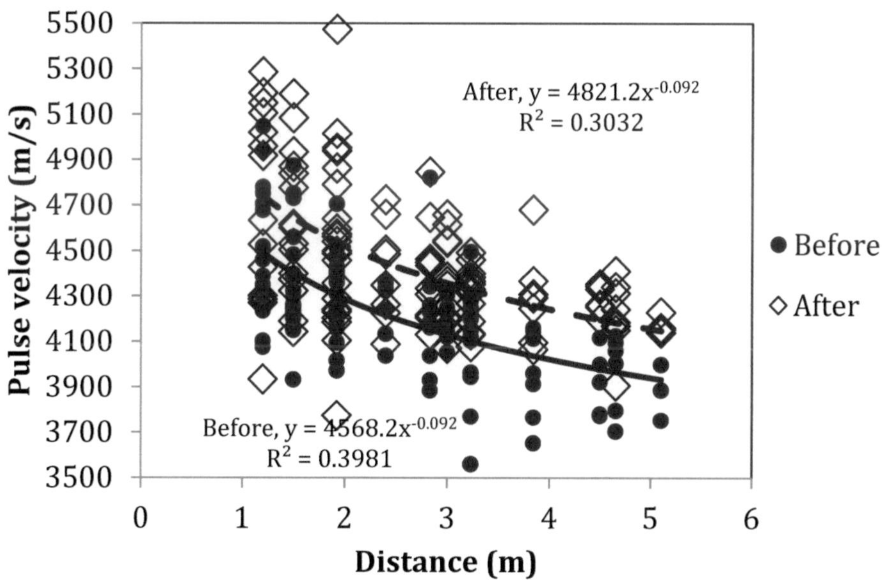

Fig. 4. Pulse velocity vs. propagation distance for different stages of repair.

As to Rayleigh waves, the dependence on the distance is much weaker both before and after repair, as seen in Fig. 5. The reason is connected to the Rayleigh wave measurement, since it uses a strong reference peak (see Fig. 3(b)), much higher than the noise level. The Rayleigh velocity averages at 2235 m/s before and 2363 m/s after repair, being increased by more than 5%, similarly to the longitudinal velocity.

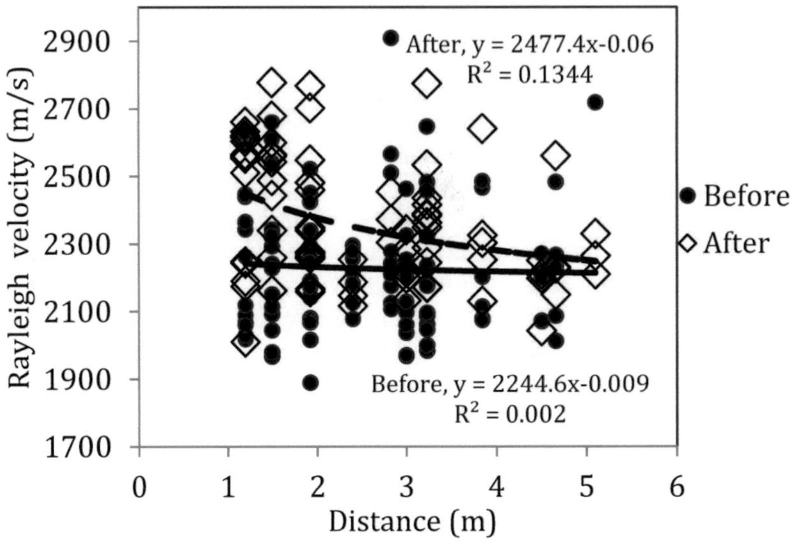

Fig. 5. Rayleigh velocity vs. propagation distance for different stages of repair.

It is mentioned that the average increase is the result of the velocity change for the individual travel paths which can differ considerably from point to point. Fig. 6 shows the velocity change for all individual measurement positions. The large majority of points showed an increase, while a smaller population exhibited decrease. Therefore, while the overall increase is considered satisfactory, examination of the change point by point enables further investigation of specific points that exhibited velocity decrease unexpectedly.

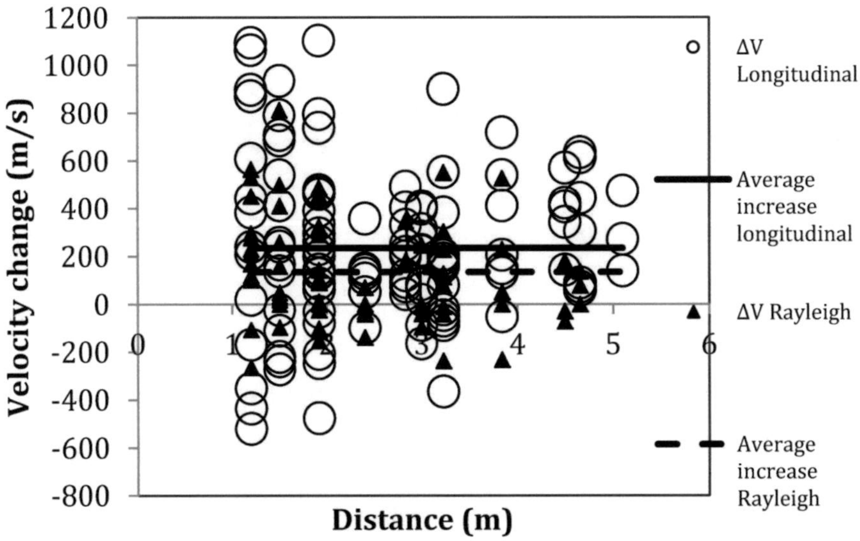

Fig. 6. Difference of velocity between before and after repair for all measurement points.

The average increase of longitudinal and Rayleigh velocities can be translated to elastic modulus improvement by Eqs. (1) and (2). Before repair the effective elastic modulus is calculated at the value of 34.5 GPa, while after repair it is calculated at 37.9 GPa, corresponding to an improvement of approximately 10%. This shows the efficiency of the repair action. It is possible that after full hydration of the injected cement, this increase would be even higher. However, one point needs attention. This is the environmental temperature that influences the hydration rate. In cold environments hydration is delayed and thus the material is gaining rigidity with a slow rate. It has been seen that elastic wave measurements immediately after injection and before the injection material hardens can show the opposite of the desired result (Aggelis & Shiotani 2009, Shiotani et al. 2009). This is due to scattering on the soft pockets of grout which lowers the velocity for a limited time period. Although this phenomenon is normal, and has been recently explained, it contradicts the common knowledge that velocity should always rise after repair. Therefore, it could be misleading and therefore, it is suggested that measurements for repair evaluation are conducted after a sufficient period of time for the cementitious repair agent to cure.

4.2 Pulse frequency

As mentioned in the experimental part, the impact excited a waveform with major frequency components below 20 kHz. As any pulse propagates through inhomogeneous and possibly damaged concrete, the higher of its frequency components are influenced more severely. This change of spectrum can be well monitored by the "central frequency", C (Aggelis & Philippidis 2004, Shiotani and Aggelis 2009). In the specific case the central frequency is calculated as the centroid of the FFT of the waveforms up to 40 kHz:

$$C = \frac{\int_0^{40} fM(f)df}{\int_0^{40} M(f)df} \tag{2}$$

where f is the frequency, and $M(f)$ the magnitude of the FFT.

In Fig. 7, the dependence of central frequency on the distance is depicted. Despite the experimental scatter of the points which is expected due to the inhomogeneity of the material, a certain decreasing trend is exhibited both before and after repair. This demonstrates the cumulative effect of the inhomogeneity of the travel path on the frequency content.The average frequency before repair is 12.5 kHz for short propagation distance while it is reduced to 6.8 kHz for the longest paths of 5.1 m. The trend can be fitted quite well with a decaying exponential curve, and is attributed to the stronger attenuation of higher frequencies (Tsinopoulos et al. 2000, Shiotani & Aggelis 2009). Measurements on the same points after repair showed a frequency increase of approximately 2 kHz for any distance, as seen again in Fig. 7. This can be attributed to the filling of the cracks which reduces the material scattering attenuation. It should be mentioned that the frequency is not directly correlated with concrete strength or structural integrity. However, in any case, strength can be estimated only through empirical relations even for pulse velocity. The importance lies on the comparison between the two stages (before and after repair). Since a frequency upgrade was evident this shows that there was a certain improvement on the structure's surface layer due to void elimination.

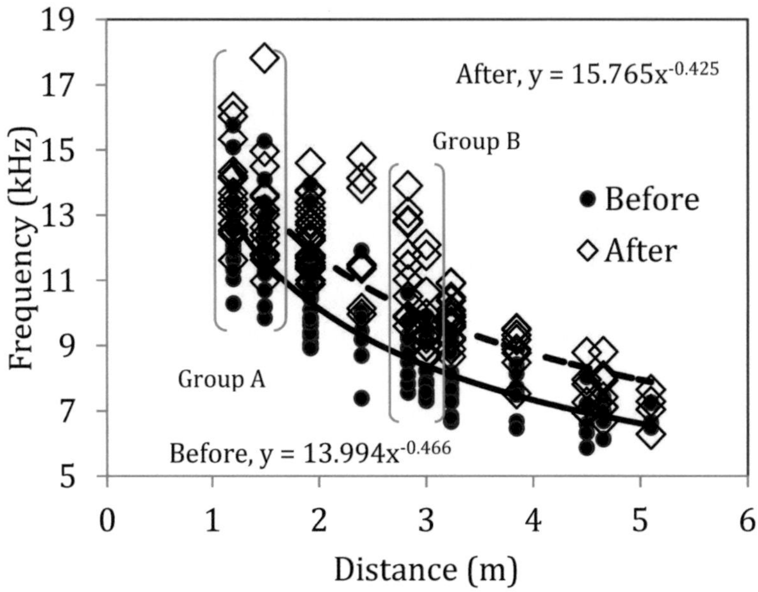

Fig. 7. Central pulse frequency vs. propagation distance for different stages of repair.

4.3 Reliability aspects

At this point it is worth to mention the importance of reliability of the signal acquisition. Strong attenuation due to inhomogeneity and spreading due to long distances decrease the signal level very effectively. This is crucial especially for the pulse velocity measurement, which is conducted by the first detectable disturbance of the received waveform. In case the signal is weak, the first cycle may be of the same level or even lower than the noise level and therefore, the velocity is underestimated. This is a phenomenon that is not widely considered in practical applications. The strength of the signal compared to the noise is measured by the "signal to noise ratio", S/N, which is simply the ratio of the waveforms peak amplitude divided by the average noise level of each waveform. The noise level is calculated by the amplitude of the "pre-trigger"period, the period after each sensor starts acquisition and before the actual wave arrives at the sensor point. In order to see the effect of noise on the measurement of wave parameters, the dependence of pulse velocity, Rayleigh velocity and central frequency are plotted vs. S/N in Fig. 8(a), (b) and (c) respectively. Concerning Fig. 8(a) before repair, an increasing trend of velocity can be found with increase of S/N. It is obvious that for S/N lower than 1000, velocities average around 4000 m/s, while for higher S/N the velocities reach 4500 m/s in average. After repair, as already discussed, the "cloud" of points is elevated to higher velocity (due to elastic modulus restoration) and higher S/N values due to the decrease of scattering attenuation. At the same time, the correlation between velocity and S/N is sufficiently weakened (from

R^2 of 0.26 to 0.07). It can be concluded that when S/N is low (e.g. less than 1000) the velocities are certainly underestimated, while the measurement becomes almost independent for S/N ratios higher than 3000. This is an aspect that should always be considered in field measurements, especially for long distance measurements.

As aforementioned the Rayleigh velocity is calculated by possibly the strongest peak of the waveform (the first peak of the Rayleigh burst) and therefore, it does not crucially depend on the S/N ratio. This is shown in Fig. 8b where the correlation coefficient between velocity and S/N is close to zero. Similar results with pulse velocity with even higher correlation coefficients can be seen in Fig. 8c for the central frequency of the pulse. Before repair there is a very clear increasing trend exhibiting also a quite high correlation coefficient R^2 of 0.61. This shows that when the signal is strongly attenuated and its strength decreases with respect to noise, the higher frequencies are the first to be influenced. The increasing trend holds for the measurements after repair, though in this case the correlation is much weaker.

The discussion of S/N is concluded with its dependence on distance. In any case it is normal that the S/N ratio decreases with distance in any material, since the effect of attenuation is accumulated for longer travel paths. However, it is remarkable to see that the S/N undergoes a change from approximately 6000 for short wave paths (1.2 m) to less than 500 for longer (5.1 m) as depicted in Fig. 9. This change of more than 12 times indicates that the data should be divided in smaller groups in order to exclude the effect of attenuation, when discussing elasticity modulus or dispersion effects, as will be analyzed in the next section. Additionally, after repair, it is evident that the S/N increased and the correlation to distance was certainly reduced from 0.59 to 0.25. It is also characteristic that for the shortest distance (1.2 m) the S/N is quite similar for both conditions (before and after repair) with values 5900 and 6600 respectively. Larger differences are observed for longer paths (>4 m) where the effect of attenuation is accumulated, i.e. less than 400 before and 1250 after repair.

4.4 Dispersion relation

Both longitudinal velocity and frequency decrease exponentially with distance; thus they are correlated with each other, as seen in Fig. 10. The clusters consist of 130 points which is the total number of possible paths between the sensors. Due to inhomogeneity and locality effects there is a certain experimental scatter. However, it is clear that both before and after repair there is a positive correlation showing a certain dependence of velocity on frequency. Nevertheless the average velocity and frequency both increase after repair. Apart from the average longitudinal velocity increase of more than 5%, central frequency increases from 9.6 kHz to 11.1 (change of 15.6%). This kind of simultaneous examination of different features can enhance repair characterization since frequency increase seems to be more sensitive to repair than velocity of longitudinal and Rayleigh waves.

It is understood that the correlations in Fig. 10 include the influence of distance since the measurements are taken for any different distance between 1.2 m and 5.1 m. In order to examine the frequency effect on the propagation velocity excluding the influence of attenuation, the information was processed in small groups of data collected over close distances. Two cases will be indicatively discussed. One group (group A of Fig. 7) includes the data collected at the shortest distances of 1.2 m and 1.5 m (totally 34 points), while the other (group B of Fig. 7) includes the approximately double distances of 2.83 m and 3 m

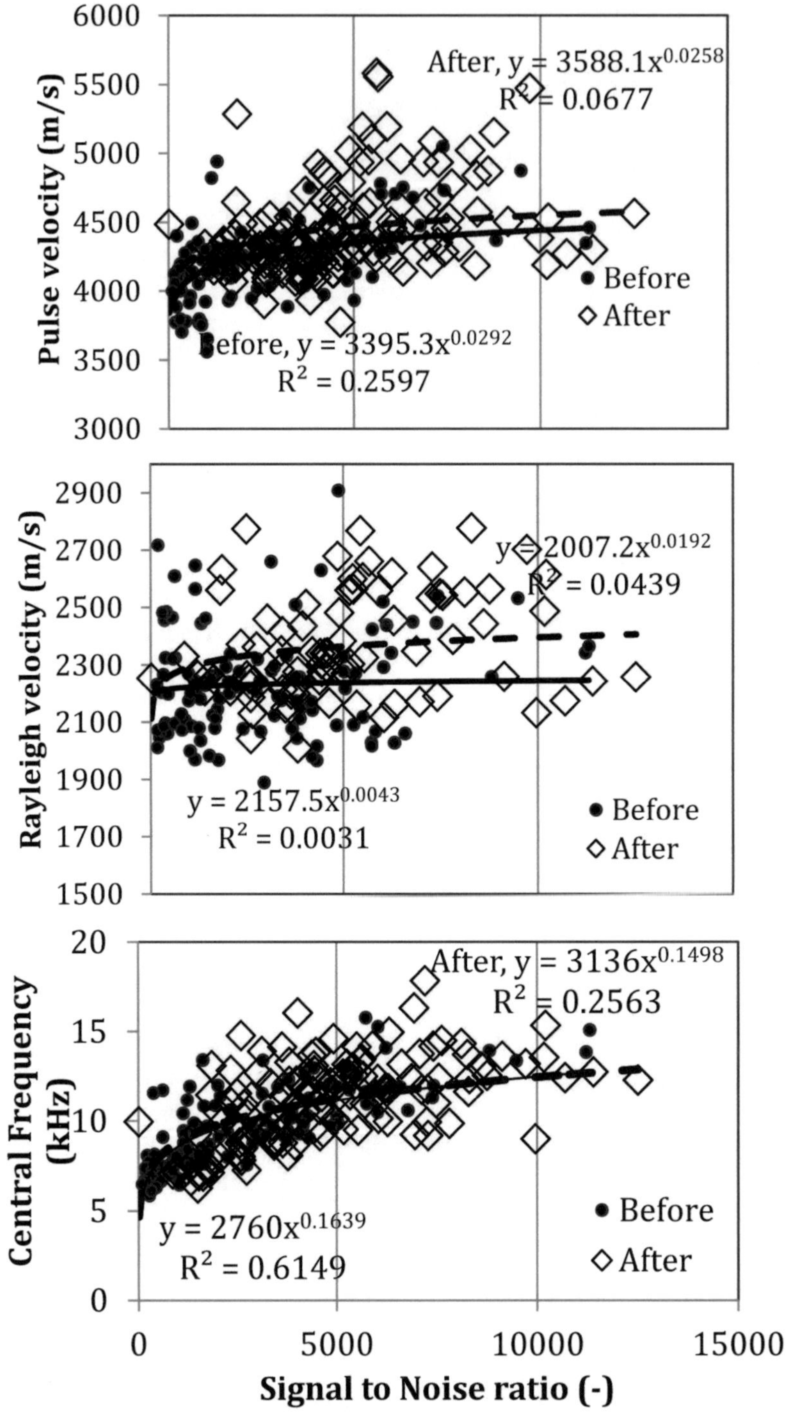

Fig. 8. Wave parameters vs. signal to noise ratio (a) pulse velocity, (b) Rayleigh velocity, (c) central frequency.

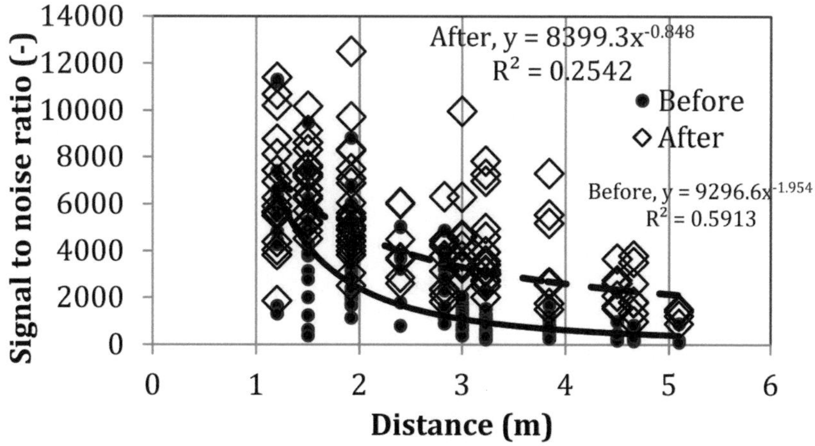

Fig. 9. Signal to noise ratio vs. propagation distance for different repair stages.

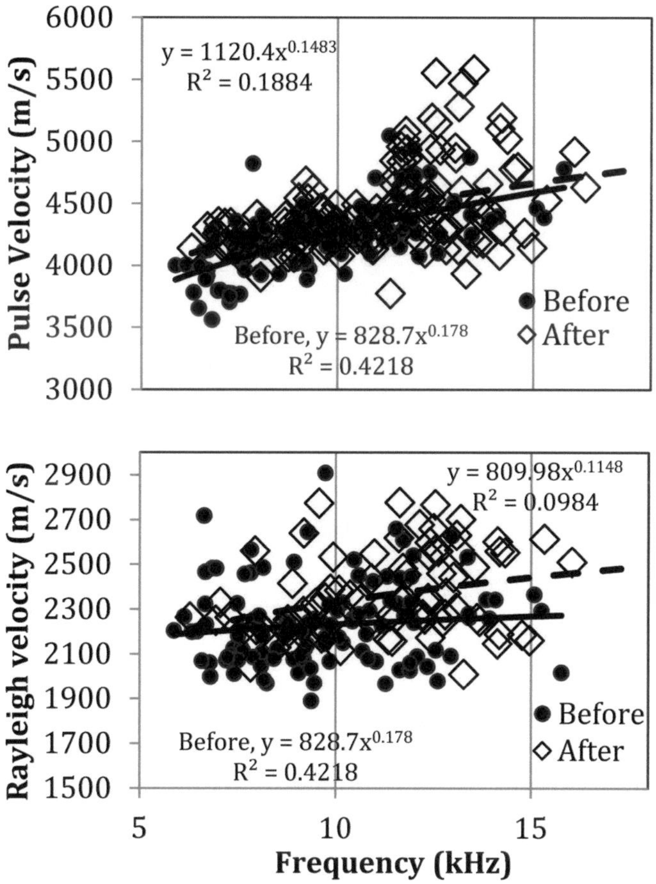

Fig. 10. Pulse (a) and Rayleigh (b) velocity vs. frequency for the total population of measurements.

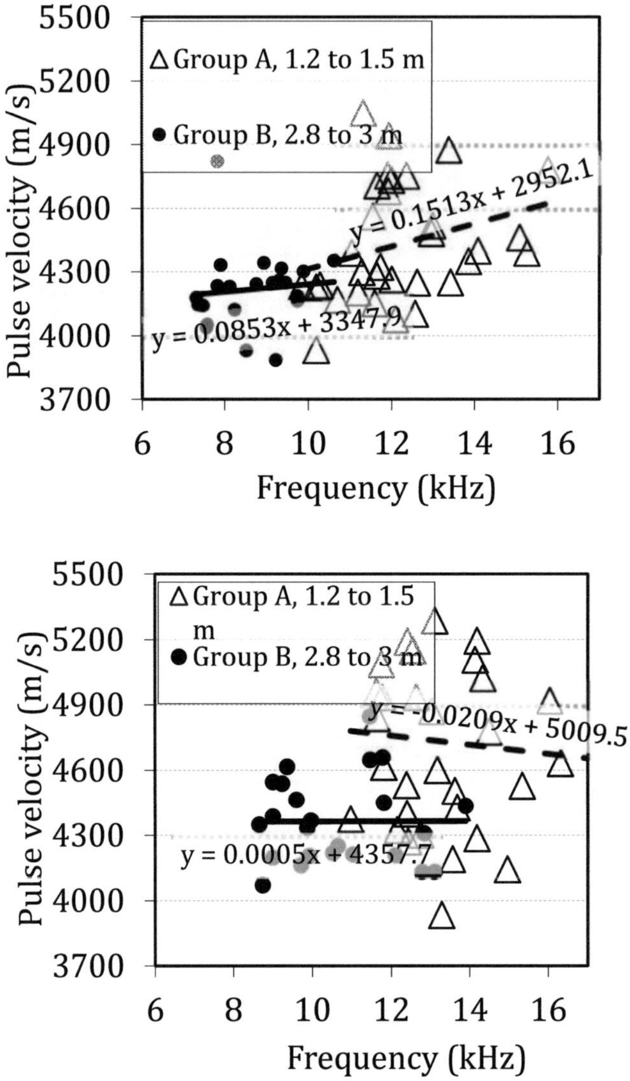

Fig. 11. Pulse velocity vs. Frequency for different length of wave paths (a) before and (b) after repair.

(totally 24 points). For each group, the velocity vs. frequency correlation is depicted in Figs. 11a and b. Since the data in each group are taken from similar distances, the effect of attenuation is eliminated allowing examination of pure dispersion effects within each group. The experimental scatter is high since each travel path in damaged concrete is unique due to the randomness of inhomogeneity. The total population shows a trend that should not be ignored; there is a positive correlation between velocity and frequency before repair. This implies that an increase in frequency content results in a slight increase of pulse velocity. This trend is in agreement with recent dispersion studies on concrete (Chaix et al. 2006, Philippidis & Aggelis 2005, Aggelis and Shiotani 2007, 2008, Aggelis 2009). They reveal that below 100 kHz the increasing slope of the dispersion curve is steep, while for frequencies around or above 100 kHz the curve reaches a plateau. This dispersion is stronger as the inhomogeneity of the material increases and has been measured for wide frequency bands up to almost 1 MHz (Philippidis & Aggelis 2005, Chaix et al. 2006). In general in all typical particulate composites or porous materials (Kinra & Rousseau 1987, Sayers & Dahlin 1993, Aggelis et al. 2004), phase velocity is lower than the velocity of the matrix for low frequencies, but rises with frequency until reaching approximately the velocity of the matrix. For higher frequencies the velocity does not exhibit serious alterations according to the above mentioned studies.

It is interesting to examine the same correlation after repair. Apart from the general translation of the points to higher velocity and frequency levels for any propagation distance, the slight positive correlation is eliminated for both groups. Cement injection filled cavities and cracks eliminating inhomogeneity to a large extend. Thus, the structure behaved in a less dispersive way. This repair-dependent dispersion should not be ignored as it shows the potential to enhance the rough characterization performed so far in concrete. It is mentioned that all individual distance groups of data exhibited the velocity-frequency dependence. Fig. 11 contains the groups with the largest population. Same behavior concerning the frequency dependence of pulse velocity was observed for through the thickness measurements of elastic waves in concrete pier before and after repair. A certain dispersive trend was exhibited initially, while after the repair material was properly hydrated into the structure, this trend was weakened (Aggelis et al. 2011).

The dependence of Rayleigh velocity on distance and frequency is much weaker than in the case of longitudinal as was seen in Figs. 5 and 10. One possible explanation is the long wavelength compared to the depth of surface-opening cracks. Some surface cracks with wide openings (i.e. 0.2 - 0.4 mm) were targeted using longitudinal and Rayleigh waves for depth measurement (Aggelis et al. 2009). The depth of cracks never exceeded 100 mm while an average value for the depth of the cracks was 35 mm. Rayleigh waves propagating through a depth of more than 200 mm (similar to their wavelength) can "dive" below the shallow defects. Therefore, their dispersion in this case is weaker, since the highest degree of damage is concentrated near the surface, while the propagation of Rayleigh takes place also through deeper, more homogeneous zones. On the other hand, longitudinal waves traveling along the surface are influenced by each surface crack as it poses a discontinuity on their path. This makes their dependence on frequency stronger, as was seen in Figs. 4 and 10. Dispersion should be further studied for material characterization since in a general case, limited dispersion of Rayleigh should imply that defects are shallower than the Rayleigh wavelength offering a means of fast estimation for a large surface.

5. Rayleigh wave tomography

Analysis of elastic wave data certainly offers insight on the structural condition of the material, as was shown above by the changes in wave properties and dispersion. However, one way to further enhance the understanding of the results is visualization. This is one global goal concerning the inspection. Techniques like X-rays, radar, thermography and elastic waves aim to produce "tomograms" of the structure (Aggelis et al. 2011, Diamanti et al. 2008, Ito et al. 2001, Ohtsu and Alver 2009) in order for the results to be more easily understood by a general audience (owner of structure or user). For the specific case of near surface damage, the establishment of Rayleigh wave-based tomography technique is beneficial since it materializes single-side assessment, which is particularly useful for certain structures with open access only on one surface, such as tunnels and bridge deck panels. Also using the tomography technique, examination of in-situ structures becomes more realistic because it provides general indications on problematic spots within a relatively short time of execution. These spots could then be assessed by a pin-point approach in a detailed manner to characterize the defect.

5.1 Analytical study of Rayleigh waves propagating in concrete with horizontal crack

To clarify the behaviour of Rayleigh waves impinged by a horizontal crack lying within the concrete sub-surface, a series of wave motion simulations have been conducted. The simulations were conducted using commercial software developed for solving two-dimensional elastic wave propagation based on finite difference method. As illustrated in Fig. 12, the analytical model was generally composed of the concrete medium with four sensors located on the top side of the model at a uniform distance of 100 mm. A point source for generating elastic waves was located 50 mm away from the trigger sensor to the left. The trigger sensor initiate simultaneous recording of waveforms once the incoming of wave was detected. Sensors R1, R2 and R3 were placed to record waveforms at different locations as the generated waves propagate further from the source. In the simulations, the concrete was modeled with the first and second lame constants of 10 GPa and 15 GPa, respectively, and density of 2300 kg/ m^3. Damping of elastic waves in concrete was appropriately defined to simulate waveforms reasonably close to those observed in the experimental measurement. The configuration resulted in a corresponding longitudinal wave velocity of approximately 4200 m/s, a typical value for homogeneous concrete of normal strength. The left, right and bottom sides of the concrete model were configured with infinite boundary conditions in order to avoid reflections and simulate a larger structure geometry. Beside the model of homogeneous concrete, models with an additional 150 x 2 mm void placed parallel to the concrete surface at varying depths from the top side of concrete were also prepared to constitute delamination in concrete.

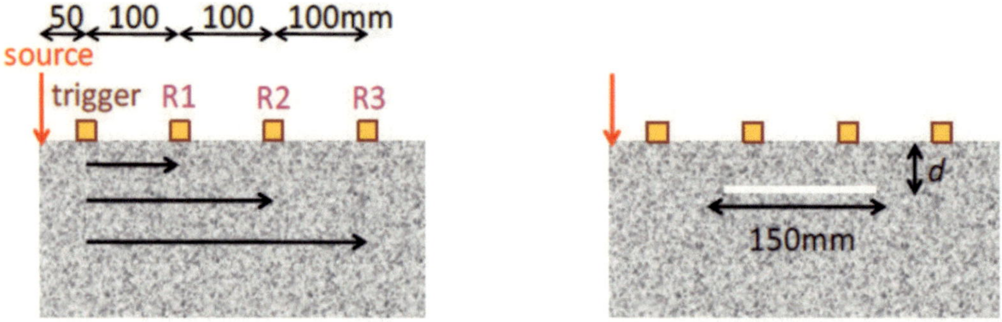

(a) (b)

Fig. 12. Wave motion simulation models, (a) Homogeneous concrete and (b) Inhomogeneity in the form of a horizontal crack

A single-cycle excitation of elastic waves from the point source was generated for all the analytical models. To study the relation between wavelength of Rayleigh waves and depth of delamination that causes distortion, excitations at different frequencies were performed. Hence the simulations produced waveform data for different combinations of excitation frequency and crack depth to serve further analysis purposes.

Fig. 13 presents snapshots of typical simulated cases of elastic waves propagation for the homogeneous model. Knowing that velocity of Rayleigh waves is approximately 57% that of the primary waves for a normal strength concrete (Sansalone and Streett 1997), in the case for 150 kHz excitation (Fig. 13(a)), in which the wavelength was relatively shorter, the separation of Rayleigh wave energy component (with the larger amplitudes) from the primary waves was evident from sensor R1 onwards. The separation became greater as the waves propagated further. For low frequency excitations, namely the 10 kHz one, the separation could not be clearly seen at propagation distance equivalent to that observed for the 150 kHz excitation, because the wavelengths of both wave modes are quite long and therefore overlap at the early distances of propagation before the different propagation velocities separate them (Fig. 13(b)). The separation of Rayleigh waves from longitudinal waves can be justified from the recorded waveforms for the homogeneous concrete model, indicating clearer separation of one full cycle for the 150 kHz excitation when reaching sensor R1. For the waveforms obtained from 10 kHz excitation, even though no complete separation can be observed, Rayleigh waves can be extracted without much difficulty since their arrival was marked by significantly large energy that initiated abrupt increase of amplitude (Qixian & Bungey 1996).

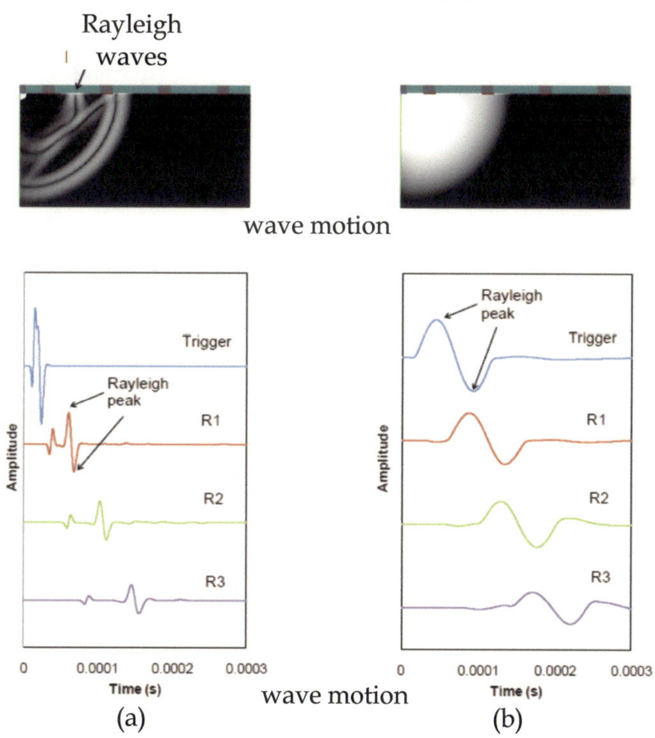

Fig. 13. Snapshots of wave motion and waveform of homogeneous concrete for (a) 150 kHz and (b) 10 kHz excitation frequencies

For the cracked case, when the elastic waves impinged on the crack tip of the delamination, as indicated by snapshots given in Fig. 14, wave diffraction and scattering occur, with some portion of wave energy being "reflected" upwards from the crack-concrete interface to eventually converge with the ones propagating on the surface. Distorted waveforms recorded by sensor R1 indicated a more apparent decrease compared to those from the receivers of the homogeneous concrete model.

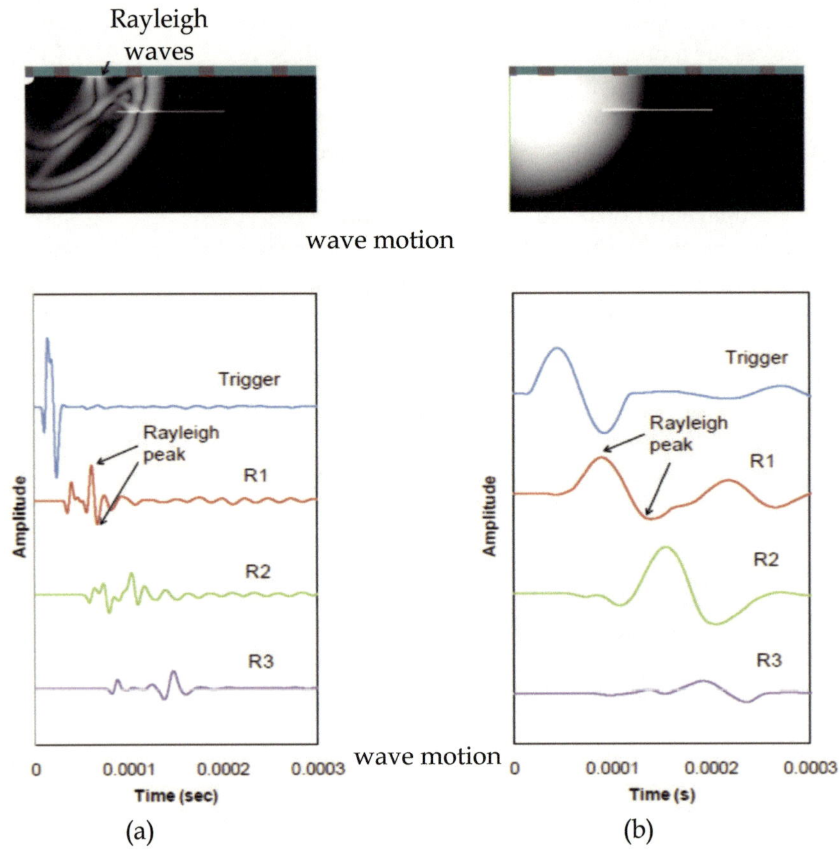

Fig. 14. Snapshots of wave motion and waveforms of concrete model with a horizontal crack located at 50 mm from top for (a) 150 kHz and (b) 10 kHz excitation frequencies

5.2 Processing for phase velocity

Since the distortion of waves due to the defect could reduce the energy of Rayleigh waves relatively to that of the body waves, which was bound to hamper accurate analysis, the characteristic amplitudes of Rayleigh waves was extracted. The extraction process was carried out simply by "muting" the other wave components so that only the one-cycle Rayleigh waves would remain. The relevant wave components of could be recognized as the maximum amplitudes in both the positive and the negative phases. The processed waveform then yielded a more clear-cut frequency response by the fast Fourier transform (FFT), indicating a characteristic peak frequency belonging to the propagating Rayleigh waves. Observation revealed that the processed waveform for the first sensor have the peak frequency almost equivalent to the frequency of excitation as configured in the simulation, suggesting that this frequency value could be regarded as the characteristic frequency of the generated Rayleigh waves.

Figure 15 plots the travel time of Rayleigh waves with excitation frequency of 20 kHz against distance of propagation for the homogeneous concrete model. The travel time was obtained by taking the time of positive peak arrival as acquired by the receivers R1, R2 and R3 with reference to the trigger. The gradient of linear regression could be inversed to yield a propagation velocity of 2398 m/s.

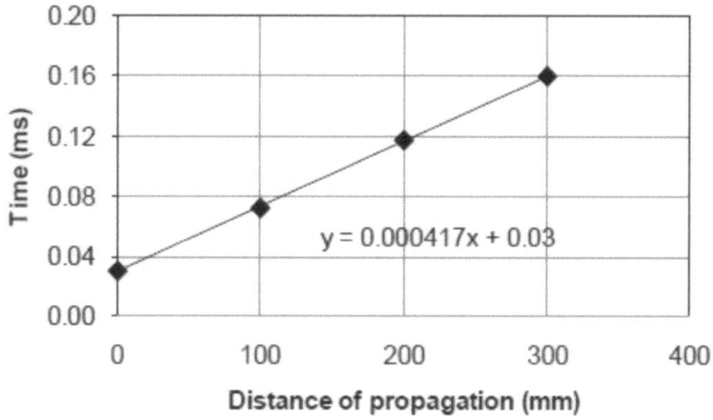

Fig. 15. Linear correlation between arrival time and propagation of Rayleigh waves with 20 kHz excitation frequency

In order to take into account the possible dependence between velocity and frequency which is expected in materials like concrete, especially when they contain inhomogeneity, the "phase velocity" curve can be calculated. The whole procedure is explained in (Sachse & Pao, 1978 and Philippidis & Aggelis, 2005). It includes Fourier transformation of the time domain signals recorded at different positions on the surface of the material. The phase of the signals is unwrapped and phase difference ($\Delta\Phi$) for each frequency is calculated. Then by the knowledge of the separation distance between the receivers, Δx, the velocity for each phase (frequency component) of phase velocity V_{ph} can be derived by the following equation:

$$V_{ph} = \lambda f = \left(\frac{2\pi\Delta x}{\Delta\emptyset}\right) f \tag{3}$$

where λ is the wavelength, f is the frequency for which the phase difference is calculated. By plotting the phase velocity versus the frequency within the bandwidth, i.e. dispersion curve, the change of phase velocity of Rayleigh waves with regards to frequency can be examined. The results can also be used to calculate theoretical wavelength of Rayleigh waves for the respective excitations. Fig. 16 shows exemplary dispersion curves for all the three trigger-receiver combinations in the case for the homogeneous concrete and that with a horizontal crack at 25 mm depth, which were acquired from data with 20 kHz of excitation frequency. The phase velocity for the homogeneous concrete corresponding to 20 kHz was averaged to a value of 2317 m/s, which was very close to that obtained by taking into account the arrival time of Rayleigh wave peak amplitude. As can be seen from the figures, the dispersion curves were seemingly uniform for the case of homogeneous concrete model, indicating that the phase velocity for each frequency component was quite similar within the bandwidth of excitation. On the other hand, for the models with horizontal crack, the dispersion curves for all the three trigger-receiver combinations exhibited considerable differences and were all translated to lower values. The average phase velocity at the excitation frequency of 20 kHz was calculated as 1723 m/s, which was lower than that obtained for the homogeneous concrete model. Considering that the penetration depth for Rayleigh waves can be equivalent up to its wavelength, it can be inferred that the generated Rayleigh waves in this particular case, which has a calculated wavelength of approximately 110 mm, was effectively distorted by the horizontal crack at 25 mm depth to exhibit translation of phase velocity to lower values as a result of the change in propagation behavior.

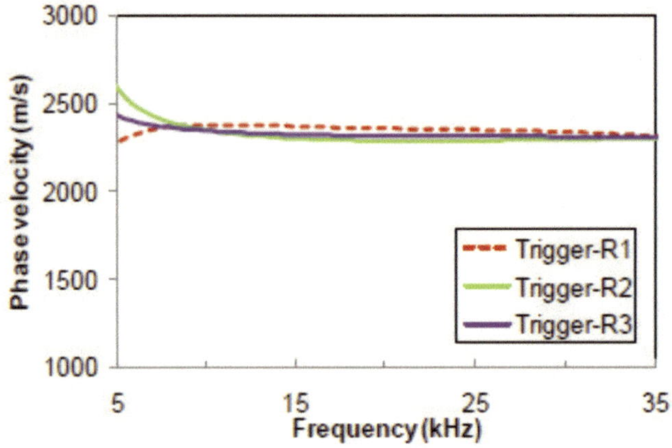

(a)

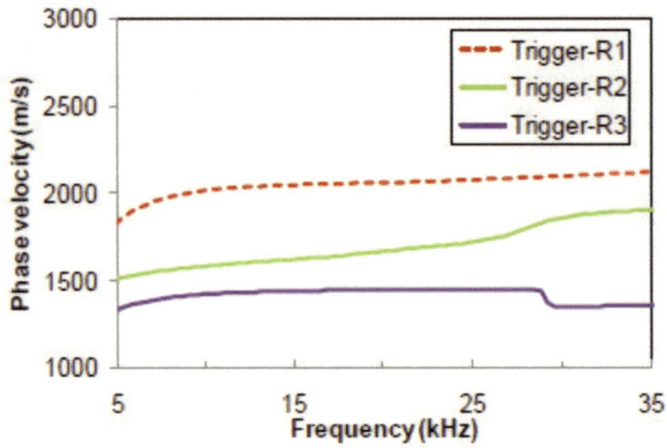

(b)

Fig. 16. Dispersion curve computed for data by excitation frequency of 20 kHz, (a) homogeneous concrete and (b) horizontal crack at 50 mm from top

5.3 Velocity change

The average phase velocity values at respective excitation frequencies were obtained from the models with delamination and normalized with those of the homogeneous concrete model to obtain the "velocity index" for facilitating comparison. The velocity indices were arranged in accordance with the depth of delamination depth, as plotted in the form of bar charts given by Fig. 17. The bars in the charts were classified under two distinct groups: data with the theoretical wavelengths equal or greater than the delamination depth ($d \geq \lambda$); and those of the opposite condition ($d < \lambda$). The data with $d \geq \lambda$ gave lower velocity index compared to the data with $d < \lambda$ in general. Among the simulated cases, the velocity index dropped most noticeably for d = 25 mm, where the soundness indices have decreased to as low as 0.68. In cases where d = 150 mm and 200 mm, although velocity indices of lower than 1.00 were also obtained for $d < \lambda$, the decrease was comparatively small. This could bear the inference that the generated waves have only been slightly disturbed by the horizontal crack although the penetration depth was greater (assuming equivalent to one wavelength) than the crack depth. In other words, the penetration depth was essentially less than the wavelength.

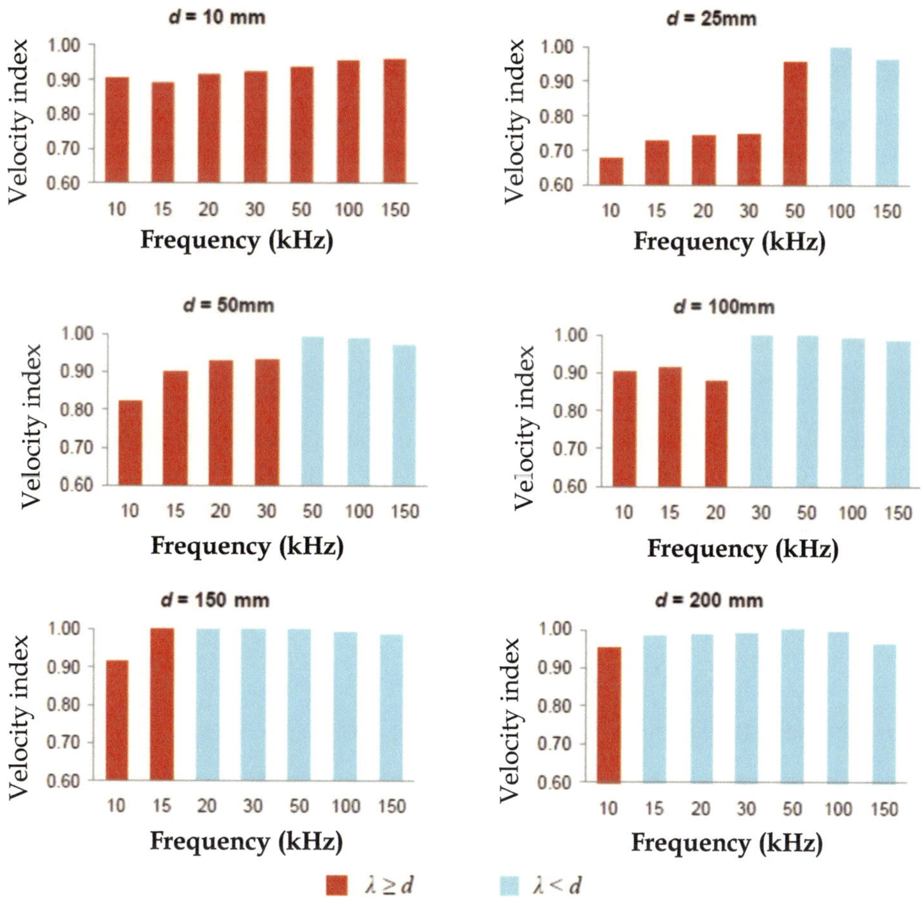

Fig. 17. Velocity indices for different cases of crack depth

Slight decrease in velocity indices in the cases of $d \geq \lambda$ was in general observed in the simulated results. The decrease could be attributed to slight change in the inherent nature of the simulated waves subject to the combinations of crack depths and excitation frequencies. As implied by the findings, inhomogeneity introduced by the presence of horizontal crack in concrete tended to produce variation of phase velocity, even though the decrease would be less significant in some cases. Further examination revealed that the wavelength should be at a minimum of 2.5 times greater than the depth of defect in order to constitute a decrease of at least 10% (velocity index = 0.9) in phase velocity.

5.4 Development of Rayleigh wave tomography

5.4.1 Principle of travel time tomography

A typical travel time tomography involves a unique algorithm that performs iterative computations to reconstruct the velocity profile of a measured object from the picked travel time of elastic waves, which can be obtained in the form of observed data through measurement at selected locations on the object. In ray theory, the travel time, t is computed from the integral of the slowness along a given ray path l

$$t = \int_l \mathrm{d}t = \int_l \frac{1}{v}\mathrm{d}l = \int_L s\mathrm{d}l \tag{4}$$

where v is the velocity, s is the slowness (reciprocal of velocity) and $\mathrm{d}l$ is the length element. If a model is subdivided into M cells of constant slowness, s_i for each corresponding segment length l_i, the integral product can be totalled as

$$t = \sum_{i=1}^{M} l_i s_i \tag{5}$$

The above equation can be expressed in a matrix-vector form:

$$\mathbf{t} = \mathbf{L}\mathbf{s} \tag{6}$$

where $\mathbf{t}$ is the travel time vector and $\mathbf{s}$ is the vector of s_i. $\mathbf{L}$ is a matrix describing the ray path segments. The matrix is generally sparse since each ray would cover only a limited number of cells. Equation (6) is solved to simulate a velocity distribution that can explain the measured travel times. Although the ray propagation itself is considered as a linear operator, the inverse problem can become non-linear, which has to be solved using iterative process. With a fundamental model $\mathbf{s}^0$, new models are computed successively based on the discrepancy of data and model response by solving

$$\mathbf{s}_{k+1} = \mathbf{s}^k + \Delta\mathbf{s}^k = \mathbf{s}^k + \mathbf{L}^{\dagger}(\mathbf{s}^k)(\mathbf{t} - \mathbf{L}(\mathbf{s}^k)\mathbf{s}^k) \tag{7}$$

with $\mathbf{L}$ recomputed and so on, whereby $\mathbf{L}^{\dagger}$ is the used inverse operator.

The computation process is commonly known as the ray tracing procedure, for which the simultaneous iterative reconstruction technique (SIRT) is adopted. The ray paths are usually restricted to the edges of the mesh and a weighted shortest path problem has to be solved. Since not every path can be used, or there is insufficient observed data, the computed travel times are always not accurately estimated in the first few iterations. The accuracy is improved following increasing refinement in the computation to reduce discrepancy.

5.4.2 Experimental measurement of concrete slab

To investigate the feasibility of Rayleigh wave properties as the observed data for tomography reconstruction to visualize defect in concrete, laboratory experiment was carried out to measure and evaluate a concrete slab specimen. The specimen was a 1500 x 1500 x 300 mm concrete slab of normal strength. Defects in the forms of horizontal cracks in concrete at different depths were modelled by including four 5 mm-thick circular Styrofoam plates of 300 mm in diameter. The Styrofoam plates were located at 30 mm, 60 mm, 100 mm and 140 mm from the top of concrete surface, respectively, as depicted in Fig. 18. Shown in the same figure also are the arrangement of 16 accelerometer sensors for sensing and recording of elastic waves. The arrangement formed a total of 240 direct ray paths. During the measurement, elastic waves were generated by manually hitting the concrete surface with a ball hammer that has a spherical head of 15 mm in diameter. A 16-channel waveform acquisition system was used to record waveforms at an interval of 5 μs for a total of 4096 samples. Each set of impact was performed by continuing hitting the specimen surface for 5 seconds, which resulted in approximately 20 impacts or waveforms to be recorded by each sensor. In the preliminary processing of signal, the multiple waveforms recorded by the

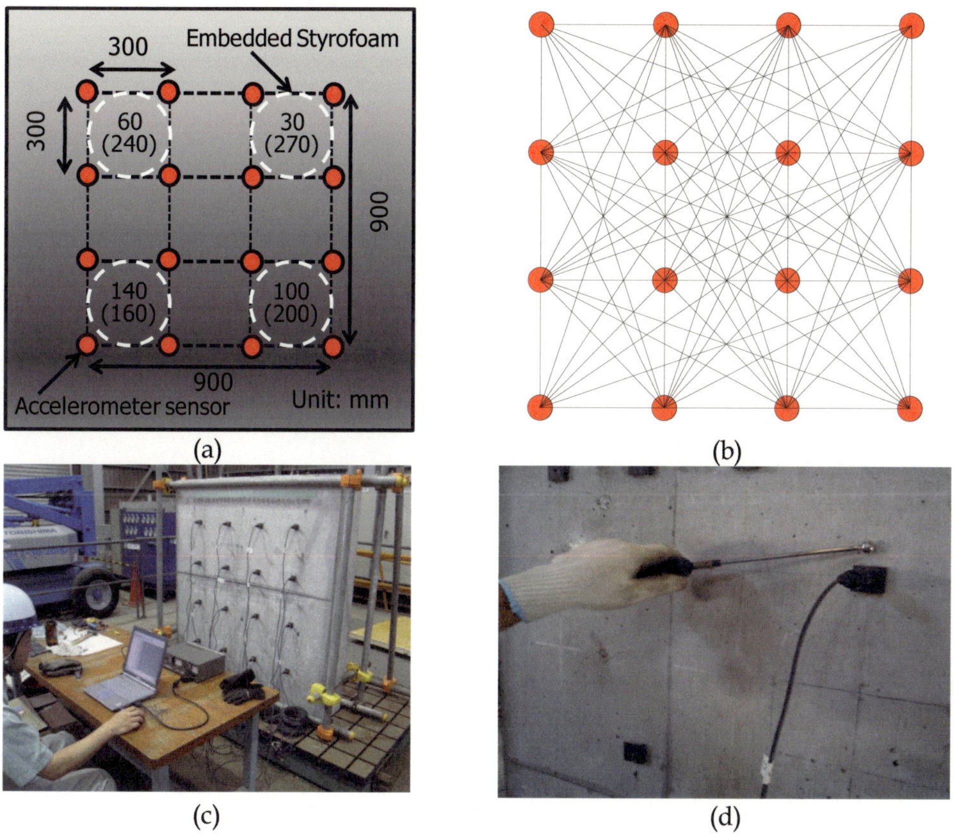

Fig. 18. Tomography measurement on concrete slab: (a) Crack depth and arrangement of accelerometer sensors (() denotes depth from the opposite surface)), (b) Ray path lay-out, (c) Snapshot during measurement, and (d) Generation of elastic waves by hammer with spherical head.

respective sensors would be stacked accordingly to eliminate random noise that could be influential in hampering proper processing in the later stage. Each time the impact was made adjacent to one sensor which has been configured as the trigger, with the others as receivers to detect incoming waves. The acquisition was synchronized in such manner that all 16 channels would start recording at the same time once the trigger channel was excited due to the arrival of elastic waves from the hitting point. The impact and recording process was repeated by subsequently setting the next sensor as the trigger in a pre-determined order until all the sensors have been assigned.

In order to measure the propagation velocity of Rayleigh waves on sound concrete, a separate measurement utilizing nine accelerometer sensors at 150 mm distance was carried out on the sound concrete portion of the same specimen. From the recorded waveforms, the peak amplitudes could be identified because of the large increase in relation to the preceding ones that belonged to the body waves. Taking into account the time of Rayleigh wave arrival, a typical travel time versus distance plot can be drawn as given in Fig. 19. The velocity could be obtained as the inverse of gradient from linear regression, giving an approximate value of 2356 m/s for the sound concrete.

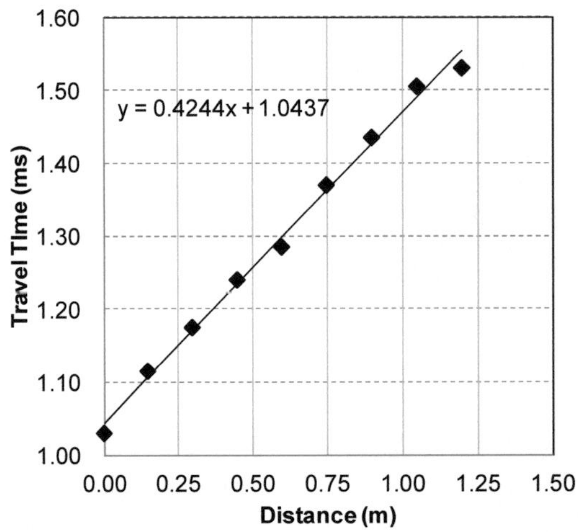

Fig. 19. Time vs. distance plot of Rayleigh waves to estimate propagation velocity

In Fig. 20, an example of time series data and its corresponding FFT frequency spectrum is shown. The frequency spectrum presented two frequency peaks, suggesting the existence of more than one dominant frequency in the waves. By using the similar procedure adopted for the processing of simulation data as discussed earlier, the Rayleigh wave amplitudes was extracted, based on the identification of time window containing the main Rayleigh wave components as anticipated based on the typical propagation velocity. This resulted in a frequency spectrum as given in Fig. 21, clearly indicating only one peak at approximately 9 kHz. Using the measured velocity and peak frequency, the dominant wavelengths was calculated as 240 mm.

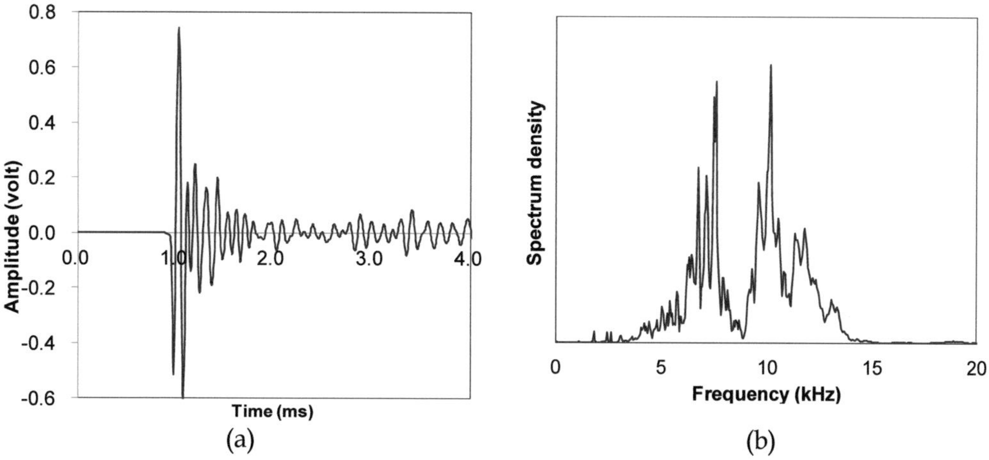

Fig. 20. Typical measurement results: (a) Time series of original waveforms, (b) Corresponding frequency spectrum

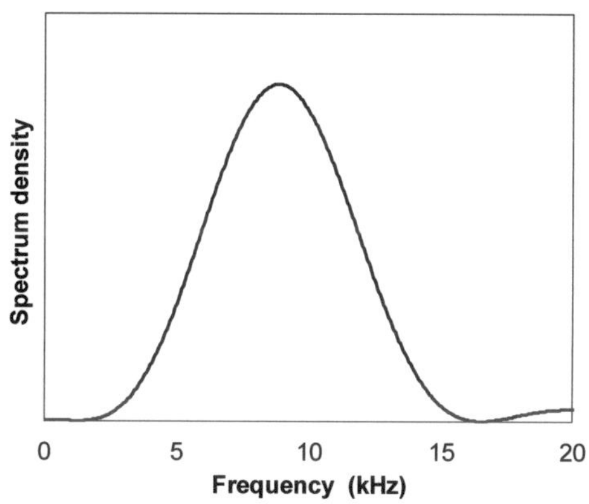

Fig. 21. Frequency spectrum of the proccessed waveform, giving characteristic peak frequncy of Rayleigh waves

5.5 Tomography reconstruction using phase velocity

All the recorded waveforms were processed for Rayleigh wave amplitudes, followed by transformation to frequency response to compute phase difference, which were used to calculate the phase velocity by utilizing Equation (1). The results were plotted against the frequency, as exemplified in Fig. 22 for the case of Rayleigh waves propagating directly from one trigger to its surrounding eight receivers. In the figure, three out of the eight dispersion curves were obtained from propagation paths that contained artificial horizontal crack in between trigger and receiver. The crack was located 30 mm from the surface. All the three paths apparently yielded reduced phase velocity values for the entire effective frequency range of excitation, which was estimated at 5-15 kHz. The phase velocity has been decreased to less than 1900 m/s in general, registering an average difference of approximately 500 m/s compared to the propagation paths containing no crack. It was interesting to note that the wavelength of Rayleigh waves as calculated based on the peak frequency and propagation velocity was greater than the crack depth. The finding was in agreement with that of the simulation work in confirming that when the waves was effectively impinged by a horizontal crack that lied within the region of penetration, distortion was bound to occur, resulting in the change of propagation behaviour to gave lower phase velocity and translation of dispersion curve in the effective frequency range. The decrease in phase velocity due to the horizontal crack was approximately 500 m/s compared to the sound concrete and was found to be greater than that observed in the simulation work.

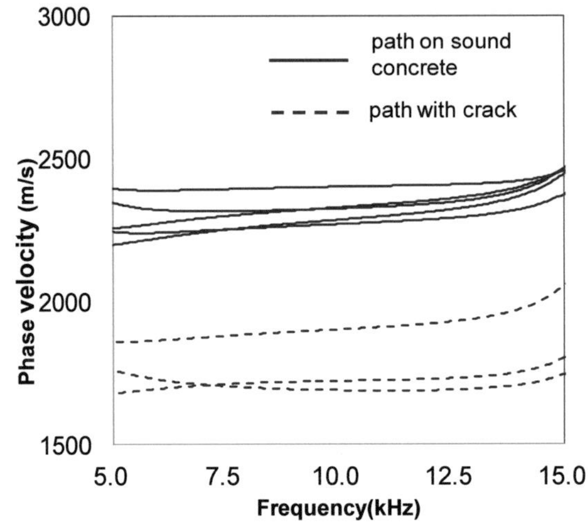

Fig. 22. Frequency spectrum of the proccessed waveform, giving characteristic peak frequncy of Rayleigh waves

For the purpose of tomography reconstruction, the measured area was divided into 6 x 6 square cells of 150 mm. The phase velocity data acquired from the peak frequency of excitation were used as the observed data and the computation process was carried out for 50 iterations to achieve satisfactory convergence. Fig. 23 presents the results of tomography reconstruction expressed in the form of phase velocity distributions for measurements from the surface with relatively small crack depths (30 mm ~ 140 mm) and large crack depths (160 mm ~ 270 mm). For measurement from the surface with small crack depths, all the four crack location were successfully detected with good visualization effect; while from the surface where the crack depths became greater, the visualization for cracks was less convincing except for the crack with 160 mm depth, which is the shallowest among the four. Based on the setting for graphical display, the cracks were indicative when the phase velocity was lower than 2000 m/s. It is also to be noted that the circular shape of the horizontal cracks was not accurately represented by the low phase velocity region and in most cases the centre of defect was indicated with the lowest value, and the change of value from the centre to the edge of defects has developed in a non-uniform way. Cracks which were not successfully visualized were due to the fact that the generated Rayleigh waves did not penetrate sufficiently deep into the concrete to be effectively impinged by these cracks. From the results it was demonstrated that the accuracy of assessment by the Rayleigh wave-based tomography was essentially dependable upon the depth of crack as well as the penetration depth.

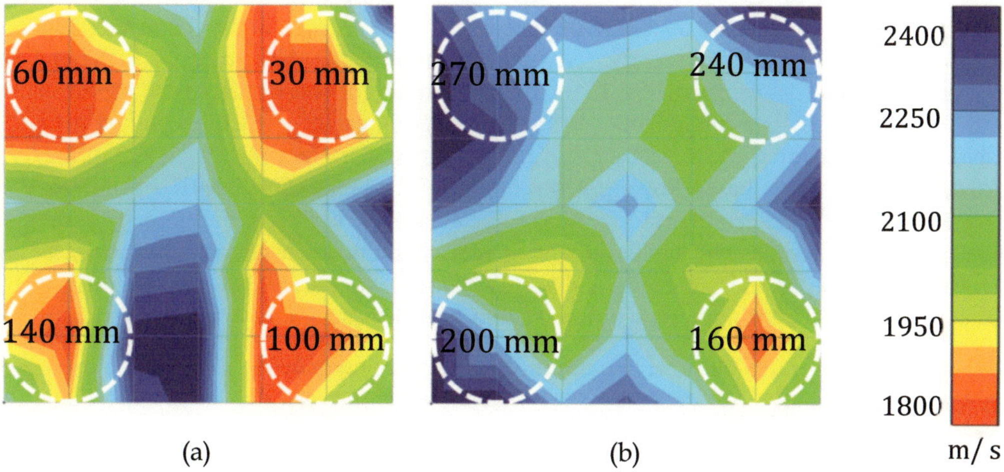

Fig. 23. Rayleigh wave tomography resutls for (a) the side with relatively small crack depth, and (b) with greater crack depth from top surface of concrete

Based on literature study, however, it is found that the penetration depth varied in accordance to material as well as structural form. For example, the was a study reporting satisfactory estimation for elastic properties of layered concrete slabs by assuming the effective Rayleigh wave penetration depth to be equivalent to half a wavelength (Wardany et. al. 1997). On the other hand, another study using concrete beams concluded that at vertical distances greater than half the beam depth, Rayleigh waves did not formed because of the fundamental modes diverged at longer wavelengths and energy was consumed in flexural mode (Zewer et. al. 2005). Also, in determining the near-surface profile of soil structures (Long & Kacaoglu 2001), it was reported that the Rayleigh wave velocity was determined primarily by the shear wave (S-wave) velocity of material in a depth range of 1/4 wavelength. Based on the findings of the current experimental study, the wavelength of Rayleigh waves has to be greater than 1.5 times the depth to crack for successful detection. The estimated wavelength, was much less than that concluded from the simulation work discussed earlier herein (2.5 times the crack depth). Nevertheless, the feasibility of utilizing the phase velocity for tomography reconstruction to detect defect in concrete has been demonstrated. It is considered further work has to be directed to refining quantitative relation between the wavelength and the depth of crack so that the vertical location of defect can be estimated in more accurately way. In any case however, it is worth mentioning that even if a crack is not visualized at its actual dimensions using the proposed method, successful detection for the fingerprint of the defect would help provide useful information for minimizing the labour and increasing the effectiveness of other pinpoint NDT techniques to characterize in detail.

6. Conclusion

The present chapter discusses elastic wave propagation of low frequency on large concrete surfaces. The objective is characterization of damage and effectiveness of repair by cement injection. A pattern of piezoelectric transducers was mounted on a large part of the surface allowing a quick and reliable scanning of the velocity of the surface of the structure. Repair was connected to a velocity increase of the level of 5% to 6% for both longitudinal and Rayleigh waves. Furthermore, the frequency content of the pulses surviving long propagation distances increased by 15% exhibiting stronger sensitivity to the repair action. Attenuation and dispersion effects impose a dependence of measured velocity on the distance, which is more clear for longitudinal waves that travel on a shallow depth on the surface. On the other hand Rayleigh waves have a propagation depth similar to their wavelength which enables them to fly below the shallow cracks and limits their dispersion. After repair the longitudinal wave dispersion was weakened offering an additional feature sensitive to the repair effectiveness. Reliability considerations that are not normally taken into account are discussed which seem to influence the measured wave parameters.

Additionally, the feasibility of Rayleigh waves for developing a single-side access tomography technique of concrete was investigated numerically and experimentally. Tomography reconstructions results indicated the suitability of the measurement method and data analysis procedure for visualizing subsurface defects. It was demonstrated that by increasing the dominant wavelength of Rayleigh waves, wave penetration could be improved and deeper defect could be detected. Studies on the quantitative relation between dominant wavelength and depth of defect, in addition to clarifying the dispersion characteristic of Rayleigh waves are imperative in future in order to realize accurate assessment of concrete structures using the proposed tomography technique. Numerical simulations are expected to greatly enhance knowledge concerning the both the interaction between dominant wavelength and depth of defect that can be identified as well as the dependence of velocity on frequency. Nowadays, since waveform acquisition is standard to almost all ultrasonic equipment the rough characterization based on pulse velocity can be easily enhanced by features like frequency and dispersion of each wave mode. So far these have been applied successfully in laboratory conditions for material characterization but it is expected that their application in real structures will definitely improve NDT capabilities.

7. References

Aggelis, D.G., Hadjiyiangou, S., Chai, H.K., Momoki, S., Shiotani, T. (2011). Longitudinal waves for evaluation of large concrete blocks after repair, NDT&E International 44 (2011) 61–66

Aggelis D.G., Momoki S., Chai H. K. (2009). Surface wave dispersion in large concrete structures", NDT&E International 42, 304–307.

Aggelis D. G., Philippidis, T. P. (2004). Ultrasonic wave dispersion and attenuation in fresh mortar. NDT & E International, 37(8), 617-631.

Aggelis D.G., Shiotani T. (2007a). Experimental study of surface wave propagation in strongly heterogeneous media. J Acoust Soc Am; 122(5):EL151–7.

Aggelis, D.G., Shiotani, T. (2007b). Repair evaluation of concrete cracks using surface and through-transmission wave measurements, Cement & Concrete Composites 29; 700–711.

Aggelis, D. G., Shiotani, T. (2008a). Effect of inhomogeneity parameters on the wave propagation in cementitious materials. American Concrete Institute Materials Journal, 105(2), 187-193.

Aggelis, D. G., Shiotani, T. (2008b). Surface wave dispersion in cement-based media: inclusion size effect. NDT&E INT, 41, 319-325.

Aggelis, D.G., Shiotani, T., Polyzos, D. (2009). Characterization of surface crack depth and repair evaluation using Rayleigh waves. Cement and Concrete Composites, 31(1), 77-83.

Aggelis, D.G., Tsimpris, N., Chai, H.K., Shiotani, T., Kobayashi, Y. (2011). Numerical simulation of elastic waves for visualization of defects, Construction and Building Materials 25 1503–1512.

Aggelis, D. G. and T. Shiotani. (2009). An experimental study of wave propagation through grouted concrete. American Concrete Institute Materials Journal, 106(1) 19-24

Anderson, D.A., Seals, R.K. (1981). Pulse velocity as a predictor of 28- and 90-day strength, ACI J. 78- 79, 116- 122.

Chai, H.K., Aggelis, D.G., Momoki, S., Kobayashi, Y., Shiotani, T. (2010). Single-side access tomography for evaluating interior defect of concrete, Construction and Building Materials 24 2411–2418.

Chaix, J. F., Garnier V., Corneloup G. (2006). Ultrasonic wave propagation in heterogeneous solid media: Theoretical analysis and experimental validation, Ultrasonics 44, 200–210.

Diamanti, N., Giannopoulos, A., Forde, M.C. (2008) Numerical modelling and experimental verification of GPR to investigate ring separation in brick masonry arch bridges. NDT&E Int ;41:354–63.

Doyle, P.A., Scala, C.M. (1978). Crack depth measurement by ultrasonics: a review. Ultrasonics 16(4):164–70.

Graff, K.F. 1975. Wave motion in elastic solids. New York: Dover Publications.

van Hauwaert, A.; Thimus, J. F.; and Delannay, F. (1998). Use of Ultrasonics to Follow Crack Growth. *Ultrasonics*, 36 209-217.

Hevin, G., Abraham, O., Pedersen, H.A., Campillo, M. (1998). Characterisation of surface cracks with Rayleigh waves: a numerical model. NDT&E Int ;31(4): 289–97.

In, C-W., Kim, J.Y., Kurtis, K.E.,Jacobs, L.J. (2009). Characterization of ultrasonic Rayleigh surface waves in asphaltic concrete, NDT&E International 42 610–617.

Issa, C.A., Debs, P. (2007). Experimental study of epoxy repairing of cracks in concrete. Constr Build Mater;21:157–63.

Ito, F., Nakahara, F., Kawano, R., Kang, S-S., Obara, Y. (2001). Visualization of failure in a pull-out test of cable bolts using X-ray CT. Construct Build Mater 15:263–70.

Jacobs, L. J., Owino, J. O. (2000). Effect of aggregate size on attenuation of Rayleigh surface waves in cement-based materials, J. Eng. Mech.-ASCE. 126 (11) 1124-1130.

Jian, X., Dixon, S., Guo, N., Edwards, R.S., Potter, M. (2006). Pulsed Rayleigh wave scattered at a surface crack. Ultrasonics 44:1131–1134.

Jones R (1953) Testing of concrete by ultrasonic-pulse technique, Proceedings of the thirty-second annual meeting. Highway Res Board 32:258–275

Kaplan, M.F. (1959). The effects of age and water/cement ratio upon the relation between ultrasonic pulse velocity and compressive strength, Mag. Concr. Res. 11 (32) 85- 92.

Kaplan, M. F. (1960). The relation between ultrasonic pulse velocity and the compressive strength of concretes having the same workability but different mix proportions, Magazine of Concrete Research, 12 (34) 3-8.

Kase, E.J., Ross, T.A. (2003). Quality assurance of deep foundation elements, Florida Department of Transportation. In: Proceedings of the 3rd international conference on applied geophysics - geophysics 2003, Orlando, Florida, December 8–12.

Keating, J.; Hannant, D. J.; and Hibbert, A. P. (1989). Correlation between Cube Strength, Ultrasonic Pulse Velocity and Volume Change for Oil Well Cement Slurries, *Cement and Concrete Research*, 19(5); 715-726.

Kepler, W.F., Bond, L.J., Frangopol, D.M. (2000). Improved assessment of mass concrete dams using acoustic travel time tomography, part II—application. Construct Build Mater 14:147–56.

Kheder, G.F. (1999) A two stage procedure for assessment of in situ concrete strength using combined non-destructive testing. Mater Struct 32:410–417

Kinra V. K., Rousseau, C. (1987). Acoustical and optical branches of wave propagation," J. Wave Mater. Interaction 2, 141–152 .

Kobayashi, Y.; Shiotani, T.; Aggelis, D. G.; and Shiojiri, H. (2007). "Three- Dimensional Seismic Tomography for Existing Concrete Structures,"*Proceedings of the Second International Operational Modal AnalysisConference*, IOMAC 2007, (April 30-May 2, Copenhagen), V. 2.pp. 595-600.

Long, L. T., Kacaoglu, A. (2001). Surface-wave group-velocity tomography for shallow structures. J Env Eng Geophys 6(2); 71-81.

Landis, E. N., Shah, S. P. (1995). Frequency-dependent stress wave attenuation in cement-based materials, J. Eng. Mech.-ASCE 121 (6) 737-743.

Liu, P.L., Lee, K.H., Wu, T.T., Kuo, M.K. (2001). Scan of surface-opening cracks in reinforced concrete using transient elastic waves. NDT&E Int 34:219–26.

Mikulic, D., Pause, Z., Ukraincik, V. (1999). Determination of concrete quality in a structure by combination of destructive and non-destructive methods. Mater Struct 25:65–69

Monteiro, P. J. M., Helene, P. R. L., Kang, S. H. (1993). Designing concrete mixtures for strength, elastic modulus and fracture energy, Materials and Structures, 26 443-452

Naik TR, Malhotra VM, Popovics JS. (2004).The ultrasonic pulse velocity method. In: Malhotra VM, Carino NJ, editors. Handbook on nondestructive testing of concrete. Boca Raton: CRC Press.

Ohtsu M, Tomoda Y. (2008). Phenomenological model of corrosion process in reinforced concrete identified by acoustic emission. ACI Mater J. 105(2):194–9.

Ohtsu, M., Alver, N. (2009). Development of non-contact SIBIE procedure for identifying ungrouted tendon duct. NDT&E Int 42:120–7.

Ono, K. (1988) Damaged Concrete Structures in Japan due to Alkali Silica Reaction," *The International Journal of Cement Composites and Lightweight Concrete*, 10;4 247-257.

Owino, J. O., Jacobs, L. J. (1999). Attenuation measurements in cement-based materials using laser ultrasonics, J. Eng. Mech.-ASCE. 125 (6) 637-647.

Pecorari, C. (2001). Scattering of a Rayleigh wave by a surface-breaking crack with faces in partial contact. Wave Motion ;33:259–70.

Philippidis, T. P., Aggelis D. G. (2003) An acousto-ultrasonic approach for the determination of water-to-cement ratio in concrete. Cement and Concrete Research, 33(4), 525-538.

Philippidis, T. P., Aggelis, D. G. (2005). Experimental study of wave dispersion and attenuation in concrete. Ultrasonics, 43(7), 584-595.

Popovics, S. (2001). Analysis of the Concrete Strength versus Ultrasonic Pulse Velocity Relationship," *Materials Evaluation*, 59(2) 123-130.

Punurai W, Jarzynski J, Qu J, Kurtis KE, Jacobs LJ. (2006). Characterization of entrained air voids in cement paste with scattered ultrasound. NDT&E Int 2006;39(6):514–24.

Qasrawi HY (2000) Concrete strength by combined nondestructive methods simply and reliably predicted. Cem Concr Res 30:739–746

Qixian L., Bungey J.H. (1996). Using compression wave ultrasonic transducers to measure the velocity of surface waves and hence determine dynamic modulusof elasticity for concrete. Construct Build Mater 4(10):237–42.

Sachse, W., Pao, Y.-H. (1978). On the determination of phase and group velocities of dispersive waves in solids, J. Appl. Phys. 49 (8) 4320–4327.

Sansalone, M.J., Streett, W.B. (1997). Impact-echo nondestructive evaluation of concrete and masonry. Ithaca, NY: Bullbrier Press.

Sayers C. M., Dahlin, A. (1993). Propagation of ultrasound through hydrating cement pastes at early times, Advanced cement based materials, 1:12-21.

Sassa K. (1988). Suggested methods for seismic testing within and between boreholes. Int J Rock Mech Min Sci Geomech Abstr;25(6):449–72.

Shah, S.P., Popovics, J.S., Subramanian, K.V., Aldea, C.M. (2000). New directions in concrete health monitoring technology, J. Eng. Mech. – ASCE 126 (7) 754–760.

Shiotani, T., Aggelis D. G. (2009). Wave propagation in concrete containing artificial distributed damage. Materials and Structures, 42(3), 377-384.

Shiotani, T., Momoki, S., Chai, H. K., Aggelis, D.G. (2009). Elastic wave validation of large concrete structures repaired by means of cement grouting, Construction and Building Materials 23 2647–2652.

Shiotani, T., Aggelis, D. G. (2007). Evaluation of repair effect for deteriorated concrete piers of intake dam using AE activity. Journal of Acoustic Emission, 25, 69-79.

Song, W.J., Popovics, J.S., Aldrin, J.C., Shah, S.P. (2003). Measurements of surface wave transmission coefficient across surface-breaking cracks and notches in concrete. J Acoust Soc Am 113(2):717–25.

Thanoon, W.A., Jaafar, M.S., Razali, M., Kadir, A., Noorzaei, J. (2005). Repair and structural performance of initially cracked reinforced concreteslabs. Constr Build Mater 19(8):595–603.

Tsinopoulos, S. V., Verbis, J. T., Polyzos, D. (2000). An iterative effective medium approximation for wave dispersion and attenuation predictions in particulate composites," Adv. Composite Lett. 9, 193–200.

Tsutsumi, T., Wu, J., Wu, J., Huang, X., Wu, Z. (2005). Introduction to a new surface-wave based NDT method for crack detection and its application in large dammonitoring. In: Proceedings of international symposium on dam safety and detection of hidden troubles of dams and dikes, 1-3 November, Xian, China (CD-ROM).

Yokota, O., Takeuchi, A. (2004). Injection of repairing materials to cracks using ultrasonic rectangular diffraction method. In: 16th World conference on non destructive testing 2004 (WCNDT), Montreal, Canada, August 30–September 3.

Zerwer, A., Polak, M.A., Santamarina, J.C. (2005). Detection of surface breaking cracks in concrete members using Rayleigh waves. J Environ Eng Geophys;10(3): 295–306.

9

Ultrasonic Testing of HPC with Mineral Admixtures

R. Hamid, K. M. Yusof and M. F. M. Zain
Universiti Kebangsaan Malaysia
Malaysia

1. Introduction

High strength and high performance concrete (HSC and HPC) have been introduced commercially for the columns of high-rise buildings in 1970s in the United States of America and now being made worldwide, one example being the tallest building in Malaysia, the Petronas Twin Towers (Shah, 1997). Fly ash, silica fume, or slag are often mandatory in the production of high-strength and high performance concrete for the reason that the strength gain obtained with these supplementary cementing materials cannot be attained by using additional cement alone. These mineral additives are usually added at dosage rates of 5% to 20% or higher by mass of cementing material (Portland Cement Association EB001 [PCA EB001], 2002). HSC and HPC with mineral additives have different microstructure than normal concrete. The matrix is denser and has lower capillary porosity compared to ordinary concrete. In HSC at more than 105 MPa with silica fume, the matrix is perfectly homogenous, apparently amorphous and the capillary porosity is diminished and discontinuous compared to that of other concretes where it is interconnected (Malier, 1992).

There are a number of ultrasonic testing techniques applied to concrete since 40 years ago (International Atomic Energy Agency TCS 17 [IAEA TCS17], 2002). These techniques are relatively not as sophisticated as compared to other materials application such as metals, but it is found that these techniques are relevant in some cases. These techniques include ultrasonic pulse velocity (UPV), ultrasound pulse echo, the amplitude or the pulse energy measurement, impact-echo/resonance frequency/stress wave test (pulse-echo, impact-echo, impulse-response and spectral analysis of surface waves), and relative amplitude method (RA).

Ultrasonic pulse velocity measurement in concrete is applicable in assessing the quality of concrete. The UPV measurements are correlated with the concrete strength as a measure of concrete quality. For normal concrete with ordinary Portland cement, with strength range between 5 MPa up to 60 MPa, the empirical correlations between strength and UPV are recorded as in the form of exponential (Bungey, 1984; Facaoaru, 1969; Popovics, 1986; Teodaru, 1989; etc as cited in Ismail et al., 1996) with correlation equation, R^2 higher than 90%. For strength between 5 MPa to 60 MPa, the UPV varies between as low as below 2.0 km/s and as high as more than 4.7 km/s. The exponential relationship between the UPV and strength is expressed as:

$$K = ae^{bv} \quad (1)$$

where K is the compressive strength of concrete, v is the UPV and a and b are the empirical constants.

Researches on improvement on the strength-UPV correlation are done vastly either to have more accurate relationship (there are many factors that affect the UPV measured but not the concrete strength) or to make it applicable in assessing concrete structures (Carino, 1994). British Standard (BS 1881: Part 203. Testing Concrete - Recommendation for measurement of velocity of ultrasonic pulses in concrete) listed some factors that affect the ultrasonic pulse measurement in concrete. These factors are path length; lateral dimension of the specimen tested; moisture content of the concrete; temperature of the concrete and the presence of reinforcing steel. The corrections for each factor are as suggested by BS 1881: Part 203.

As the usage of mineral admixtures is inevitable in HPC, researches on influence of additional cementitious materials on strength as well as UPV have been studied. Demirbogă et al., 2004; Türkmen et al., 2010 reported the coefficient of correlation (R^2) of more than 0.90 which indicate a very good exponential relationship between UPV and compressive strength of concrete containing Portland cement, fly ash (FA), natural zeolite (NZ) and blast furnace slag (BFS) and both NZ + BFS concrete (Figure 1). They confirmed the exponential relationship of strength-UPV of concrete with these mineral admixtures (as expressed in Eq. 1). The UPV measured varies between 3250 m/s at strength of 5 MPa up to 4400 m/s at strength 55 MPa. As the concrete strength increase, as in HPC and HSC, the exponential relationship still applied (Khatib, 2008; Hamid et al., 2010), but for strength range between 60 to 100 MPa, the variation of UPV is between 4.5 to 5.0 km/s, making the UPV measurement insensitive towards the variation of strength of HPC and HSC in which their strengths are in this range (Figure 2).

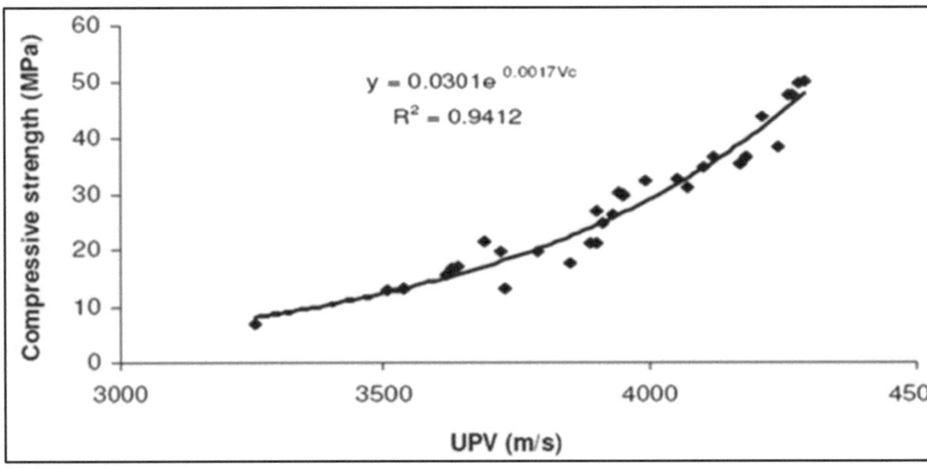

Fig. 1. Relationship between compressive strength and UPV for all results between 3 and 90 days of curing periods for Portland cement, natural zeolite (NZ) and blast furnace slag (BFS) and both NZ + BFS concrete (Türkmen et al., 2010)

Relative amplitude (RA) method is an alternative method to assess the concrete strength (IAEA TCS17, 2002). The principle of this method will be discussed further in this chapter.

Ismail et al., 1996 described the relationship between RA and strength for normal concrete up to 60 MPa. The RA-strength relationship of HPC with silica fume is also described (Hamid et al., 2010). Meanwhile, researches on combined ultrasound method were done to have more accurate estimation of concrete strength (Galan, 1990). The strength of normal concrete has been successfully estimated with only 5% error using the combination of ultrasonic pulse velocity-strength and relative amplitude-strength correlation curves (Ismail et al., 1996).

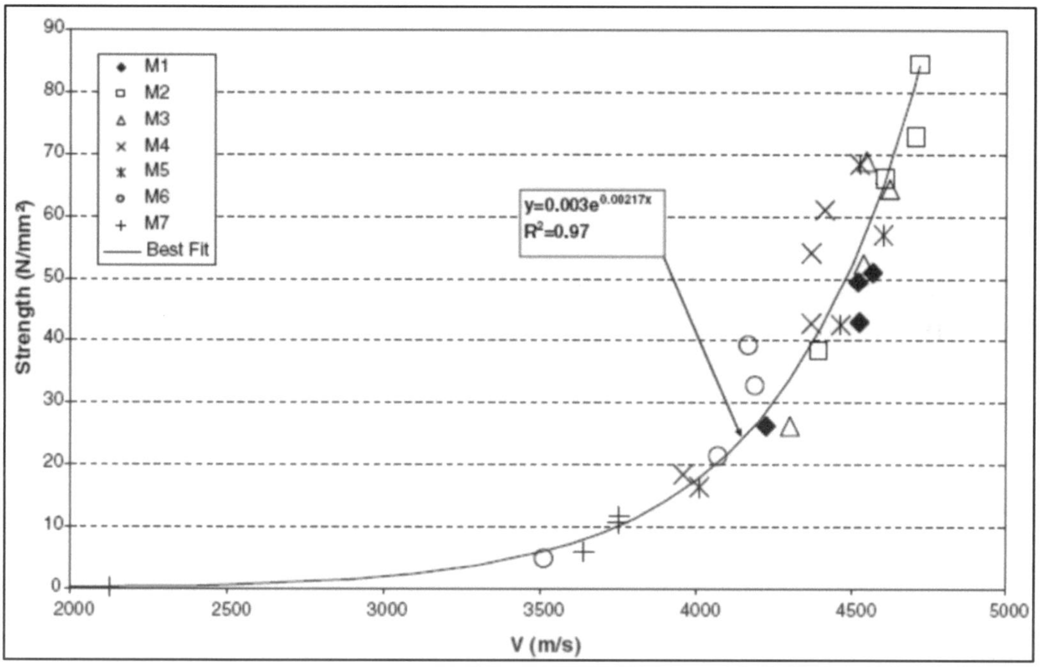

Legend: M1, M2, M3 - 0% FA; M4 – 20% FA; M5 – 40% FA; MA6 – 60%FA; MA7 – 80%FA

Fig. 2. Relationship between strength and UPV for concrete with different proportion of fly ash (FA) (Khatib, 2008)

This chapter emphasizes on ultrasonic pulse velocity (UPV) method in assessing the strength of high performance concrete (HPC) with mineral admixtures particularly fly ash and silica fume. Adding mineral admixtures to replace certain percentages of cement content increase the strength of HPC compare to the same concrete combination containing only ordinary Portland cement (OPC). This chapter discusses further in detail results of experiments on the UPV and RA method in strength estimation of HPC with fly ash.

2. Experimental program, results and discussion

2.1 Experimental program

About 300 specimens, representing 32 trial mixes were prepared and tested in this study. Fabrication, curing and testing of specimens were done based on ASTM C 192, ASTM C618, ACI Committee 211, BS 1881: Part 116 and BS 1881: Part 203.

2.1.1 Materials

The cement used is Type I Portland cement produced locally and conforms with the Malaysian Standard. Its specific gravity is 3.15 (SI unit) and the moisture content is 0.6%. The finess modulus of the mine sand used is 2.48, with its maximum size at 4.75 mm; specific gravity 2.25 (SI unit) and total moisture content 2.1%. The maximum size of granite aggregate used is 20 mm, its specific gravity is 2.29 (SI unit) and the total moisture content is at 0.7%. Superplasticizer (SP) used in this study is a naphthalene formaldehyde sulfate based condensed superplasticizer with brand name Super-20 supplied by Ready Mixed Concrete Malaysia. Its specific gravity is 1.21 (SI unit) with 40% of solid content and 0% chloride content.

2.1.2 Mix design

To reduce the number of variants, the ratio of sand to aggregate is held constant at 0.4. Any admixtures added in the form of liquid, is calculated as part of the mixing water. Three series of HPC were designed with 10, 20 and 30% replacement with class F Malaysian fly ash plus one series of mixes with 0% fly ash content. Each series consisted of 10 different water/binder ratios (7 for mixes with 30% FA and 5 for mixes with 0% FA) to get variation of strength, with the percentage of SP added were changed for each w/b ratio in consideration for workability, but maintained within the series. The w/b ratios are from 0.200 to 0.380 with 0.02 intervals or otherwise stated. Total cementitious materials are 600 kg/m^3 and the ratio of fine aggregate to coarse is 0.4. Table 1 shows the detailed mixes in this study.

2.1.3 Fabrication, curing and testing

Mixing is done in a rotating mixer and enough time is allowed until the mix shows consistency visually. The actual time to achieve the mix consistency depends on the slump/workability of the mixes. Slump test is done on each water/binder ratio mix in each series. The test specimens are 150 mm × 150 mm × 150 mm cubes. The specimens are demoulded after 24 hours. One third of the total samples are immersed in water for 28 days, the other one third are immersed in water for 7 days and the last one third is let dried in open air inside an uncontrolled temperature and humidity building. The temperature inside the building is on average 32°C. Compression test are done with a compression machine with the capacity of 3000 kN after the ultrasonic testing were done. All tests are done on day 91 after casting with 3 samples tested for each test. The concrete specimens were tested in wet condition (more than 48 hours immersed in water before testing) as required by ASTM C 39.

2.1.4 Method of ultrasonic testing

Ultrasonic pulse velocity (UPV) is measured in materials by the equation:

$$v = \frac{L}{T} \text{ m/s} \tag{2}$$

where v is the longitudinal (pressure) pulse velocity, L is the path length and T is the time taken by the pulse to traverse that length.

A set up of the ultrasonic pulse velocity measurement on laboratory specimen is as shown in Figure 3.

Mix	Water/ binder ratio	Binder (kg/m³)		SP (Liter)	Sand (kg/m³)	Aggregate (kg/m³)
		Cement	Fly ash			
I	0.220	600	0	17.4	653	980
	0.260	600	0	14.9	630	945
	0.300	600	0	12.4	607	910
	0.340	600	0	9.9	584	876
	0.380	600	0	7.4	561	841
II	0.198	540	60	17.4	648	973
	0.216	540	60	17.4	638	957
	0.234	540	60	14.9	628	942
	0.252	540	60	14.9	618	926
	0.270	540	60	12.4	607	911
	0.288	540	60	9.9	597	895
	0.306	540	60	9.9	587	880
	0.324	540	60	7.4	576	865
	0.342	540	60	7.4	566	849
	0.360	540	60	7.4	556	834
III	0.200	480	120	19.8	643	965
	0.220	480	120	17.4	632	948
	0.240	480	120	17.4	620	930
	0.260	480	120	14.9	609	913
	0.280	480	120	14.9	597	896
	0.300	480	120	12.4	586	879
	0.320	480	120	9.9	575	862
	0.340	480	120	9.9	563	845
	0.360	480	120	7.4	552	828
	0.380	480	120	7.4	540	810
IV	0.260	420	180	14.9	602	903
	0.280	420	180	14.9	590	885
	0.300	420	180	12.4	579	868
	0.320	420	180	9.9	567	851
	0.340	420	180	9.9	556	834
	0.360	420	180	7.4	545	817
	0.380	420	180	7.4	533	800

Table 1. Details of Mix Proportions

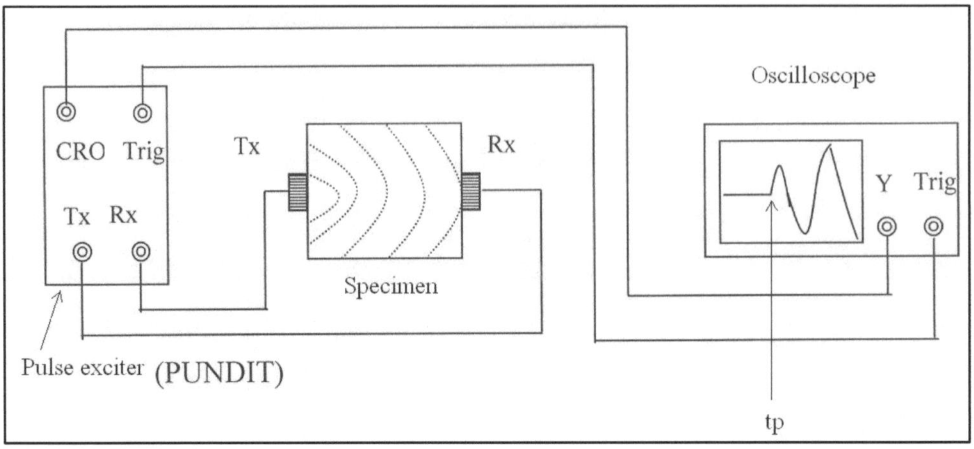

Tx - Transmitter
Rx - Receiver
CRO - Cathode Ray Oscilloscope
Trig - Trigger

Fig. 3. Ultrasonic pulse velocity measurement set up

The piezoelectric transducers (lead zirconate-lead titanates ceramic type) with frequencies of 54 kHz, 82 kHz and 180 kHz as the ultrasonic pulse transmitter and receiver are employed. The ultrasonic pulse is generated by alternate current using commercial ultrasonic-scope (brand name PUNDIT) with digital display of transit time in microsecond. Figure 1 shows a digital storage oscilloscope connected to the pulse exciter to trace the transmitted wave – from where the pulse amplitude is obtained. The oscilloscope display shows that the pulse amplitude is represented as displacement in one axis (normally y-axis) and the ultrasonic pulse transit time is represented by the displacement at the other axis (x-axis). The arrival time of the pressure wave (transit time) can be located as t_p shown in Figure 3. In linear amplification system, vertical displacement is proportional to the signal amplitude. By using logarithmic scale amplification, y-axis is represented by logarithmic scale.

The relative amplitude method is basically an attenuation method, which measures the ratio of the wave amplitudes. Ultrasonic waves are attenuated as they pass through the materials. The attenuation of the sound pressure propagating in solid consists of two components, namely the geometric divergence and energy dissipation. For quality evaluation of materials, the dissipation factor is critical since it is related to the properties of the material. The dissipation effect consists of two processes, namely scattering and absorption. Scattering is due mainly to the non-homogenous composition of the medium and absorption is the actual loss of sound energy. When the properties of material vary, the intensity of the lost energy will also vary, thus the attenuation characteristic can be used to determine the properties of the material. For low amplitude sinusoidal wave, the attenuation of wave due to both conversion processes is in an exponent form and the incident amplitude (either the particle displacement amplitude or the pressure amplitude) can be expressed as [Krautkramer & Krautkramer, 1983] :

$$P_a = P_0 e^{-\alpha x} \tag{3}$$

where P_0 and P_a are the initial pressure wave amplitude and pressure wave amplitude at a distance x in a cross section with length d and attenuation coefficient, a (alpha). The natural logarithm of Eq. (3) is:

$$\alpha x = ln\frac{P_a}{P_0} \tag{4}$$

articulated as the attenuation at a distance x, a dimensionless number expressed in Nepers (Np). Attenuation coefficient is in Np/cm or dB/m. Since $\log x = \ln x/\ln 10 = 0.434 \ln x$ and 1 Np = 8686.0 dB, Eq. (4) can be written as:

$$\alpha x = 20log\frac{P_a}{P_0}\ \text{dB} \tag{5}$$

The relative amplitude, β, is expressed as:

$$\beta = 20log\frac{A_p}{A_{ps}}\quad \text{dB} \tag{6}$$

where β is the measurement of attenuation with A_p taken as the first peak amplitude after the arrival of pressure wave (at t_p = transit time of UPV through the concrete specimen) and A_{ps} as the peak amplitude after the arrival of shear wave (at t_s = two times the transit time of UPV). The location where t_s is located is based on the fact that shear wave is slower (about two times) than the pressure wave in concrete. Figure 4 shows the typical oscilloscope display defining the transit time and wave amplitude that delineate the relative amplitude (β). As β is also a ratio between two amplitudes, it is also known as relative amplitude ratio (RAR).

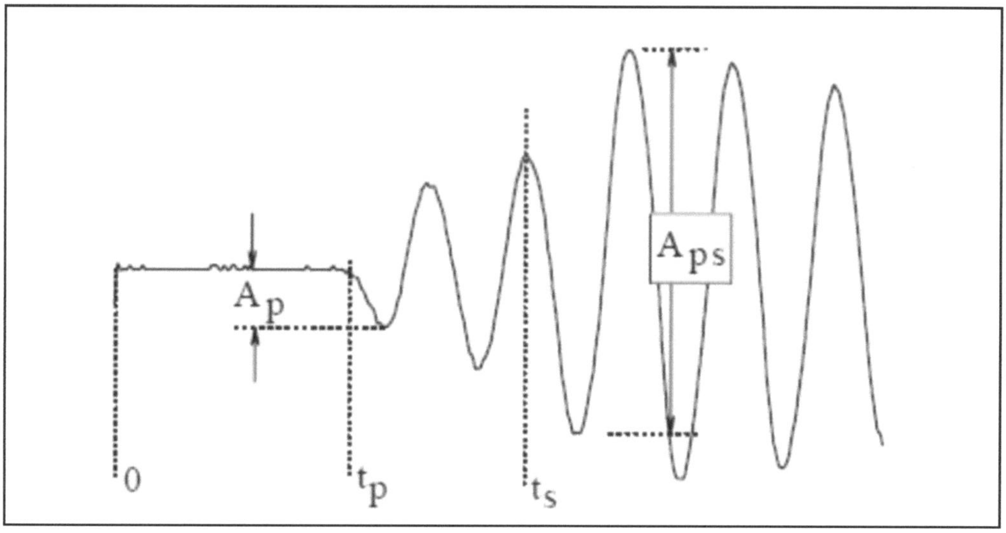

Fig. 4. Typical oscilloscope display defining the transit time and wave amplitude (IAEA TCS17, 2002)

However, consistent and constant coupling pressure for the transducers has to be maintained and will affect the attenuation values. It is suggested that a simple pressure measurement mechanism based on load cell and dial is built to check this problem, though, results discussed here did not include this mechanism.

2.2 Results and discussion

2.2.1 Ultrasonic pulse velocity-strength relationship of HPC with mineral admixtures

At high strength, the well established exponential relationship between strength and UPV (Eq. 1) is not sensitive to estimate the strength accurately (UPV between 4.5 to 5 km/s for strength range 50 to 100 MPa). The correlation between strength and UPV is also poor. Uysal & Yilmaz, 2011 reported the UPV values ranged from 4222 m/s to 4998 m/s for strength value of 55 to 105 MPa for concrete containing limestone powder (LP), basalt powder (BP) and marble powder (MP). They also found that the use of limestone powder (LP), basalt powder (BP) and marble powder (MP) as cement replacement at various contents had caused a reduction in UPV as well as the compressive strength compared to the control mixture. Increasing amounts of these kind of mineral admixtures (limestone powder, basalt powder and marble powder) generally decrease the strength. The correlation between the compressive strength and UPV is poor (R^2 = 0.75) at high strength compare to concrete with mineral admixtures (fly ash, natural zeolite and blast furnace slag) at low strength (Demirbogă et al., 2004; Türkmen et al., 2010).

Khatib, 2008 reported that concrete with 40% fly ash (FA) mix shows the largest value of UPV compared with the 20% FA mix. Generally there is a decrease in strength with the increase in FA content. The strength versus UPV values shows an exponential relationship with an R^2 of 0.97 but the strength range covered is from 5 MPa up to 85 MPa. Khatib, 2008 result also shows that for strength range 50 to 85 MPa, the UPV only varies 4.5 to 4.7 km/s; however the relationship seems to be independent of the FA content.

Silica fume is a mineral admixture that provides early strength gain and high strength gain to HPC. Hamid et al., 2010 reported that at a given strength, specimens with higher content of silica fume (30%) shows lower readings of UPV compared to specimens with lower content of silica fume (10% and 20%) (Fig. 5). The argument was that the mixture with lower content of silica fume is denser than the mixture with higher content of silica fume at the same water/binder ratio, even though they bear the same strength. The strength-UPV relationship is also reported as exponential form which resulted in at strength higher than 85 MPa, the variation of UPV reading reduced. The coefficient of correlation, R^2, for strength-UPV curve for HPC with 10%, 20% and 30% silica fume are good which are 0.93, 0.92 and 0.96 respectively. Hamid et al, 2010 also reported that the sensitivity of the UPV measurement decreased with the increased on HPC with silica fume strength.

For HPC with FA at strength range 50 to 90 MPa, Hamid, 2004 found that the correlation between strength and UPV of FAHPC (strength range 50 – 90 MPa) is poor (Figure 6). Results of experiments show that using different transducer's frequency gave the same best-fitted line of strength-UPV curve, but the correlation of coefficient using transducer 150 kHz is higher than using transducer 54 kHz. Measuring the UPV on saturated samples increase the value of UPV compared to measuring on dry samples. The correlations of the best-fitted line strength-UPV curve for the saturated samples are higher than the correlations of those on dry samples. The compressive strength of mixes with higher content of fly ash decrease compare to mixes with low content of FA. The increase packing density influences the measured UPV very significantly but in reverse to the effect towards strength. Higher content of FA produce higher measured UPV at the same strength level of HPC with lower fly ash content.

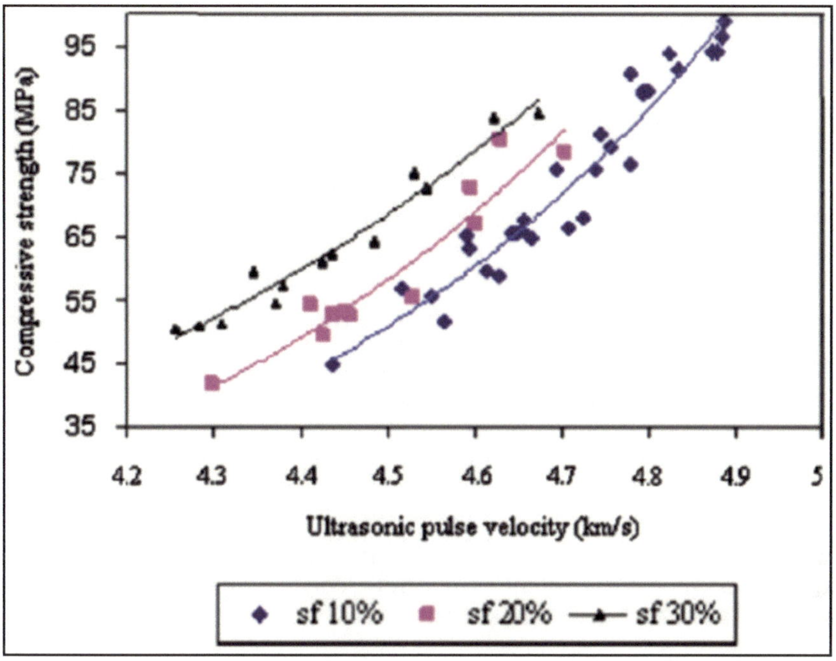

Fig. 5. Relationship between silica fume content with UPV (Hamid et al., 2010)

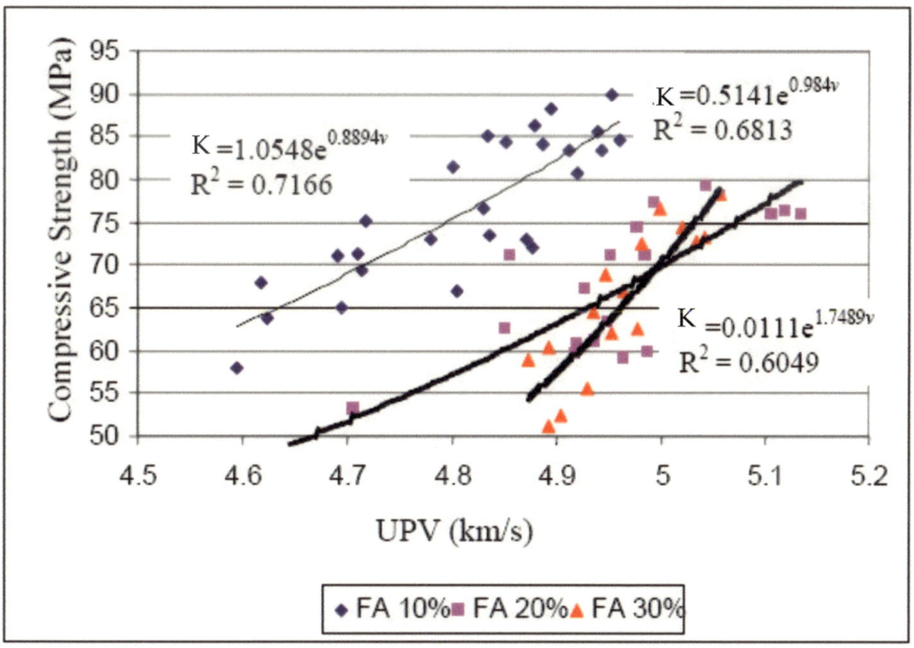

Fig. 6. Relationship between UPV and strength of HPC with fly ash (Hamid, 2004).

It can be concluded that, these phenomena are observed in the measurement of UPV and their relationship with strength of HPC with mineral admixtures:

1. Ultrasonic pulse velocity (UPV) is not sensitive to estimate the strength accurately (UPV between around 4.2 to 5.0 km/s for strength range 50 to 100 MPa).
2. The correlation of coefficient (R^2) of the UPV-strength curve is poor (below 0.8) for all mineral admixtures except silica fume.
3. The content of mineral admixtures in the HPC affects the measured UPV such that at different mineral admixtures content, but at the same strength level, the UPV measured are different. Generally, the UPV is higher in HPC with higher content of mineral admixtures.

2.2.2 Relative amplitude-strength relationship in HPC with mineral admixtures

Relative amplitude (RA) method is described in IAEA TCS17, 2002 and general relationship between strength, *K* and relative amplitude, β is an exponential form:

$$K = e^{c - d\beta} \tag{7}$$

where *c* and *d* are empirical constants.

The relative amplitude decreases as the strength increased. The factors which influence the relationship are the moisture content of concrete; concrete age; aggregate size and type; curing method; bar size and beam path length.

The RA, in contrast with the UPV, shows greater sensitivity towards the strength change, that is, 6 - 20 dB for strength range between 5 MPa to 60 MPa of normal concrete with coefficients of correlation greater than 80% (Ismail, 1996).

Fig. 7 shows that for HPC with silica fume at strength higher than 85 MPa, the variation of RA value is 14% for variation of strength of 40 MPa to 100 MPa. The coefficients of correlation, R^2, for strength-RA curve for all three series of mixes are 0.54, 0.80 and 0.48 respectively (Hamid et al., 2010).

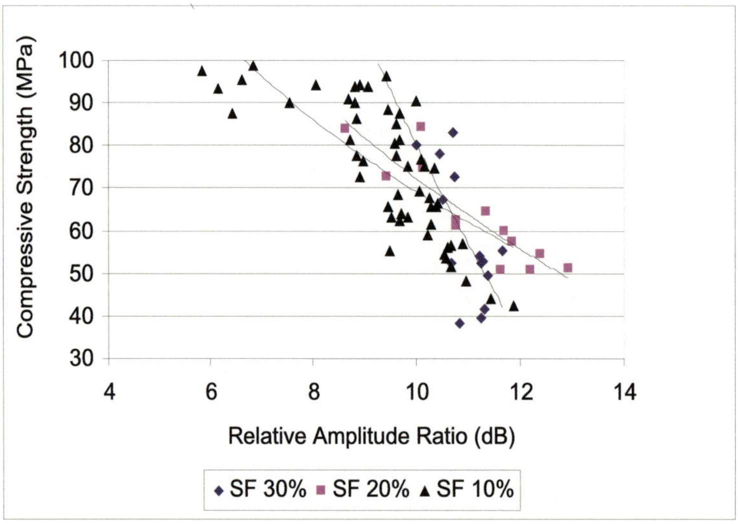

Fig. 7. The relationship between the silica fume content with the relative amplitude (Hamid et al., 2010)

Figure 8 shows that each series of HPC with different fly ash content exhibits an exponential relationship with the relative amplitude ratio (RAR), which is in the form:

$$K = ce^{d\beta} \tag{8}$$

where K is the compressive strength, β is the RAR and c and d are the empirical constant.

The RAR, in contrast with the UPV, shows greater sensitivity towards the strength change, that is, 80% (2 - 10 dB) for the same strength rage. The coefficients of correlation are greater than 80% as shown in Figure 8.

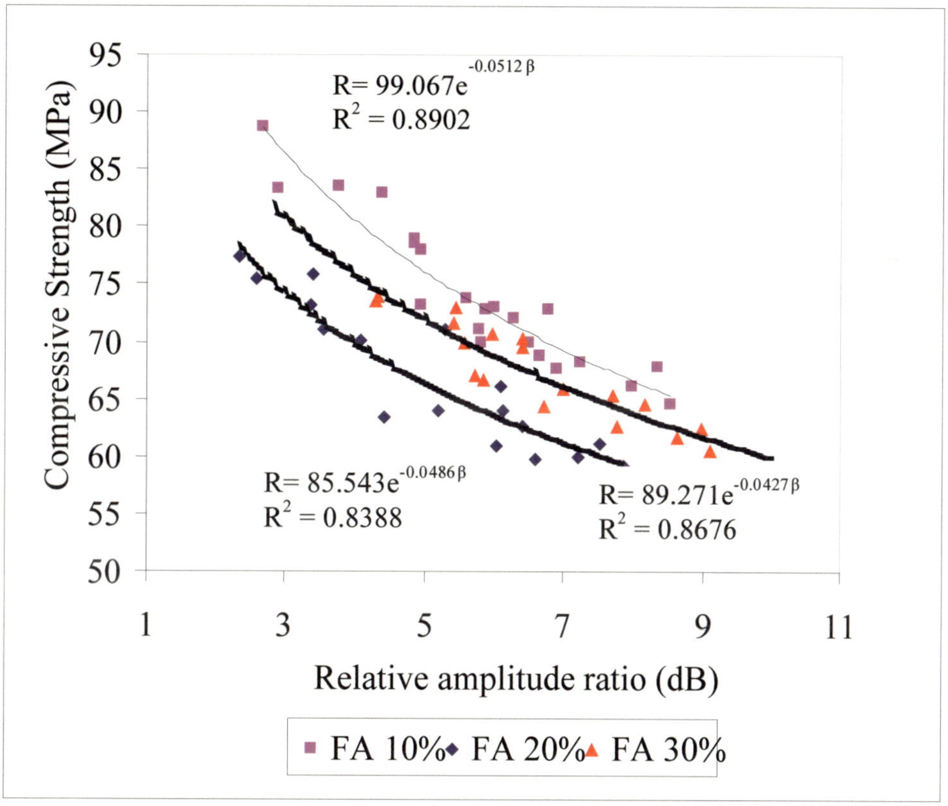

Fig. 8. The relationship between the strength of HPC with FA and the RAR (Hamid , 2004)

2.3 Factors affecting the UPV-strength relationship of HPC with fly ash

Tests were done to determine the effect of the transducer's frequency, moisture content, samples' shape, transmission type and FA content on the UPV-strength and RA-strength relationships. The frequencies of the transducers selected for the test are 54 kHz and 150 kHz. Table 2 shows the effect of different transducer's frequency on the correlation coefficient (R^2) of UPV-strength relationship of HPC with FA. It can be seen from Table 2 that higher transducer frequency gives better correlation coefficient (R^2) of UPV-strength relationship of HPC with FA.

Fly ash content	R^2 (54 kHz transducer)	R^2 (150 kHz transducer)	Empirical constant (Eq.1)		Empirical constant (Eq.1)	
			a	*b*	*a*	*b*
10%	0.8921	0.7075	2.8076	0.6873	0.8252	0.9507
20%	0.4466	0.5317	9.2439	0.4142	9.0802	0.4195
30%	0.0007	0.4205	108.42	-0.0945	0.0034	1.9929

Table 2. Effect of transducer's frequency on the correlation coefficient (R^2) of UPV-strength relationship of HPC with FA

The samples were then submerged in water for more than 72 hours and tested. The increases in moisture content recorded are between 0.4% and 0.8%. Table 3 shows the effect of moisture content on the correlation coefficient (R^2) of UPV-strength relationship of HPC with FA. Saturated samples give better correlation coefficient (R^2) of UPV-strength relationship of HPC with FA.

Fly ash content	R^2 (54 kHz, saturated)	R^2 (150 kHz, saturated)	R^2 (150 kHz, air dry)
10%	0.6253	0.8023	0.7075
20%	0.5521	0.6808	0.5317
30%	0.4503	0.4387	0.4205

Table 3. Effect of moisture content on the correlation coefficient (R^2) of UPV-strength relationship of HPC with FA

Cylindrical samples of 200 mm in height and 100 mm in diameter with the same mix proportions were tested and the resulting UPV measured are compared with those of 150 mm^3 cube samples. Results show that UPV measured in cubes are higher than in cylinders at the same strength level. Equation 9 shows the relationship between the UPV measured in cubes and cylinders:

$$v_{\text{cylinder}} = 0.6115 v_{\text{cube}} + 1.7953 \tag{9}$$

for $v_{\text{cube}} > 4.6$ km/s.

As described in IAEA TCS17, 2002, the ultrasonic pulse transmission types affect the value of the measured UPV. The transmission types are the direct, semi direct and indirect transmission. Equations 10 and 11 show the relationship between the UPV measured using indirect method, semi direct method and the direct method:

$$v_{\text{direct}} = 1.2089 v_{\text{semi direct}} \tag{10}$$

$$v_{\text{direct}} = 1.3000 v_{\text{indirect}} \tag{11}$$

It is observed than there is an increase in the indirect (surface) transmission correction factor (CF) for HPC with fly ash compared to 1.05 reported for normal concrete in BS 1881: Part 203.

2.4 Correction factors concerning UPV-strength and RA-strength relationships of HPC with fly ash

Tests were also done to determine the effect of the transducer's frequency, moisture content, samples' shape, transmission type and FA content on the RA-strength. Table 4 shows the correction factor (CF) suggested from the analysis the test results.

Fly ash content	Correction Factor (CF)	
	UPV $v_{150\,kHz} = CFv_{54\,kHz}$	RA $\beta_{150\,kHz} = CF\beta_{54\,kHz}$
10%	1.0041	1.6397
20%	0.9969	1.5463
30%	1.0211	1.5417

Table 4. Correction factors applied with different transducer's frequency for UPV and RA-strength relationship of HPC with FA

On the other hand, effect of moisture content on UP-strength and RA-strength curve are found to be insignificant as shown in Equations 12 and 13:

$$v_{\text{saturated}} = 1.0034v_{\text{dry}} \tag{12}$$

$$\beta_{\text{saturated}} = 1.0021\beta_{\text{dry}} \tag{13}$$

The effect of fly ash content on the UPV and RA-strength correlation curves for HPC with fly ash compared to the same mix design without fly ash content (FA 0%) is as shown in Table 5.

Fly ash content	Correction Factor (CF)	
	UPV	RA
10%	$v_{FA10\%}=1.4942v_{FA0\%}-2.521$	$\beta_{FA10\%} = 0.2711\beta_{FA0\%}^{1.8769}$
20%	$v_{FA20\%}=1.3250v_{FA0\%}-1.4169$	$\beta_{FA20\%} = 0.0594\beta_{FA0\%}^{2.4494}$
30%	$v_{FA30\%}=0.6587v_{FA0\%}-1.8073$	$\beta_{FA30\%} = 0.1199\beta_{FA0\%}^{2.2340}$

Table 5. Correction Factor (CF) applied with different fly ash content for UPV and RA-strength relationship of HPC with FA

3. Conclusion

This chapter emphasizes on ultrasonic pulse velocity (UPV) and relative amplitude (RA) method in assessing the strength of high performance concrete (HPC) with mineral admixtures such as fly ash. Adding mineral admixtures to replace certain percentages of cement content increase or decrease the strength of HPC compare to the same concrete combination containing only ordinary Portland cement (OPC). By using extremely fine pozzolanic material, such as fly ash, the compactness of the concrete improves because the fly ash fills the micro voids in grain packing. This feature results in higher measured UPV values in higher fly ash content concrete compared to lower fly ash content concrete at the same measured compressive strength. The strength-UPV correlation in HPC with fly ash is found to be poor (correlation coefficient, R^2, reduces with increase of fly ash content). HPC with silica fume samples produces good strength-UPV correlation but, as with the case of fly ash concrete, the silica fume content affects the measured UPV significantly. The correction factor for fly ash content is proposed. It is recommended that when establishing iso-strength curve for concrete with mineral admixtures, correction factor for each mineral admixture is included.

4. References

Carino, N.J. (1994). Nondestructive Testing of Concrete: History and Challenges, In: *Concrete Technology – Past, Present and Future*, Mehta, P.K., (Ed.), 623-678, ACI SP-144, American Concrete Institute, Detroit, MI.

Demirboğă, R., Türkmen, İ. and Karakoc, M. B. (2004). Relationship between ultrasonic velocity and compressive strength for high-volume mineral-admixtured concrete. *Cement and Concrete Research*, Vol. 34, pp.2329–2336.

Galan, A. (1990). *Combined Ultrasound Methods of Concrete Testing*. Elsevier, Amsterdam, North Holland.

Hamid, R. (2004). The Characteristics of Ultrasonic Wave in Hardened High Performance Concrete with Fly Ash and Silica Fume (in Malay), *PhD Thesis*, Universiti Kebangsaan Malaysia, Malaysia.

Hamid, R., Yusof, K.M., Zain and M.F.M. (2010). A combined ultrasound method applied to high performance concrete with silica fume, *Construction and Building Materials*, Vol. 24, No, 1, (January 2010), pp. 94–98, ISSN 0950-0618.

International Atomic Energy Agency TCS 17, *Guidebook on Non-Destructive Testing of Concrete Structures*, Vienna, 2002.

Ismail, M.P., Yusof, K.M. and Ibrahim, A.N. (1996). A combined ultrasonic method on the estimation of compressive concrete strength. *INSIGHT*, Vol. 38, No. 11, (November 1996), pp. 781-785.

Khatib, J.M. 2008. Performance of self-compacting concrete containing fly ash. *Construction and Building Materials*, Vol. 22, No. 9, pp.1963-1971. ISSN 0950-0618.

Krautkramer, J. and Krautkramer, H. (1983). *Ultrasonic Testing of Materials*. (3rd ed.). Springer-Verlag, Berlin.

Malier, Y. (1992). Introduction, In: *High Performance Concrete: From Material to Structure*, Yves Malier, pp. (xiii – xxiv), E & F Spon, ISBN 0419176004, London, UK.

PCA EB001, *Design and Control of Concrete Mixtures*, 14th edition, Portland Cement Association. 2002.

Price, W.F. and Hynes, J.P. 1996. In-situ testing of high strength concrete. *Magazine of Concrete Research*. Vol. 48, No.176, pp. 189-197.

Shah, S.P. 1997. High performance concrete: controlled performance concrete. *Magazine of Concrete Research*. Vol. 49, No: 178, pp. 1-3.

Türkmen, I., Öz, A. and Aydin, A. C. (2010). Characteristics of workability, strength, and ultrasonic pulse velocity of SCC containing zeolite and slag. *Scientific Research and Essays*, Vol. 5, No.15, pp. 2055-2064.

Uysal, M. and Yilmaz, K. (2011). Effect of mineral admixtures on properties of self-compacting concrete. *Cement & Concrete Composites*, Vol. 33, pp. 771–776, ISSN 0958-9465.

10

Imaging Methods of Concrete Structure Based on Impact-Echo Test

Pei-Ling Liu and Po-Liang Yeh
National Taiwan University, Institute of Applied Mechanics
Taiwan

1. Introduction

The impact echo method is a non-destructive test of concrete structures. This method was developed in 1980s (Sansalone & Carino, 1986) and nowadays is extensively used because of its simplicity and robustness. In the impact echo test, an impact force is applied on the surface of the target structure. Then, a transducer is used to measure to response of the structure. Transforming the time signal to the frequency domain, one can determine the depth of the reflector beneath the test point.

The impact echo test is a point-wise detection method. The non-destructive examination of a structure usually requires a large amount of tests. How to integrate the test results to get an overall picture about the condition of the structure is a challenging task. In order to simplify the interpretation, many imaging methods were proposed, providing direct information on the concrete interior.

This chapter introduces several imaging methods based on the impact echo test, including spectral B- and C-scan, spectral tomography, surface rendering, and volume rendering. These imaging methods can be used to determine the size and location of internal defects in concrete structures. Several numerical and experimental examples are given to illustrate the results of these imaging methods. The features of these methods are also compared.

2. The impact echo test and depth spectrum

2.1 Impact echo test

In the impact echo test, a steel ball or a hammer is used to produce a wave source on the surface of the structure. Consequently, stress waves are generated and propagate in the structure. A transducer is placed near the impact point to measure the response of the structure. Then, the received signals are recorded for data analysis.

If there is an interface beneath the test point, the longitudinal waves generated by the impact will bounce between the top surface and the interface. Such echo waves will form a peak in the spectrum of the signal. The peak frequency f and the depth of the interface d is related by the following formula (Sansalone & Carino, 1986):

$$d = C_p / 2f \qquad (1)$$

where C_p is the velocity of the longitudinal wave. Equation 1 is valid only when the acoustic impedance of the material on the other side of the interface is less than that of concrete. If the material is stiffer, e.g., steel, the factor 2 in Eq. 1 should be replaced by 4.

It is proposed in the literature (Gibson & Popovics, 2005; Lin & Sansalone, 1992; Sansalone & Streett, 1997; Schubert & Köhler, 2008; Zhu & Popovics, 2007) that the depth obtained by Eq. 1 be multiplied by a correction factor to eliminate the discrepancy between the actual and predicted depths. The correction factor is dependent on the geometry of the structure. Its value has been determined for a few types of structural members. When imaging method is adopted to examine structures with defects, the geometry of the target structure is complex. It is almost impossible to choose an appropriate value for correction factor. Hence, Eq. 1 is used directly in the imaging process.

The Fourier analysis is the most widely used approach to construct the spectrum of the test response. However, the Fourier spectra usually contain ripples and multiple peaks generated by the transform process. Such artificial interferences may jeopardize the interpretation of test results (Yeh & Liu, 2008). Several time–frequency techniques have been proposed in the literature to prevent the interferences, for example, the wavelet transform and the Hilbert–Huang transform (Abraham et al., 2000; Algernon & Wiggenhauser, 2005; Lin et al., 2009; Shokouhi et al., 2006; Yeh & Liu, 2008). However, there is a trade-off because none of these methods would produce echo peaks as sharp as the Fourier transform does.

2.2 Depth spectrum

No matter which transform is adopted, the horizontal axis of the spectrum is the frequency, not depth. For the imaging purpose, it is preferable to have depth as the horizontal axis. The horizontal axis of the spectrum can be easily transformed to the depth axis by applying Eq. 1, as proposed by Yeh and Liu (2009). The frequency-depth transformation procedure is as follows:

Suppose $a(f)$ is the original spectrum of a signal.

1. Select an appropriate depth interval Δz.
2. Apply Eq. 1 to $i\Delta z,\ i=1,2,\ldots$ to find the corresponding frequency f_i.
3. Determine the maximum amplitude $\hat{a}_i$ in each interval (f_i, f_{i-1}), $i=1,2,\ldots$, as shown in Fig. 1.
4. Plotting $\hat{a}_i$ versus $i\Delta z$ yields the depth spectrum of the signal.

Notice that $\hat{a}_i$, instead of $a(f_i)$, is used to plot the depth spectrum. This is to insure that no peak is left out in the depth spectrum since peaks are the most important information in the spectrum. One should also know that the transformation cannot start from z=0 because it maps to $f=\infty$.

The depth spectrum constructed as above has a constant depth interval, which is convenient for image processing. However, the data points in the Fourier spectrum are not totally retained. Therefore, some of the details in the Fourier spectrum are not reproduced in the depth spectrum. The peaks, which contain the most important information, are preserved, but they may not appear at the precise depth. The maximal depth deviation of a peak is $\Delta z/2$. Therefore, it is advisable to make the depth interval as short as possible.

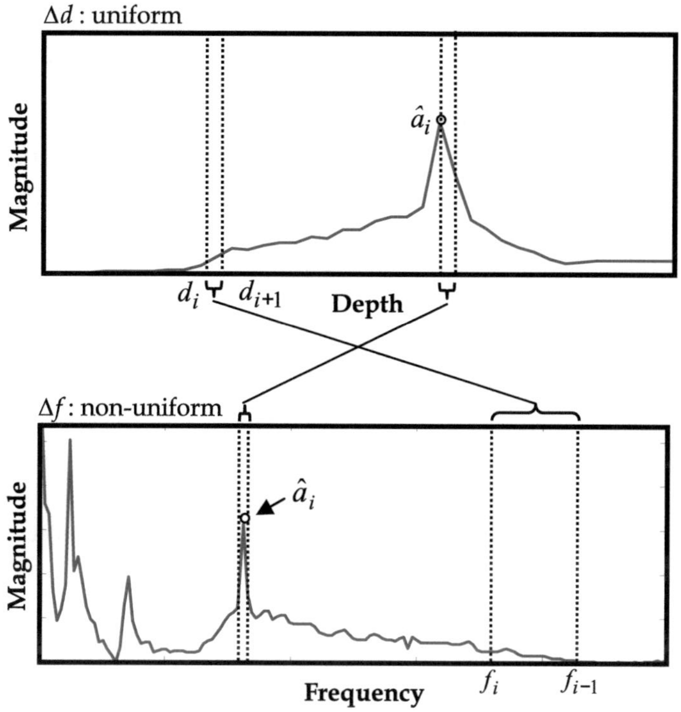

Fig. 1. Frequency-depth transformation

The depth spectrum is useful for the imaging of impact echo data. It has another advantage if the depth D of the concrete structure is known. In the impact echo test, several vibration modes of the structure will be induced. The impact echo is one of them. Unfortunately, the peaks caused by the lower-modes are often higher than the echo peak in the spectrum. This complicates the interpretation of results. The frequencies of the lower modes usually correspond to depths greater than D. Hence, if the depth spectrum is drawn only for the depth range $0 \le z \le D$, the lower-mode peaks do not appear. That simplifies the image interpretation.

3. Imaging methods

In general, the imaging of impact echo data contains three steps: data acquisition, data construction, and image rendering. The details of these steps will be introduced in the following sections.

3.1 Data acquisition and construction

To acquire data for image rendering, one has to conduct a series of impact echo tests on the target structure. Firstly, select a coordinate system such that the x-y plane coincides with surface of the concrete. Secondly, draw a mesh on the concrete surface, and perform the impact echo test at each grid of the mesh, as shown in Fig. 2(a). The location of the impact and receiver can be chosen arbitrarily inside the grid. However, their midpoint should be located at the center of a grid.

Then, apply the Fourier transform or other transforms to the test signals to obtain the frequency spectra. In tomography and 3D imaging, the spectra are further transformed into the depth spectra.

After obtaining the spectra, one can proceed with the construction of data. Assume that the numbers of test points along the x and y directions are n_x and n_y, respectively, and each spectrum contains n_z data points. All the spectra are assembled into an $n_x \times n_y \times n_z$ matrix, $V[i,j,k]$. The array $V[i,j,1 \le k \le n_z]$ is simply the amplitude of spectrum under test point (i, j).

After the volume data is constructed, one may apply any of the imaging methods to depict the interior of concrete structure.

Notice that in tomography and 3D imaging, each vector $V[i,j,1 \le k \le n_z]$ is a depth spectrum. Suppose the grid size of the test mesh is $\Delta x \times \Delta y$ and the depth interval of the depth spectrum is Δz, each element in $V[i,j,k]$ corresponds to a voxel (volume element) in the space with voxel size $\Delta x \times \Delta y \times \Delta z$. Hence, the matrix maps to a rectangular solid with side lengths L_x, L_y, and L_z, where $L_x = n_x \Delta x$, $L_y = n_y \Delta y$, and $L_z = n_z \Delta z$. In this case, $V[i,j,k]$ represents the reflection energy from voxel $[i,j,k]$ in the rectangular solid.

In spectral B- or C-scan, on the other hand, each vector $V[i,j,1 \le k \le n_z]$ is a frequency spectrum. Hence, the third axis is frequency, not depth. Therefore, the matrix does not map to a rectangular solid in the space.

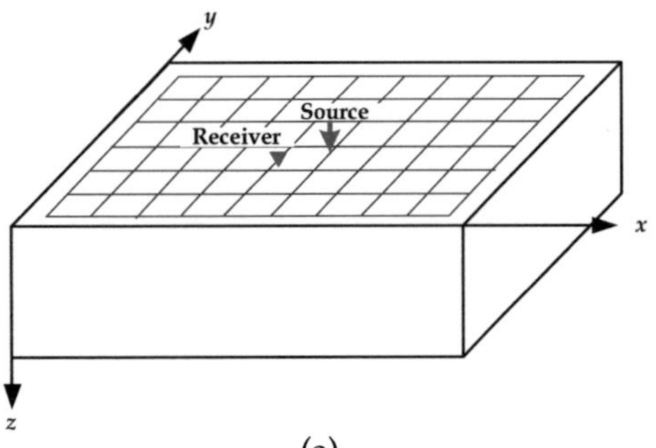

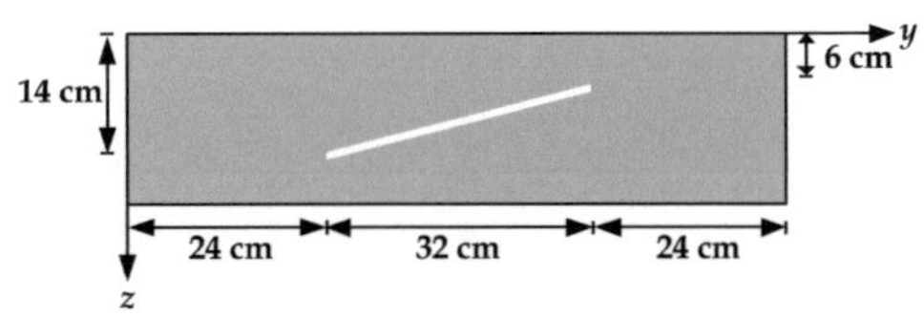

Fig. 2. (a) Test mesh on the concrete surface and (b) side view of the example specimen

In this chapter, both numerical simulations and model tests are presented to illustrate the imaging methods. A concrete specimen with an slant internal crack is considered. The dimension of the specimen is 80 cm (L) × 80 cm (W) × 20 cm (H). The depth of the internal crack ranges from 6 cm to 14 cm beneath the surface, as shown in Fig. 2(b). The coordinates of the four corners of the crack are $[x, y, z]$ = [24, 24, 14], [24, 56, 14], [56, 24, 6], and [56, 56, 6] (in cm).

The test mesh on the specimen is a 76 cm × 76 cm square with 19 × 19 grids, leaving 2 cm of margins on the four sides. In each test, the impact source and the receiver were located at the upper right and lower left corners of a grid, respectively.

In the numerical simulation, the finite element code LS-Dyna970 (Hallquist, 2003) was adopted to simulate the response of the concrete specimen due to the impact of a steel ball. Three-dimensional solid elements with side length 1cm were used in the numerical simulation. The mass density, Young's modulus, Poisson's ratio, and the longitudinal wave speed of the concrete are 2300 kg/m^3, 33.1 GPa, 0.2, and 4000 m/s, respectively.

A time-varying pressure was applied to the surface to simulate the impact of a steel ball with a diameter of 6 mm. According to Goldsmith (1960), the pressure was approximated by a half-sine function with a contact time t_c= 25 μs. The total time of simulation was 3 ms, and the time increment was 3 msec/1024 = 2.93 μs.

Non-reflecting boundary conditions were applied on the four sides of the blocks to prevent waves reflecting from the boundaries. As such, the examples could focus on the results of imaging methods, excluding the effect of specimen geometry.

In the model test, the dimension of the concrete specimen and the crack location are identical to those of the numerical model. The longitudinal wave velocity of the concrete is 3890 m/s. In the experiment, a steel ball with diameter 6 mm was dropped on the concrete surface to produce the impact source. A conical transducer, developed at the Institute of Applied Mechanics, National Taiwan University, was adopted to measure the vertical displacement on the concrete surface. The impact and receiver locations were the same as in the numerical examples. The received voltage signals were enhanced by an amplifier and then recorded by a digital oscilloscope (LeCroy WavePro940). In the model tests, the sampling rate was 1 MHz, and the total sampling time was 5 msec.

Unlike the numerical tests, the energy of the impact source in the model tests may vary from test to test. Therefore, the experimental data were normalized based on the amplitude of the surface wave.

3.2 Spectral B-scan and C-scan

The spectral B- and C-scan methods were firstly proposed by Liu and Yiu (2002) to detect the internal cracks in concrete. The spectral B-scan was adopted by Schubert et al. (2004) to measure the thickness of concrete specimens. Kohl et al. (2005) further extended the idea so that one could combine impact echo data with ultrasonic data to construct B- and C-scan images.

The concepts of spectral B- and C-scans are similar to the ultrasonic B- and C-scans. They respectively generate the images of vertical and horizontal cross-sections of a test specimen. The data acquisition and construction for spectral B- and C scans are the same as described in Section 3.1. However, in spectral B-scan one only has to perform impact echo tests along a test line. Therefore, n_y=1 in B-scan.

For the purpose of imaging, the matrix $V[i,j,k]$ needs to be transformed into a matrix of color scales $c[i,j,k]$:

$$c[i,j,k]=\begin{cases} c_{max} & V[i,j,k]>V_{max} \\ c_{max}\dfrac{V[i,j,k]-V_{min}}{V_{max}-V_{min}} & V_{min}<V[i,j,k]<V_{max} \\ 0 & V[i,j,k]<V_{min} \end{cases} \quad (2)$$

where c_{max} is the upper bound of the color scale, and $[V_{min}, V_{max}]$ defines the range in which $V[i,j,k]$ is mapped linearly to $c[i,j,k]$. If $V[i,j,k]$ exceeds V_{max}, the color scale is set to c_{max}; if $V[i,j,k]$ is less than V_{min}, the color scale is set to 0. Unless specified otherwise, V_{max} and V_{min} are chosen to be the maximum and minimum of the volume data, respectively. However, one can increase the value of V_{min} to suppress the noise and enhance the contrast of the image (Liu & Yeh, 2010). The contrast ratio V_{min}/V_{max} that yields a satisfactory result can be obtained by adjusting the ratio manually.

In spectral B-scan, the image of the vertical section under the test line can be obtained simply by using $c[i, j=1,k]$ to generate a 2D density plot. Notice that $c[i, j,k]$ reflects spectral amplitude. If a peak appears in the *i*th spectrum at frequency $k\Delta f$, the corresponding color scale is high and one can see a clear spot at location $x=i\Delta x$, $f=k\Delta f$ on the B-scan image. The clear spots will connect into a clear stripe if a crack exists. As such, one can detect the size and location of the crack in the concrete.

The spectral C-scan constructs an image for a horizontal section. Hence, one has to select the depth of the horizontal section to be examined. Suppose the depth corresponds to frequency $f=K\Delta f$, according to Eq. 1. The spectral C-scan image of the horizontal section can be obtained by using $c[i, j,k=K]$ to generate a 2D density plot.

Theoretically, the color of the image is homogeneous if no defect exists. If the section does contain a defect, there will be peaks in the spectra. Hence, one can determine if there is a defect simply by examining the color variation of the image.

In the following, numerical examples are given to illustrate the spectral B- and C-scan methods. Figure 3 shows the spectral B-scan images of the vertical sections under test line (a) x=16 cm and (b) x=40 cm. The crack occurs in the range $24 \le y \le 56$ cm. Hence, the first section contains no crack while the second section does. One can see that there are only horizontal stripes in Fig. 3(a). The stripe with the highest color scale (red) appears at *f*=10 kHz. Since the longitudinal wave velocity C_P=4000 m/s, this frequency corresponds to the depth of the bottom, 20 cm, according to Eq. 1. The bright stripe near 20 kHz is caused by the multiple peaks of the Fourier spectra. There are also red stripes at the bottom of the image. They are induced by the lower-mode vibrations of the specimen.

Figure 3(b) looks quite different from Fig 3(a). One can find an inclined red stripe occurring in the crack range. Outside that range, the image resumes the no-crack pattern, that is, red horizontal stripes appearing near *f* =10 kHz, denoting the bottom of the specimen. Although the B-scan image reveals the existence of the inclined crack, it does not exhibit the profile of the specimen under the test line. This is certainly because the vertical axis is frequency, not depth.

Figure 4 shows the spectral C-scan images of the horizontal sections at depth (a) 4 cm, (b) 10 cm and (c) 20 cm. Figure 4(a) looks just like a blue rectangle, implying that the cross-section contains no reflector. The image in Fig. 4(b) contains a red zone, depicting the crack on the cross-section. The yellow and cyan zones in the image both denote concrete, but the cyan zone is the concrete under the crack. Figure 4(c) shows the C-scan image at the bottom of the specimen. A cyan zone appears at the center of the image, surround by red zone. It can be considered as the shadow cast by the crack because the waves are blocked by the crack. The cyan zone is not a square because stress waves may go around the crack edge and reach the bottom when the test point is near the top of the edge. This zone provides a supplementary evidence for the existence of a defect above this region.

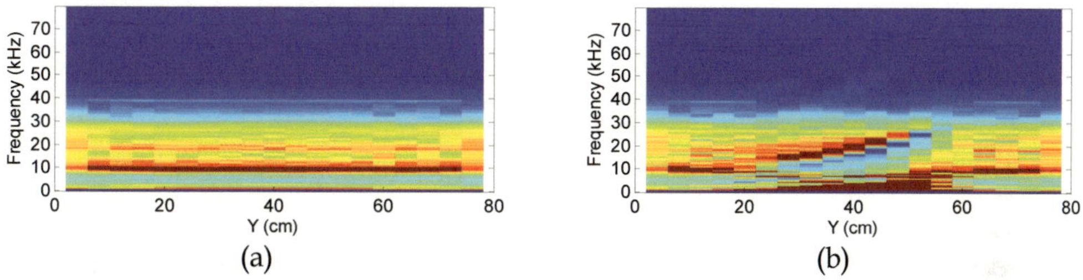

Fig. 3. Spectral B-scan of cross-sections under test lines (a) x=16 cm and (b) x=40 cm

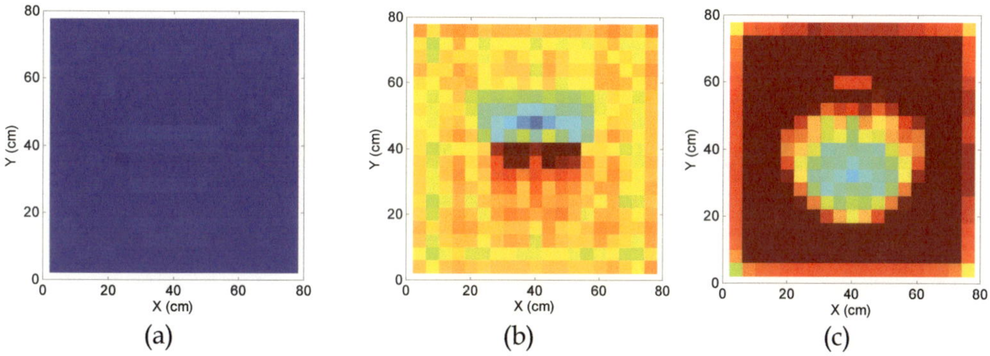

Fig. 4. Spectral C-scan of cross-sections at depth (a) 4 cm, (b) 10 cm, and (c) 20 cm

3.3 Spectral tomography

In the spectral B-scan method, the Fourier spectra of the test signals are assembled to construct an image of the test cross-section. Such image certainly provides useful information about internal defects. However, the vertical axis of the image is frequency. Hence, the spectral B-scan does not provide a "picture" of the test section.

In order to provide a more intuitive image, the vertical spectral tomography was proposed by Liu and Yeh (2010) . The imaging process of the vertical spectral tomography is the same as in spectral B-scan except that the Fourier spectra are replaced by depth spectra. Liu and Yeh (2011) extended the notion further to tomography of arbitrary cross-sections. As such, the inspector could examine the interior of a structure from various angles to get better understanding of its condition. The spectral C-scan is only a special case of spectral tomography.

In spectral tomography, the volume data is constructed using the depth spectra of the impact echo test. Similar to the spectral B- and C-scan, the matrix $V[i,j,k]$ is transformed into a matrix of color scales $c[i,j,k]$ according to Eq. 2. Then, the matrix $c[i,j,k]$ is used to produce tomograms for designated cross-sections.

Consider a cross-section defined by $\mathbf{n}^T\mathbf{x}+b=0$, where $\mathbf{n}$ is the outward normal of the cross-section, and $\mathbf{x}=[x,y,z]$ is the position vector. The spectral tomogram can be generated as follows:

1. Define a new coordinate system $\mathbf{x}'$ such that the $x'-y'$ plane coincides with the cross-section, as shown in Fig. 5(a). The coordinates in the new and old systems are related by

$$\mathbf{x}'=\mathbf{Q}^T(\mathbf{x}-\mathbf{t}) \tag{3}$$

where $\mathbf{Q}$ is the rotation matrix with component $Q(i,j)=\mathbf{e}_i\cdot\mathbf{e}'_j$, and $\mathbf{t}$ is the translation vector.

The coordinate transformation defined in Eq. (3) is not unique. $\mathbf{t}$ can be easily defined by choosing a point on the cross-section as the origin of the new coordinate system. $\mathbf{Q}$ can be constructed as follows. Suppose $\mathbf{n}=[p,\ q,\ r]^T$ is the outward normal of the cross-section, where $p^2+q^2+r^2=1$. The new base vectors can be selected as $\mathbf{e}'_z=\mathbf{n}$, $\mathbf{e}'_x=\mathbf{e}_z\times\mathbf{e}'_z$, and $\mathbf{e}'_y=\mathbf{e}'_z\times\mathbf{e}'_x$. Hence,

$$\mathbf{e}'_x=\frac{1}{\sqrt{1-r^2}}\begin{bmatrix}-q\\p\\0\end{bmatrix}\quad \mathbf{e}'_y=\frac{1}{\sqrt{1-r^2}}\begin{bmatrix}-pr\\-qr\\1-r^2\end{bmatrix}\quad \mathbf{e}'_z=\begin{bmatrix}p\\q\\r\end{bmatrix} \tag{4}$$

and

$$\mathbf{Q}=\begin{bmatrix}\dfrac{-q}{\sqrt{1-r^2}} & \dfrac{-pr}{\sqrt{1-r^2}} & p\\ \dfrac{p}{\sqrt{1-r^2}} & \dfrac{-qr}{\sqrt{1-r^2}} & q\\ 0 & \sqrt{1-r^2} & r\end{bmatrix} \tag{5}$$

2. Obtain the orthographic projection of the $L_x\times L_y\times L_z$ solid on the $x'-y'$ plane, and find a coordinate rectangle to enclose the projection, as shown in Fig. 5(b).

The orthographic projection of the solid can be obtained by projecting the vertices of the solid on the $x'-y'$ plane. The location of a projected vertex is simply the new coordinates (x',y') of that vertex obtained by Eq. (3). Suppose (x',y') satisfies $x'_{\min}\le x'\le x'_{\max}$ and $y'_{\min}\le y'\le y'_{\max}$ for every vertex. Then, the bounding rectangle is formed by the coordinate lines $x'=x'_{\min}$, $x'=x'_{\max}$, $y'=y'_{\min}$, and $y'=y'_{\max}$.

3. Draw a mesh of square grids on the rectangle. For each grid, using Eq. (3) to transform the location of its center $\mathbf{x}'_c = [x'_c, y'_c, 0]$ back to the original coordinate system, i.e.,

$$\mathbf{x}_c = \mathbf{Q}\mathbf{x}'_c + \mathbf{t} \tag{6}$$

4. Use $\mathbf{x}_c$ to determine in which voxel the center is located. Then, determine the color scale of the grid $c(\mathbf{x}_c)$ from the volume data. If $\mathbf{x}_c$ is outside the volume, no color scale is assigned and the pixel is transparent.
5. Construct the tomogram of the cross–section by filling each grid with its color $c(\mathbf{x}_c)$.

Notice that in Step 3, each grid in the mesh corresponds to a pixel in the tomogram. It is advisable to adopt a fine mesh so that one can obtain a high-quality image. The resolution of the tomogram is certainly limited by the voxel size of the volume data. However, a pixel in the tomogram may pass through more than one voxel in the solid. A fine mesh helps to display the boundary between adjacent voxels more precisely. Furthermore, it helps to better delineate the borderline of the cross-section in the tomogram. Take the cross-section in Fig. 5 for example. The borderline of the cross-section is a hexagon. Since only pixels with its center located inside the volume are colored, the tomogram of the cross-section appears as the colored polygon in Fig. 5(b). It is seen that some segments of the borderline become zigzag. Apparently, the zigzag borderline approximates the true boundary better if a finer mesh is adopted.

The proposed spectral tomography does not provide the velocity profile of a test section as most conventional nondestructive techniques do. Instead, the spectral tomogram should be considered as a profile of reflection energy due to the impact.

In the following, a numerical example is presented to illustrate the spectral tomography. Figure 6 shows the horizontal tomograms constructed at depth (a) 20 cm, (b) 12 cm, and (c) 9 cm, respectively. Similar to Fig. 4, the crack casts a shadow on the bottom tomogram in Fig. 6(a). Bright stripes appear in Figs. 6(b) to (c), depicting the crack on each cross-section. Since this is a slant crack, the bright stripes shift along the y direction as the cross-section moves up. In this case, it is difficult to get an overall picture of the crack using horizontal tomograms alone.

Figure 7 shows the vertical tomograms along x= (a) 18 cm and (b) 40 cm, respectively. No crack presents on the first cross-section. Hence, one only finds bright stripes at the bottom of Fig. 7(a). In contrast, the slant crack is clearly depicted on the tomogram in Fig. 7(b). One can use this tomogram to determine the length and inclination of the crack.

It should be mentioned that the thickness of the bright stripe in Fig. 7(b) does not represent the thickness of the crack. In fact, the impact echo test cannot provide information about the thickness of the crack. The thickness of the bright stripe results from the width of the echo peak in the depth spectra. Hence, the true depth of the crack should be determined based on the location of the brightest pixels in the tomogram.

One may construct oblique tomograms to get a picture of the whole crack, as shown in Fig. 8. The slopes of the cross-sections vary from 0° to 90°. One can see that the area of the crack image is maximal when the slope is around 14°, coincident with the crack orientation. Hence, Fig. 8(c) provides the best picture of the crack among the tomograms. The crack image is narrower near the shallow edge of the crack. Nevertheless, one can use the bright zone to estimate the size of the crack. This is quite difficult if only vertical or horizontal tomograms are available.

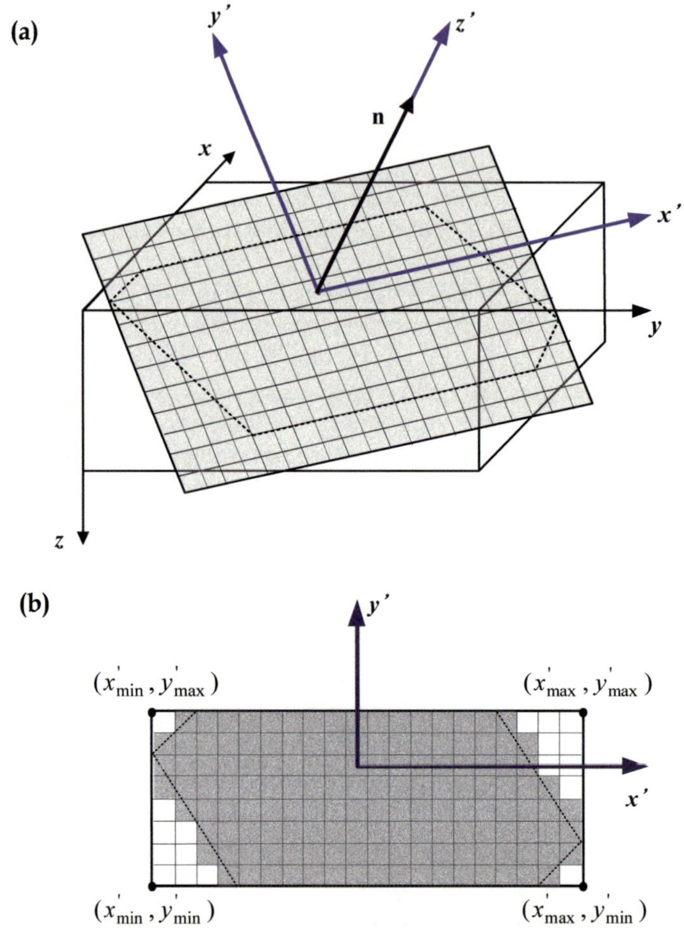

Fig. 5. (a) Coordinate transformation in tomography and (b) orthographic projection of specimen on the $x' - y'$ plane

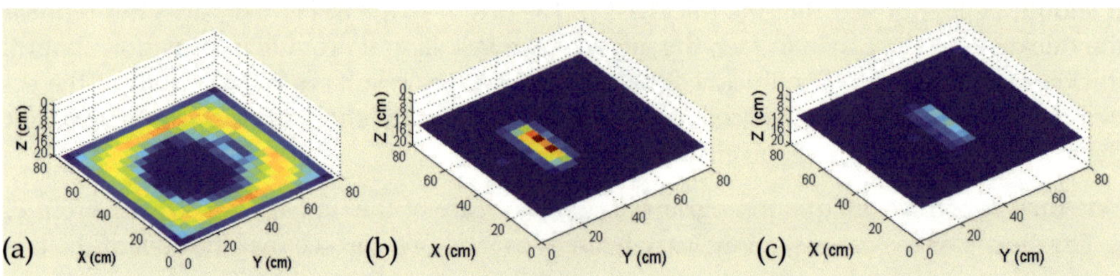

Fig. 6. Horizontal tomograms at depth (a) 20 cm, (b) 12 cm and (b) 9 cm

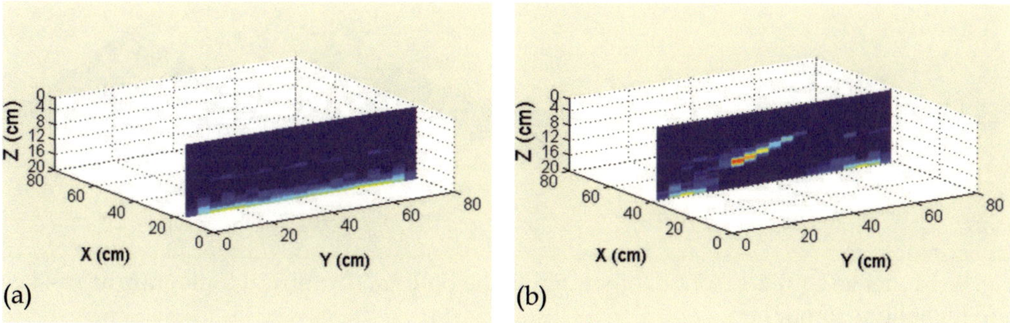

Fig. 7. Vertical tomograms along x = (a) 18 cm and (b) 40 cm

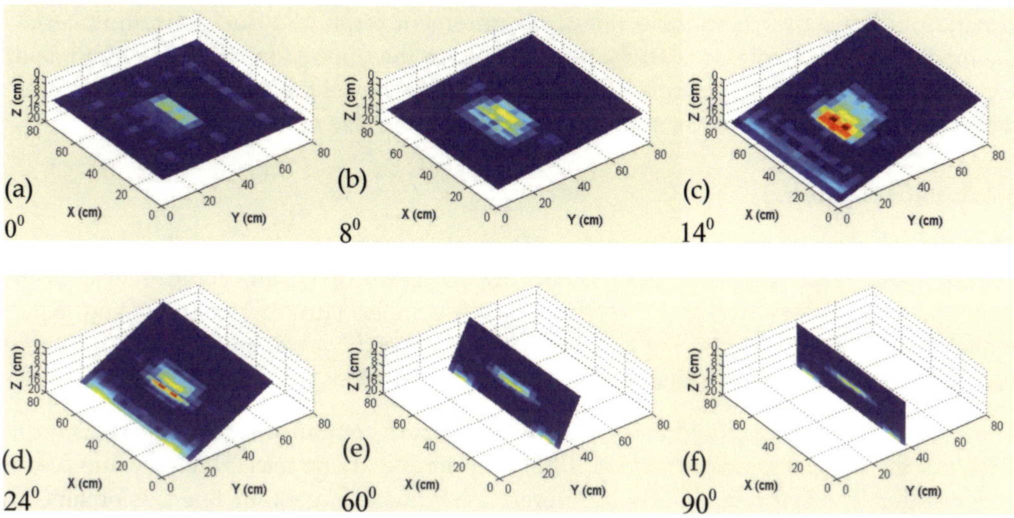

Fig. 8. Oblique tomograms with slopes (a) 0°, (b) 8°, (c) 14°, (d) 24°, (e) 60°, and (f) 90°

This numerical example demonstrates that the inspector can use spectral tomography to examine any cross-section of a specimen. To get an overall assessment of the interior condition, it is advisable to examine the specimen in a systematic way rather than scan randomly. An inspection procedure is proposed herein: Firstly, construct a horizontal tomogram at the bottom of the specimen to find the defect zone. Then, construct a series of horizontal or vertical tomograms to find the location, size, shape, and orientation of the defect. Finally, based on the scanning results, perform oblique tomography to get a better image of the defect if necessary.

Figure 9 shows the results of spectral tomography in the model test. From Fig. 9(a), one can see the dark zone also forms beneath the crack. Figure 9(b) is the vertical tomogram along x=44 cm. The result is similar to that of numerical test. Bright zones occur at the crack and bottom. Despite noise, one can easily detect the inclination of the crack using this tomogram. Figure 9(c) shows the oblique tomogram of the crack plane. A bright zone appears in the crack area.

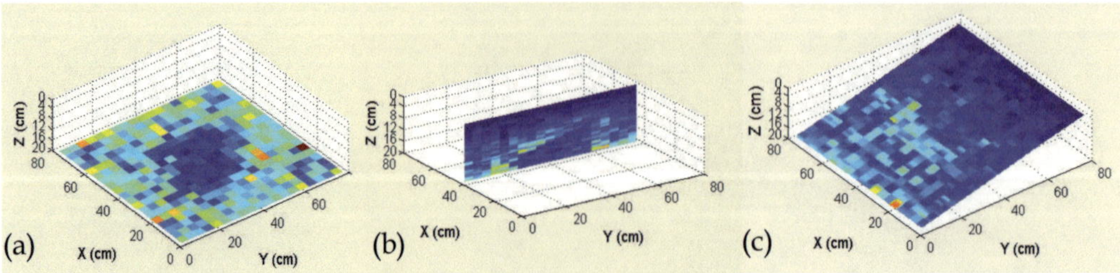

Fig. 9. Model test, (a) horizontal tomogram at the bottom, (b) vertical tomogram at x=44 cm, and (c) oblique tomogram.

Although the test data have been normalized by the amplitude of the surface wave, the tomograms still look mottled. The deterioration of tomogram quality may come from variation of impact force, random noise, measurement error, and non-uniform material of the model test. Therefore, it is advisable to compare the tomograms of different angle and sectioning before making judgment. Regardless of the noise, the bright and dark zones in these tomograms still reveal the location, orientation, and size of the crack.

3.4 Surface rendering

With the progress of computer graphics, 3D display becomes a trend in the processing of volume data. The methods that render 3D images are called volume visualization techniques. There are two branches techniques, namely, surface rendering and volume rendering. The surface rendering technique was proposed by Yeh and Liu (2009) to depict the internal cracks in concrete structures.

The surface rendering method is equivalent to drawing contour lines in a 2D density plot. Consider the vertical tomogram in Fig. 10(a) for example. If one selects an iso-value 104, one gets contour line 1; if one chooses a different iso-value 208, contour line 2 is obtained, as shown in Fig. 10(b). Apparently, if the iso-value is chosen properly, the contour line will depict the location of an interface. The same idea can be extended to the 3D case.

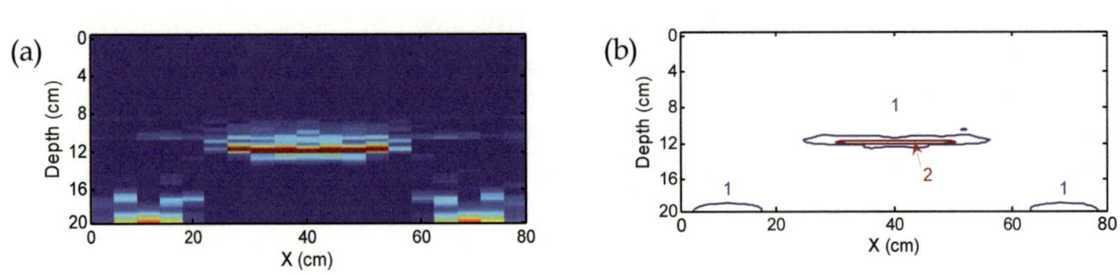

Fig. 10. 2D analogy of surface rendering (a) vertical tomogram and (b) contour plot

The main idea of surface rendering is to abstract the iso-surface from the volume data, i.e., to find a surface with the same spectral amplitude. Let $V(x,y,z)$ denote the volume data. The iso-surface corresponding to iso-value C can be represented as:

$$\{(x,y,d) : V(x,y,d) = C\} \tag{7}$$

Once an iso-value is assigned, one can use triangular patches to generate an iso-surface. Similar to the contour lines in Fig. 10, if the iso-value is chosen properly, the iso-surface will depict the location of an interface in the specimen.

After the iso-surface is generated, it is projected to a 2D view plane. Notice that when a 3D object is projected to a 2D plane, a sphere will appear as a circle and a cube is turned into a hexagon. In order to obtain a stereograph, shading and lighting are necessary.

The Phong reflection model is a popular and effective approach for this end (Angel, 2006). This model assumes three types of light-material interactions, namely, ambient, diffuse, and specular reflection. The intensity of the reflected light is dependent on four vectors: the normal vector of the surface **N**, the viewer vector **V**, the light source vector **L**, and the reflected ray vector **R**.

The ambient light has the same intensity in the space. When it encounters a surface, it is absorbed and reflected. The intensity of the ambient reflection I_a is as follows:

$$I_a = K_a L_a \quad 0 \le K_a \le 1 \tag{8}$$

where K_a is the ambient reflection coefficient and L_a is the intensity of the ambient light.

The diffuse reflection is characterized by the roughness of the surface. The intensity of the reflected light depends on the material and incident direction of the light. Since each point on the surface has a different normal vector, it will reflect different amount of light. The intensity of the diffuse reflection I_d is as follows:

$$I_d = K_d L_d \mathbf{L} \cdot \mathbf{N} \quad 0 \le K_d \le 1 \tag{9}$$

where K_d is the diffuse reflection coefficient and L_d is the intensity of the incident diffuse light.

The specular reflection is used to highlight the shiny part of surface. Although the ambient and diffuse reflection make the image look three-dimensional, the lack of specular reflection would make the surface look dull. The intensity of the specular reflection is as follows:

$$I_s = K_s L_s (\mathbf{R} \cdot \mathbf{V})^{\beta} \quad 0 \le K_s \le 1,\ \beta \ge 1 \tag{10}$$

where K_s is the diffuse reflection coefficient, β is the shininess coefficient, and L_s is the intensity of incident specular light. Usually, $L_s = L_d$. As β increases, the reflected light tends to concentrate on a smaller region.

The total intensity of the reflected light from an object is the sum of I_a, I_d, and I_s. The coefficients K_a, K_d, K_s, and β are taken as 0.6, 0.8, 0.5, and 150, respectively, in the following examples.

Figure 11 shows the top, side, and oblique views of the surface rendering image of the numerical model. An iso-surface denoting the crack is observed in all three images at the correct location. One can also find the bottom of the specimen in the image. However, a hole is formed in the bottom beneath the crack, as shown in Fig. 11(a). This is because the waves are blocked by the crack and cannot reach the bottom.

From the side view in Fig. 11(b), one can see that the crack is thicker than the real crack. Actually, the impact echo test cannot provide thickness of the crack. The thickness of crack in the image results from the width of echo peak in the spectra. Therefore, the true depth of the crack is around the center of the iso-surface.

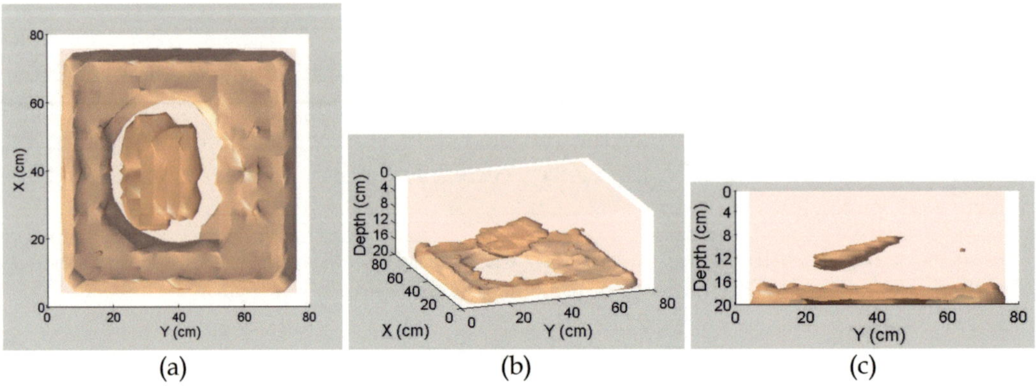

Fig. 11. Surface rendering of numerical model, (a) top view, (b) side view, and (c) oblique view

In practical applications, it is critical to choose a proper iso-value because different iso-values result in different images. The iso-value of the images in Fig. 11 is 15. Figure 12 shows a series of surface rendering images with iso-values 12, 18, and 22, respectively. Clearly, the crack iso-surface shrinks as the iso-value increases. If the iso-value is too low, the iso-surface denoting the crack becomes a large layer. On the other hand, if the iso-value is too high, the crack iso-surface becomes excessively small. Since the hole can be considered as the "shadow" of the crack, the iso-value should be chosen such that the sizes of the crack and the hole match.

The surface rendering images of the experimental model are shown in Fig. 13. Because the experimental signals are contaminated by noise, the 3D images are not as clear as in the numerical example. The crack is not enveloped in a single iso-surface. Several iso-surfaces appear around the crack location instead. This is mainly because the impact source is unsteady. Normalization of the signals may reduce the influence of the intensity of the source function, but not the shape. Therefore, it is hard to find an appropriate iso-value to surround the crack by a single iso-surface.

Nevertheless, one can still manage to find the crack by viewing the specimen at different angles. For example, the side view in Fig. 13(b) provides a clear picture of the inclined crack. Furthermore, the hole beneath the crack is still visible in Figs. 13(a) and (c).

3.5 Volume rendering

The volume rendering technique was proposed by Yeh and Liu (2008) to construct 3D images based on impact echo data. The data acquisition and construction procedures are the same as described in Section 3.1.

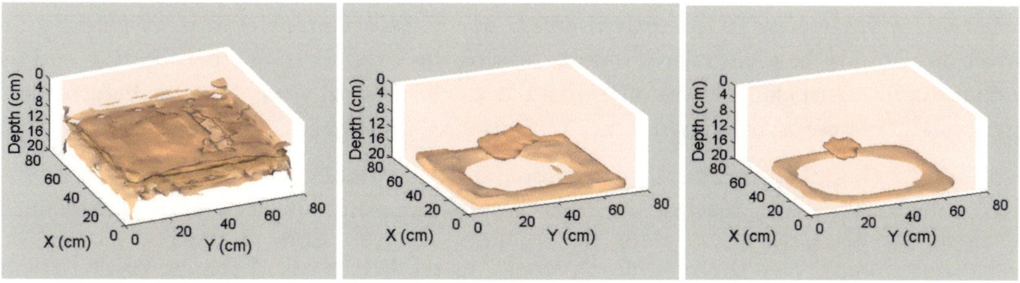

(a) (b) (c)

Fig. 12. Surface rendering of the numerical model with iso-values (a) 12, (b) 18, and (c) 22

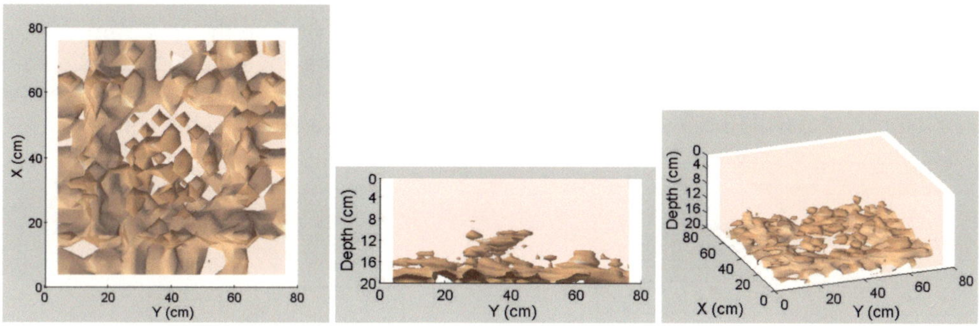

(a) (b) (c)

Fig. 13. Surface rendering of the experimental model, (a) top view, (b) side view, and (c) oblique view

In the volume rendering method, each voxel is assigned an opacity α based on the volume data such that $0 \le \alpha \le 1$. The opacity represents the level of difficulty that light goes through a voxel. If a voxel is complete opaque, $\alpha = 1$, and if it is complete transparent, $\alpha = 0$. The opacity of a voxel can be determined as follows:

$$\alpha[i,j,k] = \begin{cases} 1 & V > V_{\max} \\ \dfrac{V[i,j,k] - V_{\min}}{V_{\max} - V_{\min}} & V_{\min} < V < V_{\max} \\ 0 & V < V_{\min} \end{cases} \tag{11}$$

As in spectral tomography, the default values of $V_{\max}$ and $V_{\min}$ are the maximum and minimum of the volume data, respectively.

The process of volume rendering is analogous to X-ray examination. To construct the volume rendering image, let parallel rays emitted from a light source behind the volume transmit through the volume and reach a projection plane, as shown in Fig. 14(a). When a ray passes through the volume, it accumulates the opacity of the voxels it encounters, as shown in Fig. 14(b). The compositing operation is a recursion of opacity, as shown in the following (Angel, 2006):

$$\alpha_{out} = (1 - \alpha)\alpha_{in} + \alpha \tag{12}$$

where α is the opacity of the current voxel, α_{in} is the accumulated opacity entering the voxel, and α_{out} is the accumulated opacity leaving the voxel. If $\alpha = 1$, then $\alpha_{out} = 1$ and the light is totally obstructed. On the other hand, if $\alpha = 0$, then $\alpha_{out} = \alpha_{in}$ and the light remains unchanged. Therefore, α is an indicator of the degree that light penetrates the voxel.

Equation 12 is applied recursively to determine the accumulated opacity of a ray until it leaves the volume and reaches the projection plane. After the accumulated opacity is obtained for each ray penetrating the volume, one can draw a density plot of the accumulated opacity on the projection plane. This method is called the ray casting (Watt, 2000). Although the ray casting method describes the concept of imaging clearly, it is time-consuming. Hence, the texture mapping technique has been proposed to speed up the imaging process (Engel et al., 2006).

Recall that in the impact echo test, an interface will induce a peak in the spectrum. Hence, if a defect appears in the volume, the spectral amplitude along the defect is large, so is the opacity. Apparently, the rays that pass through the defect will be dimmer than the rays that do not. Hence, one will find a shadow in the image if there is a defect in the volume.

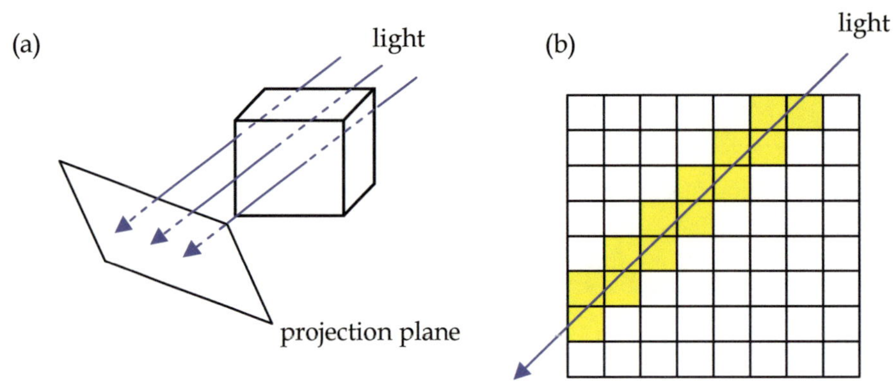

Fig. 14. The volume rendering method

The signals obtained in the impact echo tests inevitably contain noise. That may downgrade the quality of image and make the diagnosis difficult. This problem can be tackled by adjusting the relation between the volume data and the opacity, as defined in Eq. 11. It does not help to alter the value of $V_{\max}$. Hence, one can simply use its default value. The value of $V_{\min}$ can be increased to suppress noise and enhance the contrast of image.

Figure 15 shows the influence of $V_{\min}/V_{\max}$ on the volume rendering image. Generally speaking, as $V_{\min}/V_{\max}$ increases, the crack image gets clearer and the hole at the bottom gets larger. With $V_{\min}/V_{\max}$ below 20%, the image looks blurry, as seen in Figs. 15(a) and (b). When $V_{\min}/V_{\max}$ is increased to 30%, the crack and the hole at the bottom become visible. When $V_{\min}/V_{\max}$ = 40%, one gets a very good image of the specimen.

However, the contrast ratio should not be over raised. As $V_{\min}/V_{\max}$ reaches 50%, the crack starts to shrink and the size of the hole exceeds that of the crack. The situation is even worse as $V_{\min}/V_{\max} = 60\%$.

Notice that the optimal value of V_{min}/V_{max} may change from case to case. The inspector has to try different value to see which value yields the best result.

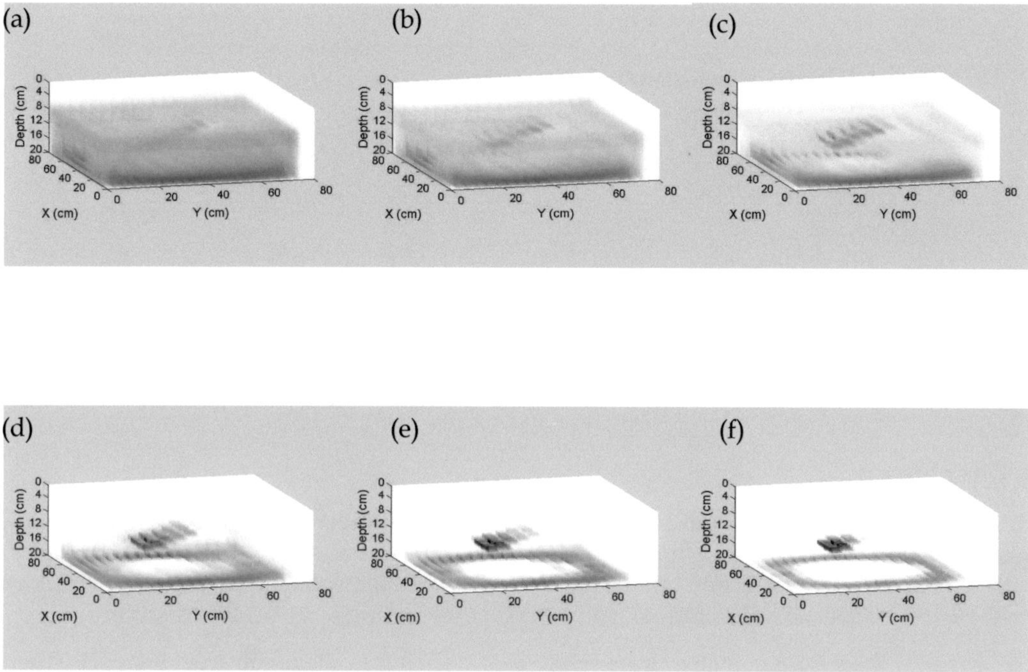

Fig. 15. Volume rendering of numerical model using V_{min}/V_{max} = (a) 10 %, (b) 20 %, (c) 30 %, (d) 40 %, (e) 50 %, (f) 60 %

Figure 16 show the volume rendering images of the experimental model with various view angles. The contrast of image is V_{min}/V_{max} =30%. The test data contain a lot of noise. Hence, the crack appears as a cluster of dark patches in the image.

In Fig. 16(a), one can find a cluster of dark patches in the central area of the image, denoting the crack. As one rotates the model, the hole at the bottom becomes visible, indicating the existence of a defect. The side view in Fig. 16(d) provides a clear view of the crack. The size and location of the crack can be estimated based on this image.

Although these images are not as clear as in the numerical examples, one can still locate the crack by viewing the specimen from different angles. This is best done with an interactive imaging program that allows the inspector to interactively adjust the view angle and the contrast ratio. With the aid of such program, one can easily manipulate the image. Furthermore, as one adjusts the view angle gradually, the image becomes stereoscopic. That helps the inspector to interpret the image. Unfortunately, such effect cannot be demonstrated in this book.

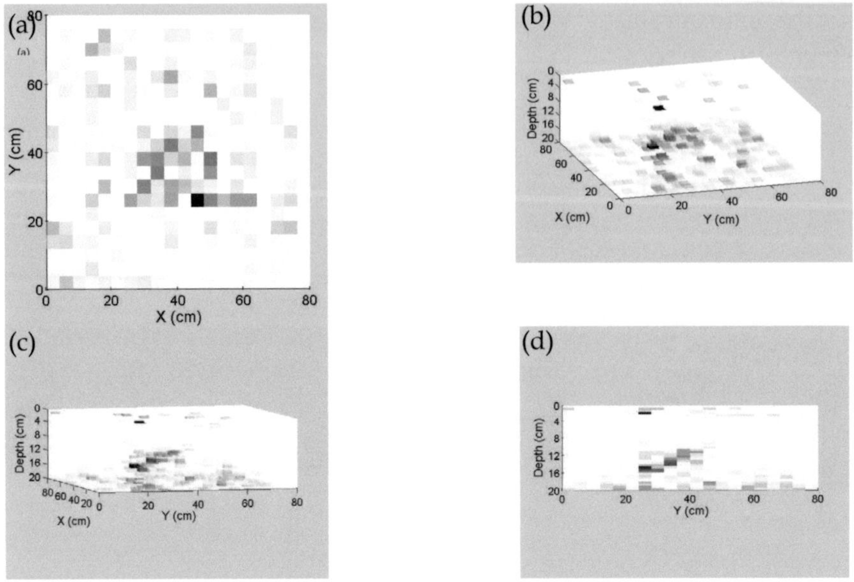

Fig. 16. Volume rendering of experimental model test with various view angles

4. Conclusion

This chapter introduces several methods to construct the image of concrete interior using impact echo data, including the spectral B-scan, spectral C-scan, spectral tomography, surface rendering, and volume rendering. With these imaging methods, the inspector may examine the interior of a structure to get better understanding of its health condition.

The imaging procedure contains three steps: data acquisition, data construction, and image rendering. Basically, the first two steps are the same for all the imaging methods. Firstly, a series of impact echo tests are performed at the grids of a mesh on the surface of the concrete. Then, the time signals are transformed into frequency spectra or depth spectra. Assembling the spectra into a 3D matrix yields the volume data, which could be used to construct images.

The spectral B- and C-scan are derived from ultrasonic scan. The spectral B-scan constructs a 2D density plot of the spectral amplitude on the vertical section under a test line; while the spectral C-scan constructs the plot for a horizontal section. Because the vertical axis of the spectral B-scan is frequency, it does not provide the profile of a vertical section. One may judge whether there is an internal defect by examining the discontinuity of horizontal stripes in B-scan. However, the size and location of the defect cannot be determined from the image directly.

The spectral C-scan, on the other hand, provides the profile of a horizontal section. Thus, one can use the image to determine the size and location of an internal defect. However, it is sometimes difficult to get an overall picture of the concrete interior by viewing horizontal sections alone.

The spectral tomogram can be considered as an extension of the spectral C-scan. It can be used to construct the profile image for arbitrary cross-sections. The inspector can observe the interior of a structure from different angles and by different sectioning. As such, the internal defects in concrete structures can be easily located. Through the numerical and experimental examples, it is seen that the spectral tomography can depict the internal crack of the concrete specimen successfully. When a crack exists in the concrete, it appears as a bright zone in the tomogram. Furthermore, no bright stripes appear beneath the crack at the bottom of the tomogram because the waves are blocked by the crack. This provides supplementary information about the size and location of the crack.

Surface rendering and volume rendering are 3D imaging techniques. The idea of surface rendering is to abstract the iso-surface from the volume data. In surface rendering, a defect is represented by one or several iso-surfaces, which can be used to estimate its size and location. Same as in spectral tomogram, a hole, approximate the size of defect, appears at the bottom beneath the defect.

Volume rendering is analogous to X-ray examination: Parallel rays are generated, transmit through a specimen, and reach a projection plane. If a defect exists in the specimen, the voxels covering the defect would have high opacity. Hence, a dark zone forms in the volume rendering image. If the view angle is chosen properly, one could also find a hole at the bottom beneath the defect.

Comparing these methods, one can see that each method has its own strength and weakness. Volume rendering is a robust technique; it is not sensitive to the interferences in spectra. Surface rendering is sensitive to noise, but it can depict the details of a defect. Spectral tomography is robust and insensitive to noise. However, it does not provide a 3D image and one can only view the specimen by sectioning.

To maintain good balance between robustness and precision, the inspector should take advantage of the strength of each method. One may apply volume rendering to get an overall picture of the specimen and to find the approximate location of the defect, if any. Then, use surface rendering or spectral tomography to observe the details.

It is seen in the experimental examples that the presence of noise downgrades the quality of images, no matter which method is adopted. Unfortunately, the test data is always noisy and the impact source is unsteady in real applications. The images obtained are sometimes difficult to interpret. In that situation, an interactive imaging program is indispensable. With the interactive graphic interface, one can adjust the imaging parameters or view angle arbitrarily and get an updated image instantly. As such, one may attain a better view of the specimen easily. More importantly, the 3D image appears stereoscopic as one changes the view angle gradually. Therefore, one can tell which object is in the front and which is in the back. This is useful especially when the quality of the image is poor.

The imaging methods presented in this chapter may provide the most direct information about the defects in concrete structures. However, its practical applications are hindered by two issues. Firstly, it is very time-consuming because a vast amount of tests need to be conducted. Secondly, the unsteadiness of the impact source deteriorates the quality of the image. It seems that an automatic test system is the solution to these problems. It is hope that such system can be developed in the near future so that the imaging techniques can be widely applied in the inspection of concrete structures.

5. Acknowledgment

The works presented in this chapter was supported by the National Science Council of Taiwan under grant NSC 98-2211-E-002-104-MY3.

6. References

Abraham, O., Leonard, C., Cote, P. & Piwakowski, B. (2000). Time-frequency Analysis of Impact-Echo Signals: Numerical Modeling and Experimental Validation. ACI Materials Journal, Vol.97, No.6, pp. 645-657

Angel, E. (2006). Interactive Computer graphics: a top-down approach using OpenGL 4th Ed., Addison Wesley, ISBN 0-321-3125-2X, MA

Engel, K., Hadwiger, M., Kniss, J. M., Rezk-Salama, C. & Weiskopf, D. (2006). Real-Time Volume Graphics, A K Peter, Ltd., ISBN 1-56881-266-3, Wellesley, MA

Gibson, A. & Popovics, J. S. (2005). Lamb wave basis for impact-echo method analysis. ASCE Journal of Engineering Mechanics, Vol.131, No.4, pp. 438-443

Goldsmith, W. (1960). Impact : The Theory and Physical Behavior of Colliding Solids, Edward Arnold Ltd., London

Hallquist, J. O. (2003). LS-DYNA Keyword User's Manual, Livermore Software Technology Corporation, Livermore

Kohl, C., Krause, M., Maierhofer, C. & Wostmann, J. (2005). 2D- and 3D-visualisation of NDT-data using data fusion technique. Materials and Structures, Vol.38, No.9, pp. 817-826

Lin, C. C., Liu, P. L. & Yeh, P. L. (2009). Application of empirical mode decomposition in the impact-echo test. NDT & E International, Vol.42, No.7, pp. 589~588

Lin, Y. & Sansalone, M. (1992). Transient Response of Thick Circular and Square Bars Subjected to Transverse Elastic Impact. J. Acoustical Society of America, Vol.91, No.2, pp. 885-893

Liu, P. L. & Yeh, P. L. (2008). Imaging of internal cracks in concrete structures using the volume rendering technique, 17th WCNDT, Shanghai, China, Oct. 25-28, 2008

Liu, P. L. & Yeh, P. L. (2010). Vertical Spectral Tomography of Concrete Structures Based on Impact Echo Depth Spectra. NDT & E International, Vol.43, No.1, pp. 45-53

Liu, P. L. & Yiu, C. Y. (2002). Imaging Of Concrete Defects Using Elastic Wave Tests, Proceedings, the 2002 Far-East Conference on Nondestructive Testing, Tokyo, Japan, Oct. 21-24, 2002

Liu, P.-L. & Yeh, P.-L. (2011). Spectral Tomography of Concrete Structures Based on Impact Echo Depth Spectra. NDT & E International, Vol.44, No.8, pp. 692~702

Sansalone, M. & Carino, N. J. (1986). Impact-Echo: A Method for Flaw Detection in Concrete Using Transient Stress Waves, Gaithersburg, MD: National Bureau of Standard

Sansalone, M. J. & Streett, W. B. (1997). Impact-echo: nondestructive evaluation of concrete and masonry, Bullbrier Press, Ithaca, N.Y.

Schubert, F. & Köhler, B. (2008). Ten Lectures on Impact-Echo. Journal of Nondestructive Evaluation, Vol.27, pp. 5-21

Schubert, F., Wiggenhauser, H. & Lausch, R. (2004). On the Accuracy of Thickness Measurements in Impact-echo Testing of Finite Concrete Specimens-numerical and Experimental Results. Ultrasonics, Vol.42, pp. 897-901

Watt, A. (2000). 3D Computer Graphics, Addison Wesley, ISBN 0-201-39855-9, Edinburgh Gate, British

Yeh, P. L. & Liu, P. L. (2008). Application of the Wavelet Transform and the Enhanced Fourier Spectrum in the Impact Echo Test. NDT & E International, Vol.41, No.5, pp. 382-394

Yeh, P. L. & Liu, P. L. (2009). Imaging of internal cracks in concrete structures using the surface rendering technique. NDT & E International, Vol.42, No.3, pp. 181-187

Zhu, J. Y. & Popovics, J. S. (2007). Imaging concrete structures using air-coupled impact-echo. ASCE Journal of Engineering Mechanics, Vol.133, No.6, pp. 628-640